STUDENT'S SOLUTIONS MANUAL

EDGAR N. REYES
Southeastern Louisiana University

TRIGONOMETRY

SECOND EDITION

Mark Dugopolski
Southeastern Louisiana University

Boston San Francisco New York
London Toronto Sydney Tokyo Singapore Madrid
Mexico City Munich Paris Cape Town Hong Kong Montreal

Reproduced by Pearson Addison-Wesley from electronic files supplied by the author.

Publishing as Pearson Addison-Wesley, 75 Arlington Street, Boston, MA 02116.

ISBN 0-321-36883-5

4 5 6 BRR 09 08 07

This Student's Solutions Manual includes worked out solutions to all odd-numbered exercises in the exercise sets, and all odd-numbered exercises in the end of the chapter "Review Exercises". Also, included are solutions to the "For Thought", "Chapter Test", and "Tying It All Together" exercises.

TABLE OF CONTENTS

For Thought

1. False, the point $(2, -3)$ is in Quadrant IV.

2. False, the point $(4, 0)$ does not belong to any quadrant.

3. False, since the distance is $\sqrt{(a-c)^2 + (b-d)^2}$.

4. False, since $Ax + By = C$ is a linear equation.

5. True

6. False, since $\sqrt{7^2 + 9^2} = \sqrt{130} \approx 11.4$

7. True

8. True

9. True

10. False, since the radius is 3.

P.1 Exercises

1. $(4, 1)$, Quadrant I

3. $(1, 0)$, x-axis

5. $(5, -1)$, Quadrant IV

7. $(-4, -2)$, Quadrant III

9. $(-2, 4)$, Quadrant II

11. $c = \sqrt{(\sqrt{3})^2 + 1^2} = \sqrt{4} = 2$

13. Since $b^2 + 2^2 = 3^2$, we get $b^2 + 4 = 9$ or $b^2 = 5$. Then $b = \sqrt{5}$.

15. Since $a^2 + 3^2 = 5^2$, we get $a^2 + 9 = 25$ or $a^2 = 16$. So $a = 4$.

17. Distance is

$$\sqrt{(4-1)^2 + (7-3)^2} = \sqrt{9+16} = \sqrt{25} = 5,$$

midpoint is $(2.5, 5)$

19. Distance is $\sqrt{(-1-1)^2 + (-2-0)^2} = \sqrt{4+4} = 2\sqrt{2}$, midpoint is $(0, -1)$

21. Distance is $\sqrt{\left(\frac{\sqrt{2}}{2} - 0\right)^2 + \left(\frac{\sqrt{2}}{2} - 0\right)^2} = \sqrt{\frac{2}{4} + \frac{2}{4}} = \sqrt{1} = 1$, midpoint is $\left(\frac{\sqrt{2}/2 + 0}{2}, \frac{\sqrt{2}/2 + 0}{2}\right) = \left(\frac{\sqrt{2}}{4}, \frac{\sqrt{2}}{4}\right)$

23. Distance is $\sqrt{\left(\sqrt{18} - \sqrt{8}\right)^2 + \left(\sqrt{12} - \sqrt{27}\right)^2} = \sqrt{(3\sqrt{2} - 2\sqrt{2})^2 + (2\sqrt{3} - 3\sqrt{3})^2} = \sqrt{(\sqrt{2})^2 + (-\sqrt{3})^2} = \sqrt{5}$,

midpoint is $\left(\frac{\sqrt{18} + \sqrt{8}}{2}, \frac{\sqrt{12} + \sqrt{27}}{2}\right) = \left(\frac{3\sqrt{2} + 2\sqrt{2}}{2}, \frac{2\sqrt{3} + 3\sqrt{3}}{2}\right) = \left(\frac{5\sqrt{2}}{2}, \frac{5\sqrt{3}}{2}\right)$

25. Distance is $\sqrt{(1.2 + 3.8)^2 + (4.4 + 2.2)^2} = \sqrt{25 + 49} = \sqrt{74}$, midpoint is $(-1.3, 1.3)$

27. Distance is $\frac{\sqrt{\pi^2 + 4}}{2}$, midpoint is $\left(\frac{3\pi}{4}, \frac{1}{2}\right)$

29. Distance is $\sqrt{(2\pi - \pi)^2 + (0 - 0)^2} = \sqrt{\pi^2} = \pi$,

midpoint is $\left(\frac{2\pi + \pi}{2}, \frac{0+0}{2}\right) = \left(\frac{3\pi}{2}, 0\right)$

31. Center$(0, 0)$, radius 4

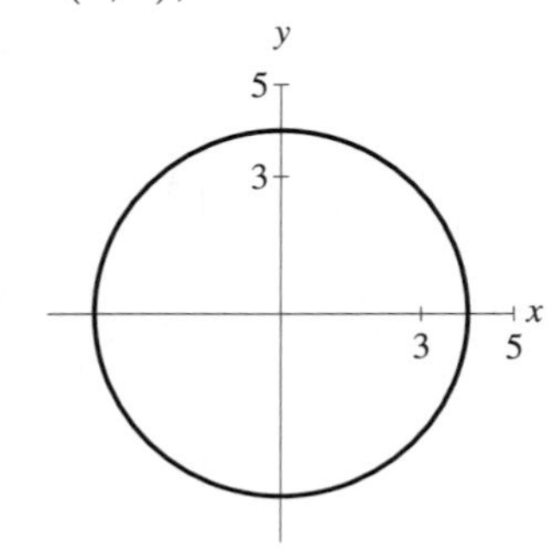

33. Center $(-6, 0)$, radius 6

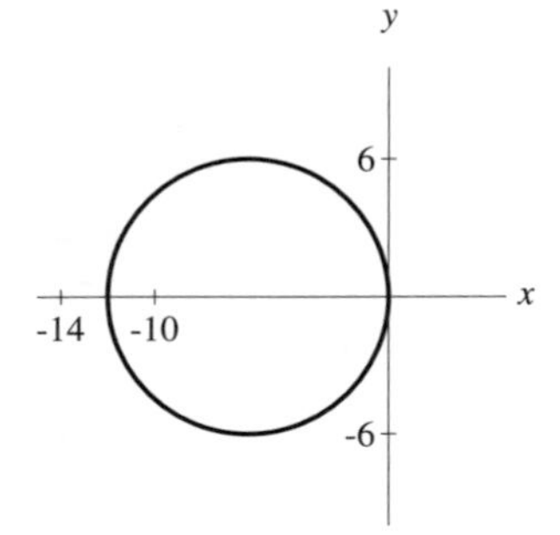

35. Center $(2, -2)$, radius $2\sqrt{2}$

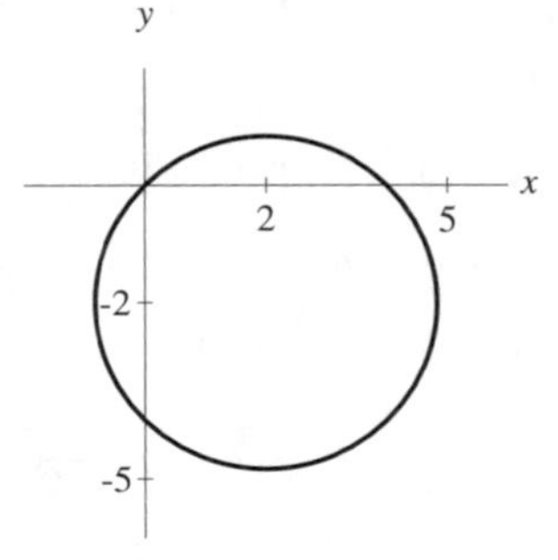

37. $x^2 + y^2 = 7$

39. $(x + 2)^2 + (y - 5)^2 = 1/4$

41. The distance between $(3, 5)$ and the origin is $\sqrt{34}$ which is the radius. The standard equation is $(x - 3)^2 + (y - 5)^2 = 34$.

43. Note, the distance between $(\sqrt{2}/2, \sqrt{2}/2)$ and the origin is 1. Thus, the radius is 1. The standard equation is $x^2 + y^2 = 1$.

45. The radius is

$$\sqrt{(-1 - 0)^2 + (2 - 0)^2} = \sqrt{5}.$$

The standard equation is $(x+1)^2+(y-2)^2 = 5$.

47. Note, the center is $(1, 3)$ and the radius is 2. The standard equation is

$$(x - 1)^2 + (y - 3)^2 = 4.$$

49. We solve for a.

$$\begin{aligned} a^2 + \left(\frac{3}{5}\right)^2 &= 1 \\ a^2 &= 1 - \frac{9}{25} \\ a^2 &= \frac{16}{25} \\ a &= \pm\frac{4}{5} \end{aligned}$$

51. We solve for a.

$$\begin{aligned} \left(-\frac{2}{5}\right)^2 + a^2 &= 1 \\ a^2 &= 1 - \frac{4}{25} \\ a^2 &= \frac{21}{25} \\ a &= \pm\frac{\sqrt{21}}{5} \end{aligned}$$

53. $y = 3x - 4$ goes through $(0, -4)$, $\left(\frac{4}{3}, 0\right)$.

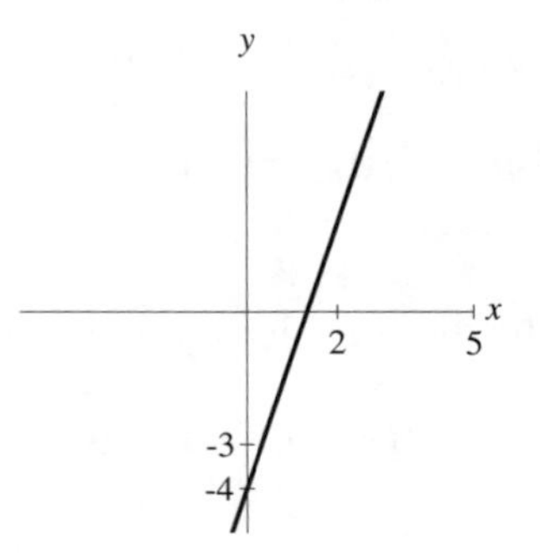

55. $3x - y = 6$ goes through $(0, -6)$, $(2, 0)$.

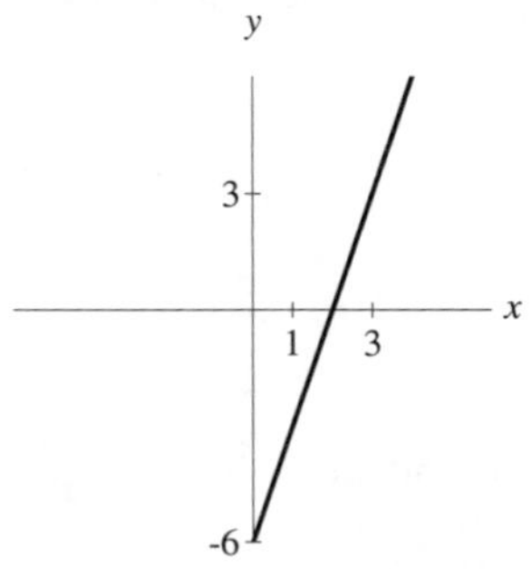

57. $x + y = 80$ goes through $(0, 80)$, $(80, 0)$.

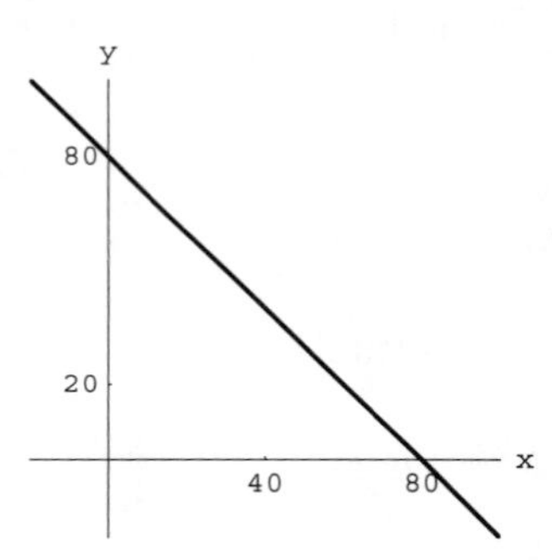

59. $x = 3y - 90$ goes through $(0, 30), (-90, 0)$.

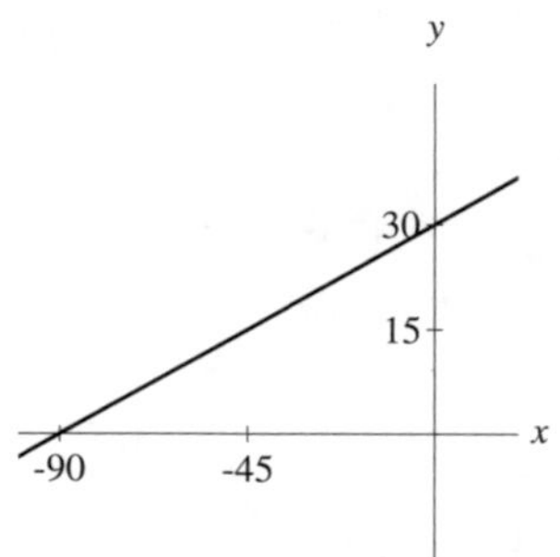

61. $\frac{1}{2}x - \frac{1}{3}y = 600$ goes through $(0, -1800)$, $(1200, 0)$.

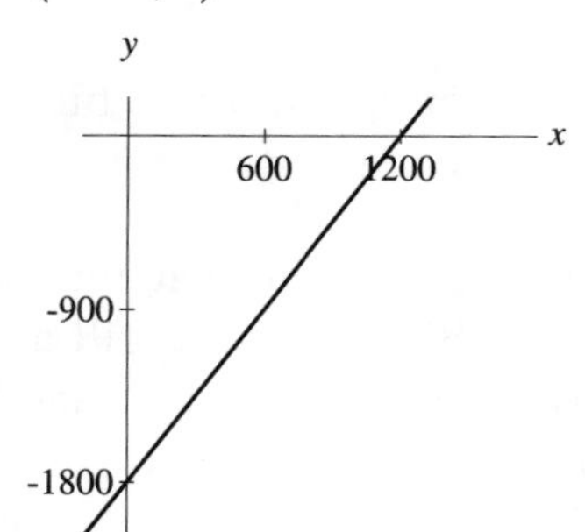

63. Intercepts are $(0, 0.0025)$, $(0.005, 0)$.

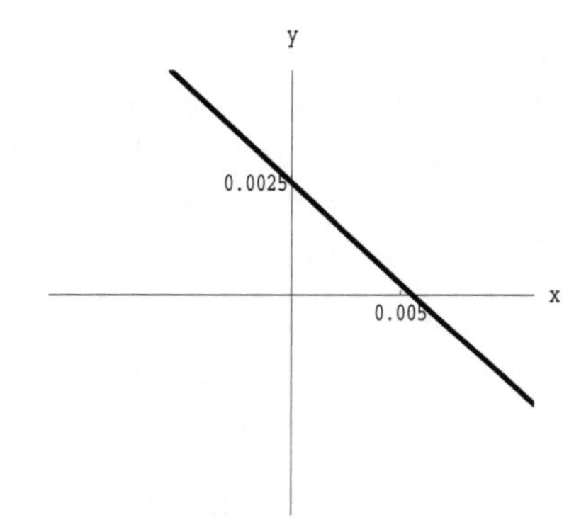

65. $x = 5$

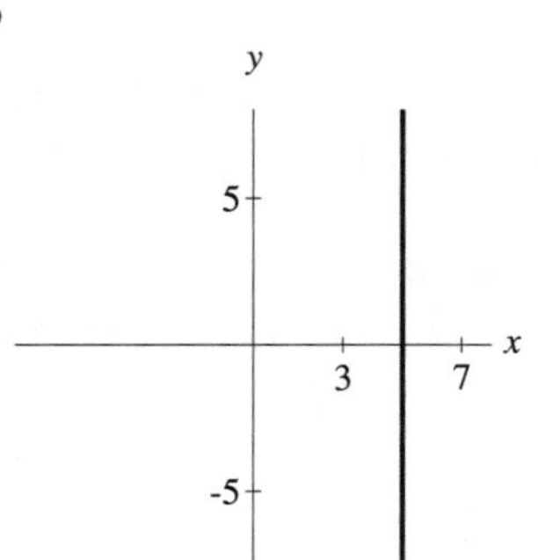

67. $y = 4$

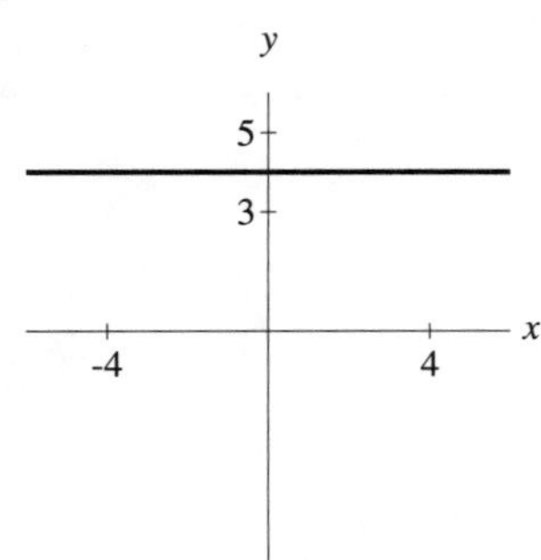

69. $x = -4$

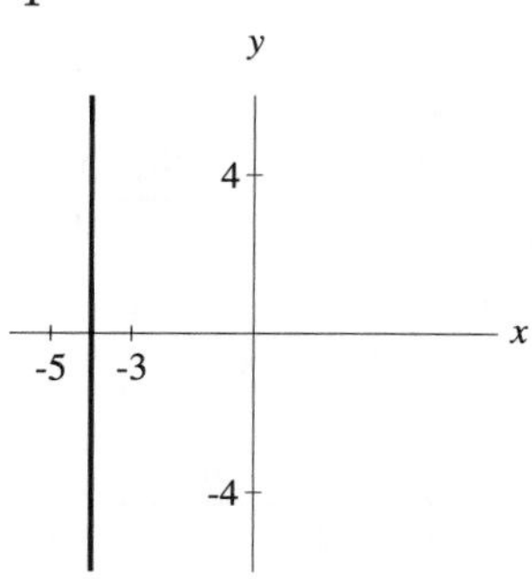

71. Solving for y, we have $y = 1$.

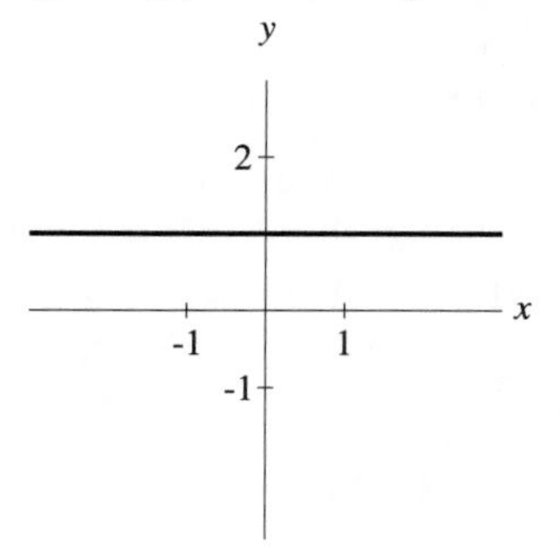

73. $y = x - 20$ goes through $(0, -20)$, $(20, 0)$.

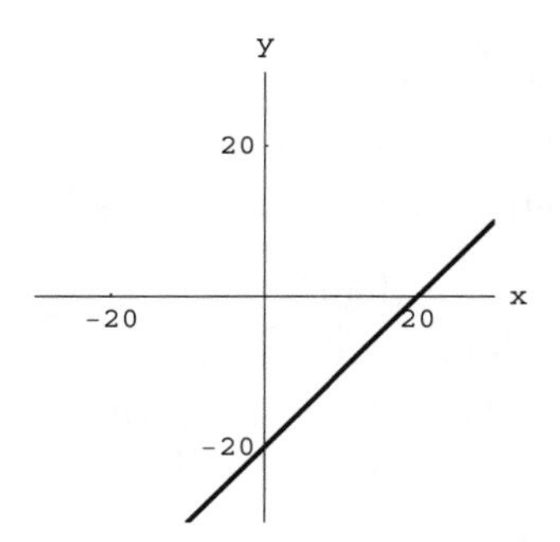

75. $y = 3000 - 500x$ goes through $(0, 3000)$, $(6, 0)$.

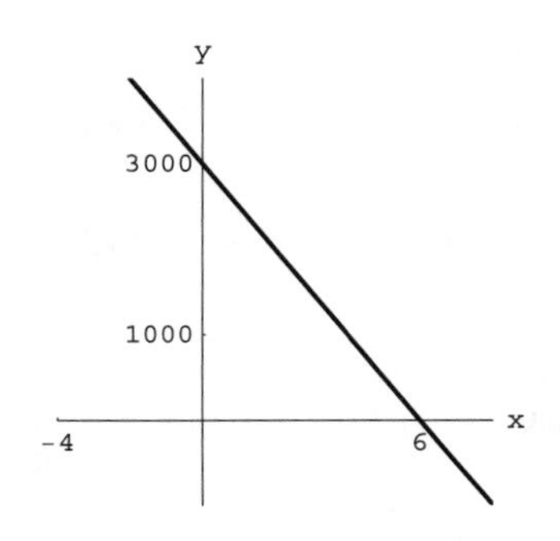

77. The hypotenuse is $\sqrt{6^2+8^2}=\sqrt{100}=10$.

79. a) The midpoint is

$$\left(\frac{1970+2000}{2}, \frac{20.8+25.1}{2}\right) = (1985, 22.95).$$

The median of age at first marriage in 1985 was 22.95 years.

b) The distance is

$$\sqrt{(2000-1970)^2+(25.1-20.8)^2} \approx 30.3$$

Because of the units, the distance is meaningless.

81. The distance between $(10,0)$ and $(0,0)$ is 10. The distance between $(1,3)$ and the origin is $\sqrt{10}$. If two points have integer coordinates, then the distance between them is of the form $\sqrt{s^2+t^2}$ where $s^2, t^2 \in \{0,1,2^2,3^2,4^2,...\} = \{0,1,4,9,16,...\}$.

Note, there are no numbers s^2 and t^2 in $\{0,1,4,9,16,...\}$ satisfying $s^2+t^2=19$. Thus, one cannot find two points with integer coordinates whose distance between them is $\sqrt{19}$.

For Thought

1 True, since the number of gallons purchased is 20 divided by the price per gallon.

2. False, since a student's exam grade is a function of the student's preparation. If two classmates had the same IQ and only one prepared then the one who prepared will most likely achieve a higher grade.

3. False, since $\{(1,2),(1,3)\}$ is not a function.

4. True

5. True

6. True

7. False, the domain is the set of all real numbers.

8. True

9. True, since

$$f(0) = \frac{0-2}{0+2} = -1.$$

10. True, since if $a-5=0$ then $a=5$.

P.2 Exercises

1. Note, $b=2\pi a$ is equivalent to $a=\dfrac{b}{2\pi}$.

Thus, a is a function of b, and b is a function of a.

3. a is a function of b since a given denomination has a unique length. Since a dollar bill and a five-dollar bill have the same length, then b is not a function of a.

5. Since an item has only one price, b is a function of a. Since two items may have the same price, a is not a function of b.

7. a is not a function of b since it is possible that two different students can obtain the same final exam score but the times spent on studying are different.

b is not a function of a since it is possible that two different students can spend the same time studying but obtain different final exam scores.

9. Since 1 in $\approx$ 2.54 cm, a is a function of b and b is a function of a.

11. Since $b=a^3$ and $a=\sqrt[3]{b}$, we get that b is a function of a, and a is a function of b.

13. Since $b=|a|$, we get b is a function of a. Since $a=\pm b$, we find a is not a function of b.

15. $A=s^2$ **17.** $s=\dfrac{\sqrt{2}d}{2}$

19. $P=4s$ **21.** $A=P^2/16$

23. $y=2x-1$ has domain $(-\infty,\infty)$ and range $(-\infty,\infty)$, some points are $(0,-1)$ and $(1,1)$

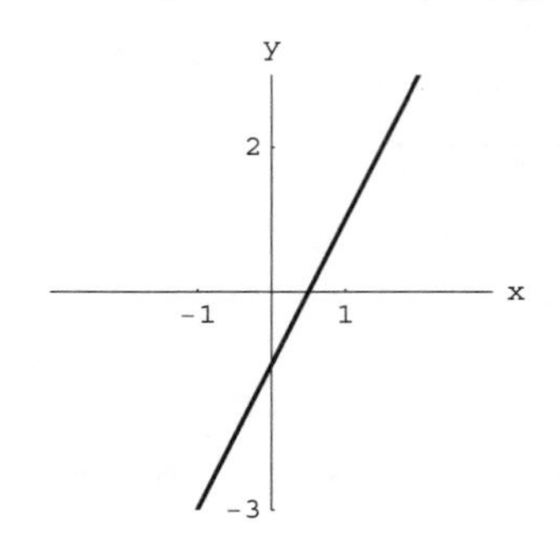

25. $y = 5$ has domain $(-\infty, \infty)$ and range $\{5\}$, some points are $(0, 5)$ and $(1, 5)$

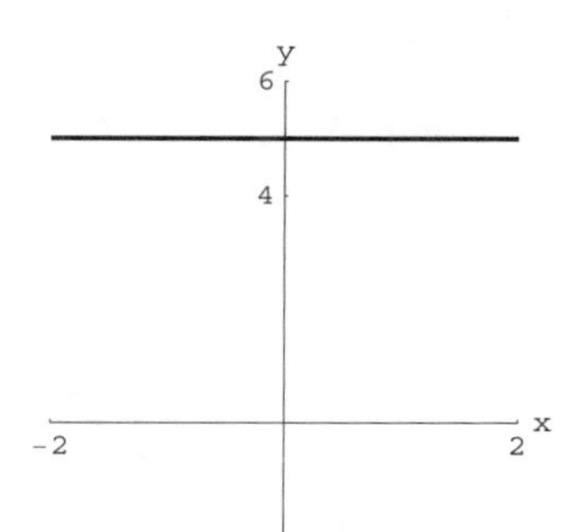

27. $y = x^2 - 20$ has domain $(-\infty, \infty)$ and range $[-20, \infty)$, some points are $(0, -20)$ and $(6, 16)$

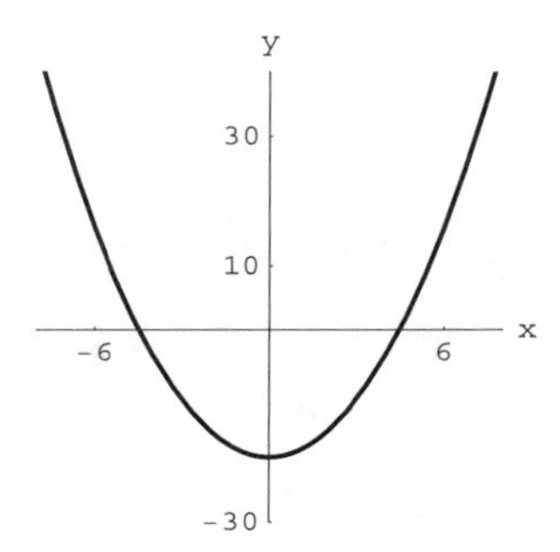

29. $y = 40 - x^2$ has domain $(-\infty, \infty)$ and range $(-\infty, 40]$, some points are $(0, 40)$ and $(6, 4)$

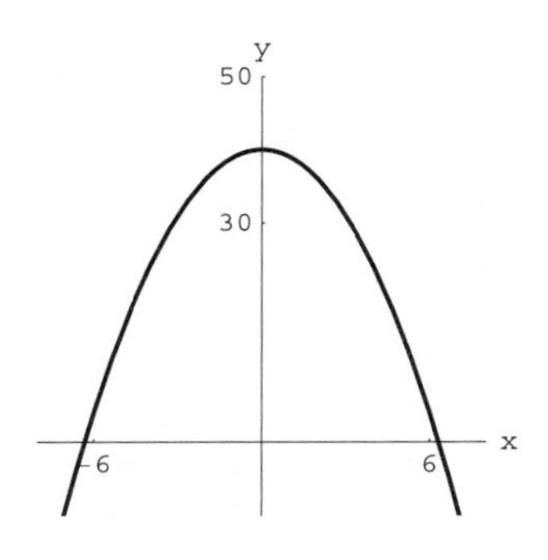

31. $y = x^3$ has domain $(-\infty, \infty)$ and range $(-\infty, \infty)$, some points are $(0, 0)$ and $(2, 8)$

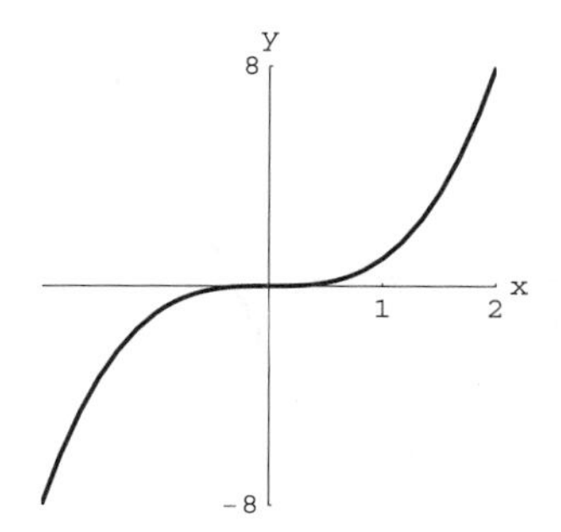

33. $y = \sqrt{x - 10}$ has domain $[10, \infty)$ and range $[0, \infty)$, some points are $(10, 0)$ and $(14, 2)$

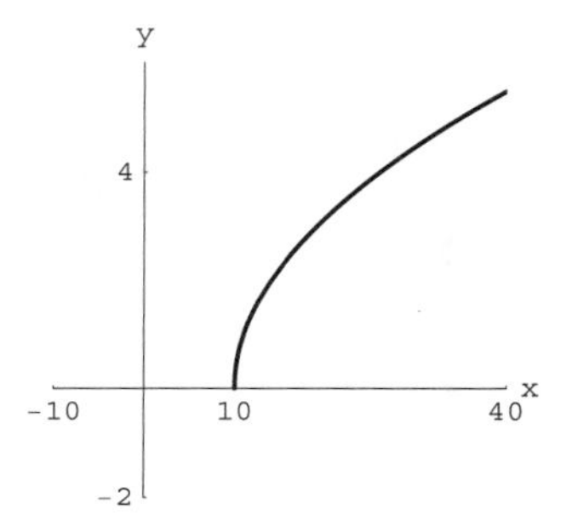

35. $y = \sqrt{x} + 30$ has domain $[0, \infty)$ and range $[30, \infty)$, some points are $(0, 30)$ and $(400, 50)$

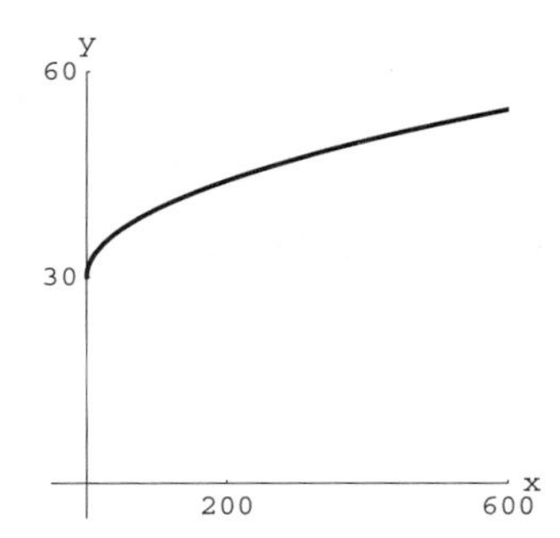

37. $y = |x| - 40$ has domain $(-\infty, \infty)$ and range $[-40, \infty)$, some points are $(0, -40)$ and $(40, 0)$

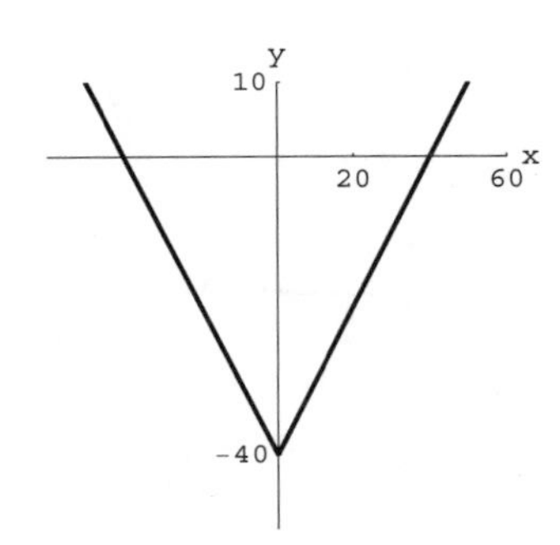

39. $y = |x - 20|$ has domain $(-\infty, \infty)$ and range $[0, \infty)$, some points are $(0, 20)$ and $(20, 0)$

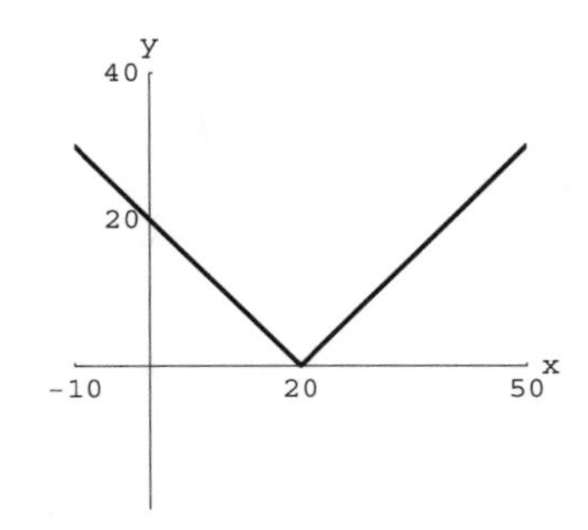

41. $3 \cdot 4 - 2 = 10$

43. $-4 - 2 = -6$

45. $|8| = 8$

47. $0.9408 - 0.56 = 0.3808$

49. $4 + (-6) = -2$

51. $80 - 2 = 78$

53. $3a^2 - a$

55. $f(-x) = 3(-x)^2 - (-x) = 3x^2 + x$

57. Factoring, we get $x(3x - 1) + 0$. So $x = 0, 1/3$.

59. Since $|a + 3| = 4$ is equivalent to $a + 3 = 4$ or $a + 3 = -4$, we have $a = 1, -7$.

61. $C = 353n$

63. $C = 35n + 50$

65. We find

$$C = \frac{4B}{\sqrt[3]{D}} = \frac{4(12 + 11/12)}{\sqrt[3]{22,800}} \approx 1.822$$

and a sketch of the graph of $C = \dfrac{4B}{\sqrt[3]{22,800}}$ is given below.

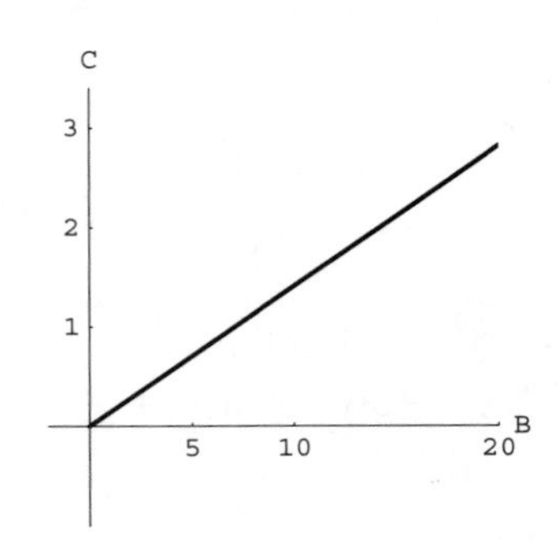

For Thought

1. False, it is a reflection in the y-axis.

2. False, the graph of $y = x^2 - 4$ is shifted down 4 units from the graph of $y = x^2$.

3. False, rather it is a left translation.

4. True

5. True

6. False, the down shift should come after the reflection.

7. True

8. False, since their domains are different.

9. True

10. True, since $f(-x) = -f(x)$ where $f(x) = x^3$.

P.3 Exercises

1. $f(x) = \sqrt{x}$, $g(x) = -\sqrt{x}$

3. $y = x$, $y = -x$

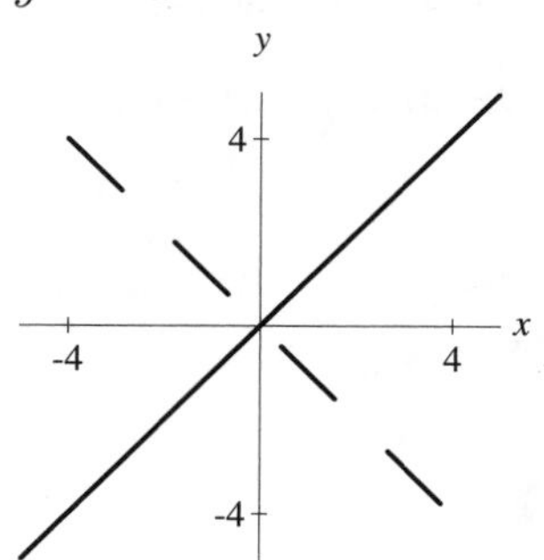

5. $f(x) = |x|$, $g(x) = |x| - 4$

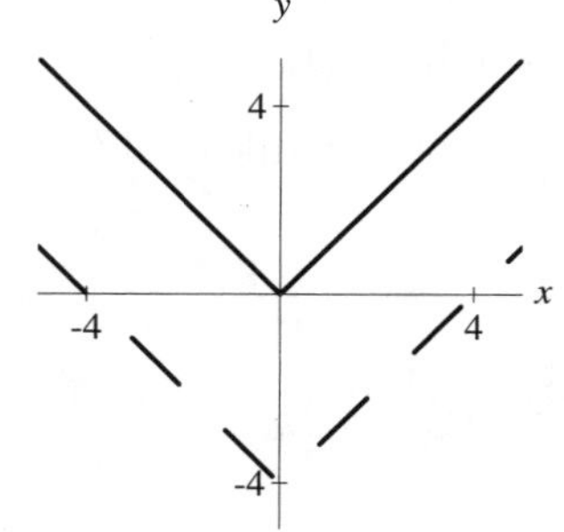

7. $f(x) = x$, $g(x) = x + 3$

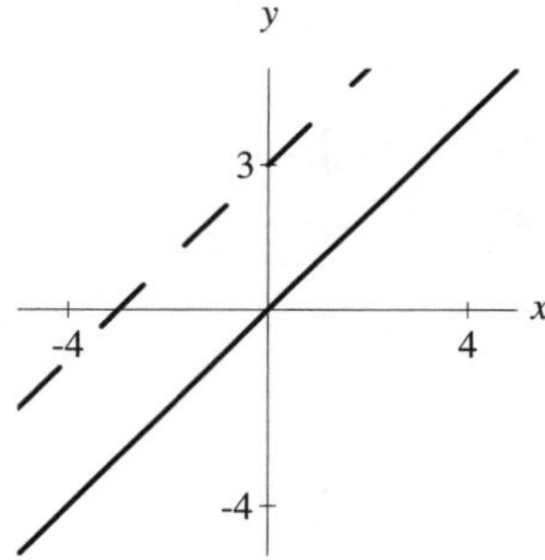

9. $y = x^2$, $y = (x - 3)^2$

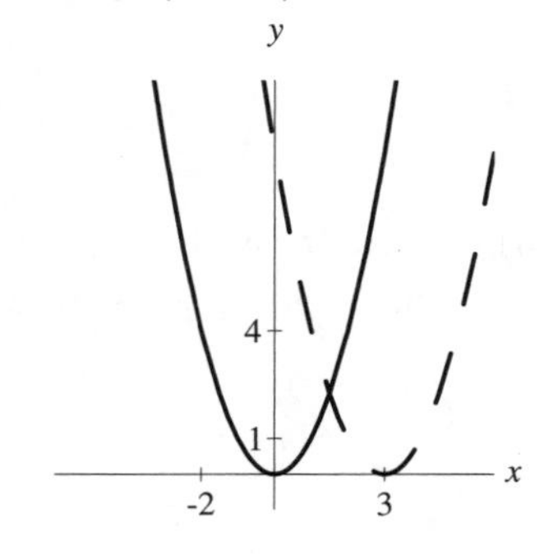

11. $f(x) = x^3$, $g(x) = (x + 1)^3$

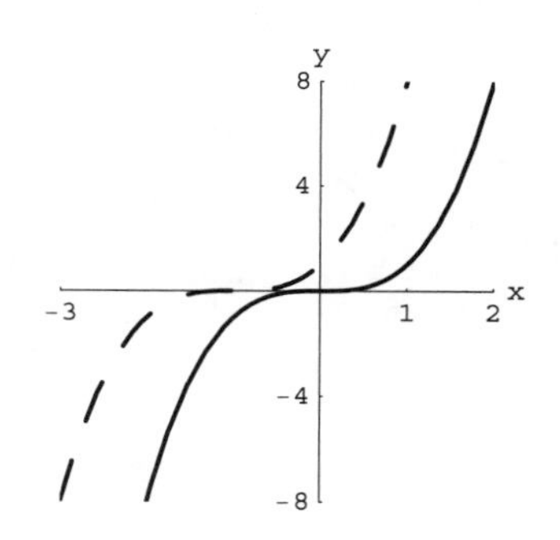

13. $y = \sqrt{x}, y = 3\sqrt{x}$

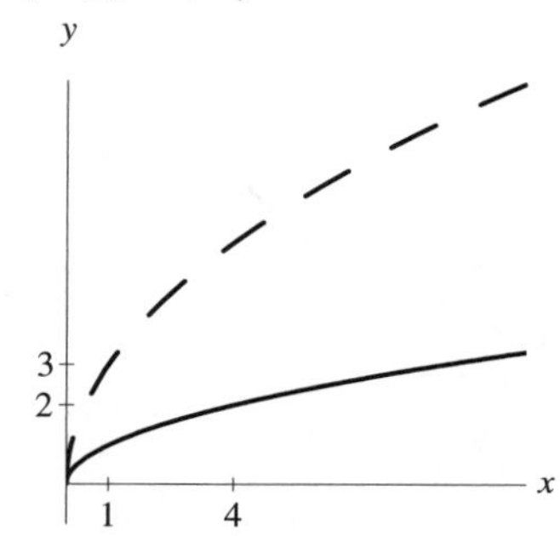

15. $y = x^2, y = \frac{1}{4}x^2$

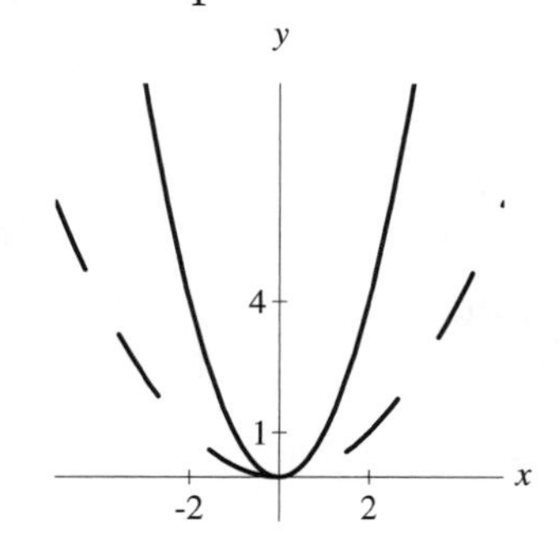

17. g **19.** b

21. c **23.** f

25. $y = (x - 10)^2 + 4$

27. $y = -3|x - 7| + 9$

29. $y = -(3\sqrt{x} + 5)$ or $y = -3\sqrt{x} - 5$

31. $y = \sqrt{x-1} + 2$; right by 1, up by 2

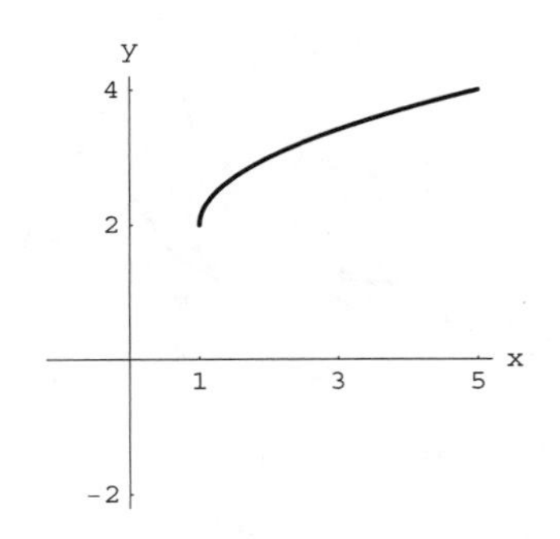

33. $y = |x-1| + 3$; right by 1, up by 3

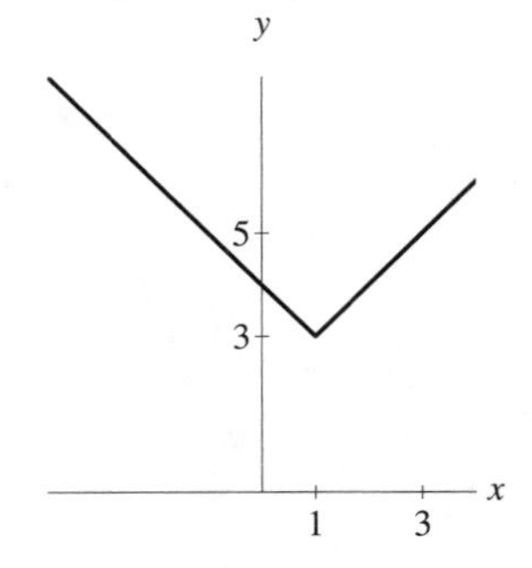

35. $y = 3x - 40$

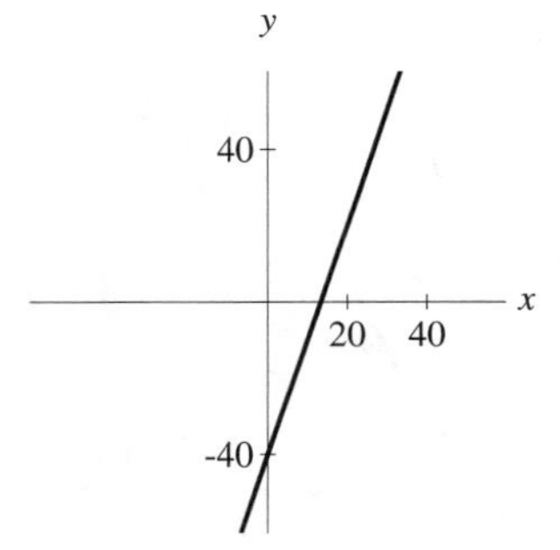

37. $y = \dfrac{1}{2}x - 20$

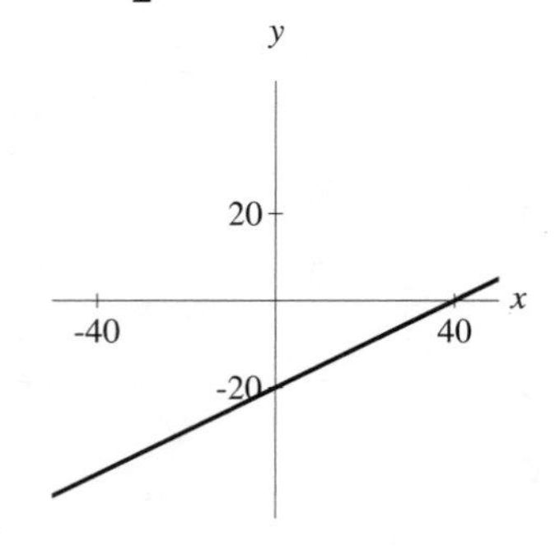

39. $y = -\dfrac{1}{2}|x| + 40$, shrink by 1/2, reflect about x-axis, up by 40

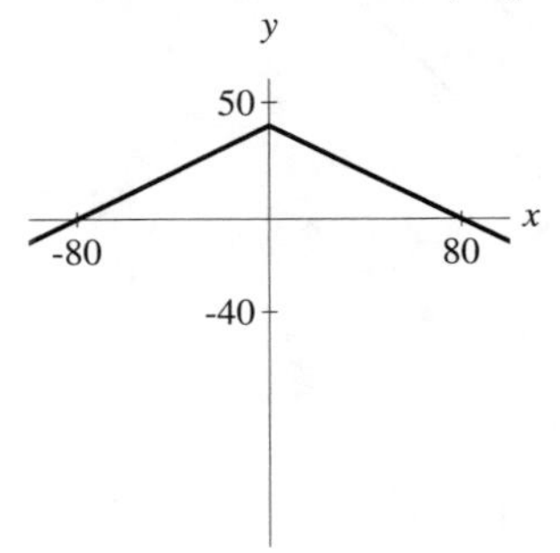

41. $y = -\dfrac{1}{2}|x+4|$, left by 4, reflect about x-axis, shrink by 1/2

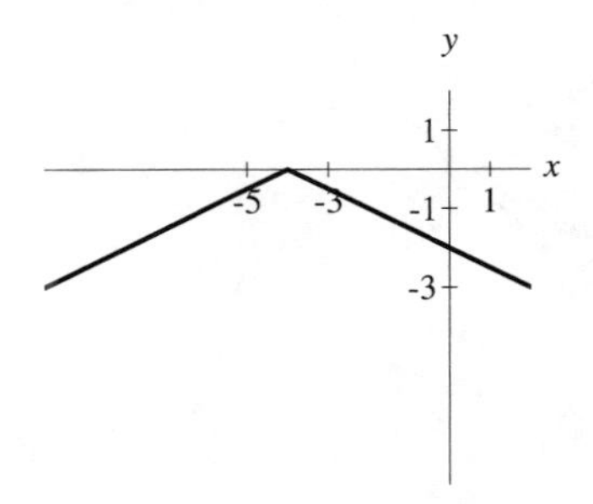

43. $y = -(x-3)^2 + 1$; right by 3, reflect about x-axis, up by 1

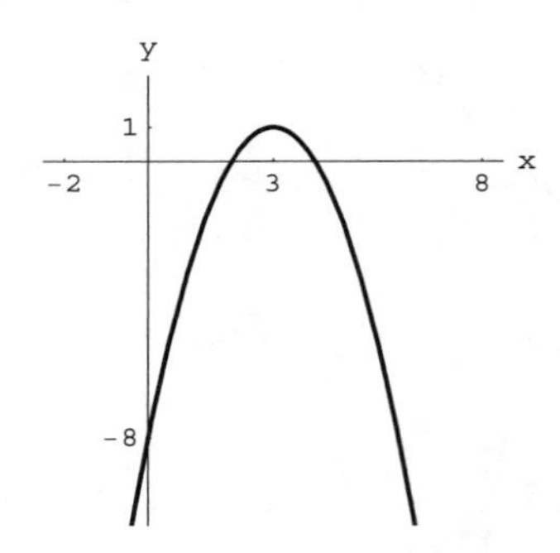

45. $y = -2(x+3)^2 - 4$; left by 3, stretch by 2, reflect about x-axis, down by 4

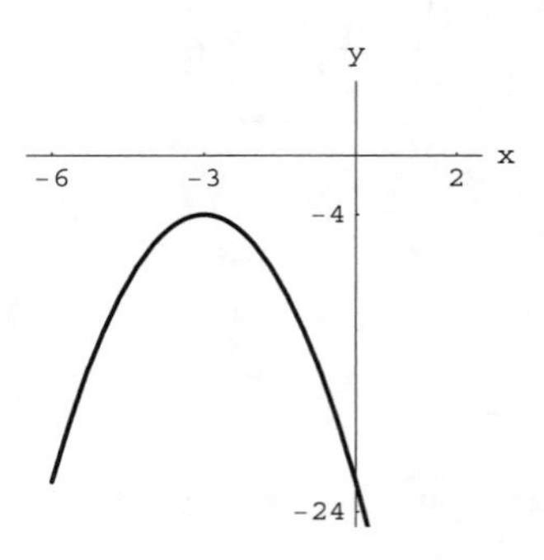

47. $y = -2\sqrt{x+3}+2$, left by 3, stretch by 2, reflect about x-axis, up by 2

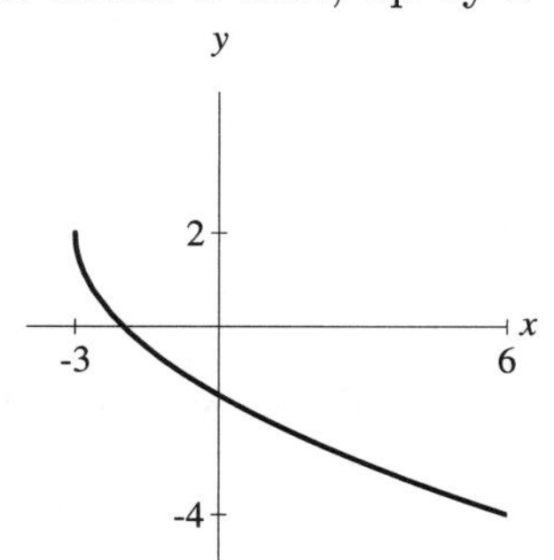

49. Symmetric about y-axis, even function since $f(-x) = f(x)$

51. No symmetry, neither even nor odd since $f(-x) \neq f(x)$ and $f(-x) \neq -f(x)$

53. Neither symmetry, neither even nor odd since $f(-x) \neq f(x)$ and $f(-x) \neq -f(x)$

55. No symmetry, not an even or odd function since $f(-x) = -f(x)$ and $f(-x) \neq -f(x)$

57. Symmetric about the origin, odd function since $f(-x) = -f(x)$

59. No symmetry, not an even or odd function since $f(-x) \neq f(x)$ and $f(-x) \neq -f(x)$

61. No symmetry, not an even or odd function since $f(-x) \neq f(x)$ and $f(-x) \neq -f(x)$

63. Neither symmetry, not an even or odd function since $f(-x) \neq f(x)$ and $f(-x) \neq -f(x)$

65. Symmetric about the y-axis, even function since $f(-x) = f(x)$

67. Symmetric about the y-axis, even function since $f(-x) = f(x)$

69. e **71.** g

73. b **75.** c

77. $N(x) = x + 2000$

79. If inflation rate is less than 50%, then $1 - \sqrt{x} < \frac{1}{2}$. This simplifies to $\frac{1}{2} < \sqrt{x}$. After squaring we have $\frac{1}{4} < x$ and so $x > 25\%$.

81.

(a) Both functions are even functions and the graphs are identical

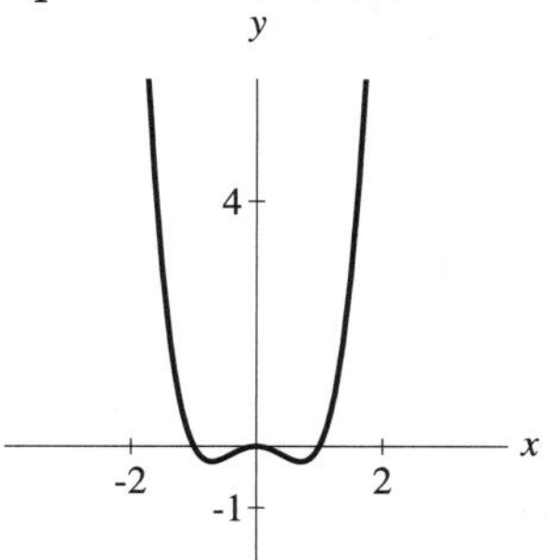

(b) One graph is a reflection of the other about the y-axis. Both functions are odd.

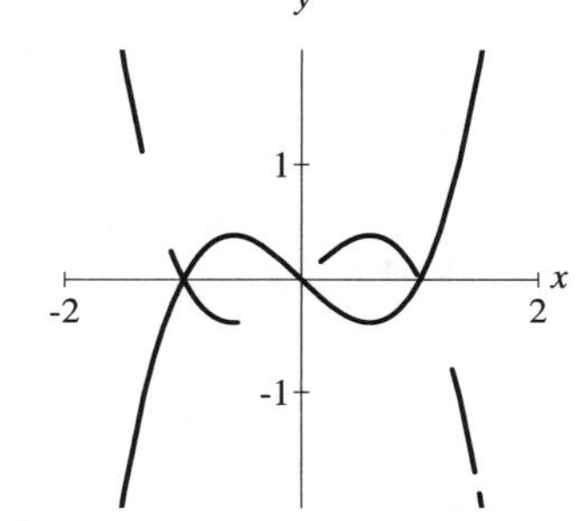

(c) The second graph is obtained by shifting the first one to the left by 1 unit.

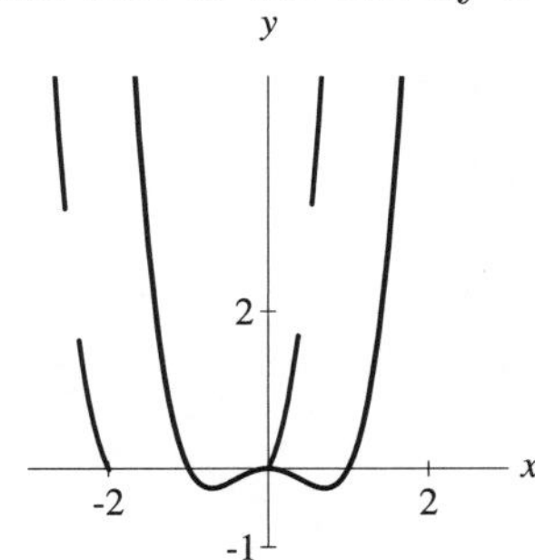

(d) The second graph is obtained by translating the first one to the right by 2 units and 3 units up.

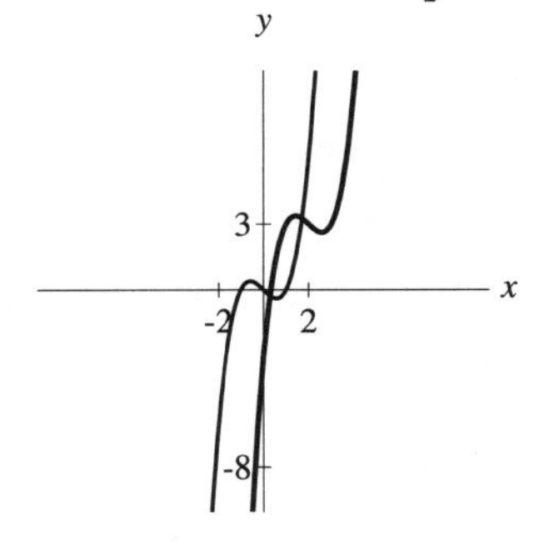

For Thought

1 True, since $A = P^2/16$.

2. False, rather $y = (x-1)^2 = x^2 - 2x + 1$.

3. False, rather $(f \circ g)(x) = \sqrt{x-2}$.

4. True

5. False, since $(h \circ g)(x) = x^2 - 9$.

6. False, rather $f^{-1}(x) = \frac{1}{2}x - \frac{1}{2}$.

7. True

8. False; g^{-1} does not exist since the graph of g which is a parabola fails the horizontal line test.

9. False, since $y = x^2$ is a function which does not have an inverse function.

10. True

P.4 Exercises

1. $y = 2(3x+1) - 3 = 6x - 1$

3. $y = (x^2 + 6x + 9) - 2 = x^2 + 6x + 7$

5. $y = 3 \cdot \frac{x+1}{3} - 1 = x + 1 - 1 = x$

7. $y = 2 \cdot \frac{x+1}{2} - 1 = x + 1 - 1 = x$

9. $f(2) = 5$

11. $f(2) = 5$

13. $f(17) = 3(17) - 1 = 50$

15. $3(x^2+1) - 1 = 3x^2 + 2$

17. $(3x-1)^2 + 1 = 9x^2 - 6x + 2$

19. $\frac{x^2+2}{3}$

21. $F = g \circ h$

23. $H = h \circ g$

25. $N = f \circ h$

27. Interchange x and y then solve for y.

$$\begin{aligned} x &= 3y - 7 \\ \frac{x+7}{3} &= y \\ \frac{x+7}{3} &= f^{-1}(x) \end{aligned}$$

29. Interchange x and y then solve for y.

$$\begin{aligned} x &= 2 + \sqrt{y-3} \quad \text{for } x \ge 2 \\ (x-2)^2 &= y - 3 \quad \text{for } x \ge 2 \\ f^{-1}(x) &= (x-2)^2 + 3 \quad \text{for } x \ge 2 \end{aligned}$$

31. Interchange x and y then solve for y.

$$\begin{aligned} x &= -y - 9 \\ y &= -x - 9 \\ f^{-1}(x) &= -x - 9 \end{aligned}$$

33. Interchange x and y then solve for y.

$$\begin{aligned} x &= -\frac{1}{y} \\ xy &= -1 \\ f^{-1}(x) &= -\frac{1}{x} \end{aligned}$$

35. Interchange x and y then solve for y.

$$\begin{aligned} x &= \sqrt[3]{y-9} + 5 \\ x - 5 &= \sqrt[3]{y-9} \\ (x-5)^3 &= y - 9 \\ f^{-1}(x) &= (x-5)^3 + 9 \end{aligned}$$

37. Interchange x and y then solve for y.

$$\begin{aligned} x &= (y-2)^2 \quad x \ge 0 \\ \sqrt{x} &= y - 2 \\ f^{-1}(x) &= \sqrt{x} + 2 \end{aligned}$$

39. $(f^{-1} \circ f)(x) = \frac{1}{2}(2x-1) + \frac{1}{2} = x$ and $(f \circ f^{-1})(x) = 2\left(\frac{1}{2}x + \frac{1}{2}\right) - 1 = x$

41. $(f^{-1} \circ f)(x) = 0.25(4x+4) - 1 = x$ and $(f \circ f^{-1})(x) = 4(0.25x - 1) + 4 = x$

43. We obtain

$$\begin{aligned}(f^{-1}\circ f)(x) &= \frac{4-\left(\sqrt[3]{4-3x}\right)^3}{3}\\ &= \frac{4-(4-3x)}{3}\\ &= \frac{3x}{3}\\ (f^{-1}\circ f)(x) &= x\end{aligned}$$

and

$$\begin{aligned}(f\circ f^{-1})(x) &= \sqrt[3]{4-3\left(\frac{4-x^3}{3}\right)}\\ &= \sqrt[3]{4-(4-x^3)}\\ &= \sqrt[3]{x^3}\\ (f\circ f^{-1})(x) &= x.\end{aligned}$$

45. We obtain

$$\begin{aligned}(f^{-1}\circ f)(x) &= \sqrt[5]{\left(\sqrt[3]{x^5-1}\right)^3+1}\\ &= \sqrt[5]{x^5-1+1}\\ &= \sqrt[5]{x^5}\\ (f^{-1}\circ f)(x) &= x\end{aligned}$$

and

$$\begin{aligned}(f\circ f^{-1})(x) &= \sqrt[3]{\left(\sqrt[5]{x^3+1}\right)^5-1}\\ &= \sqrt[3]{x^3+1-1}\\ &= \sqrt[3]{x^3}\\ (f\circ f^{-1})(x) &= x.\end{aligned}$$

47. $f^{-1}(x) = \dfrac{x-2}{3}$

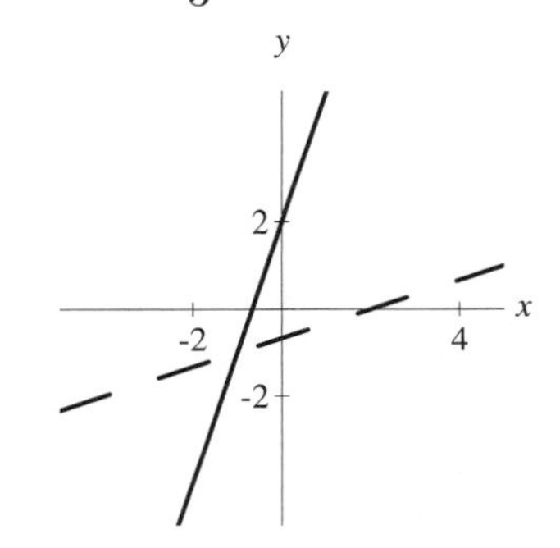

49. $f^{-1}(x) = \sqrt{x+4}$

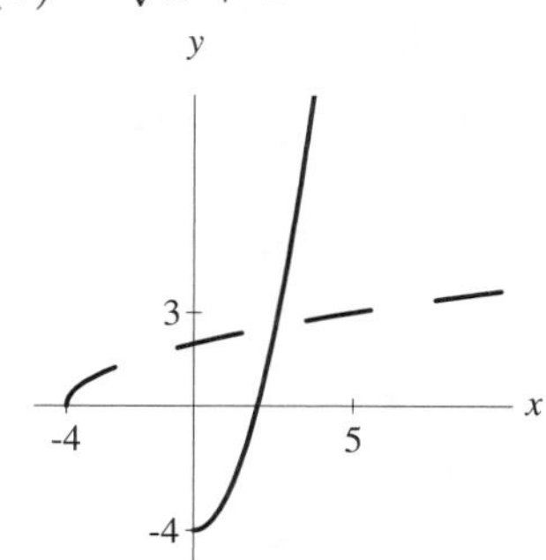

51. $f^{-1}(x) = \sqrt[3]{x}$

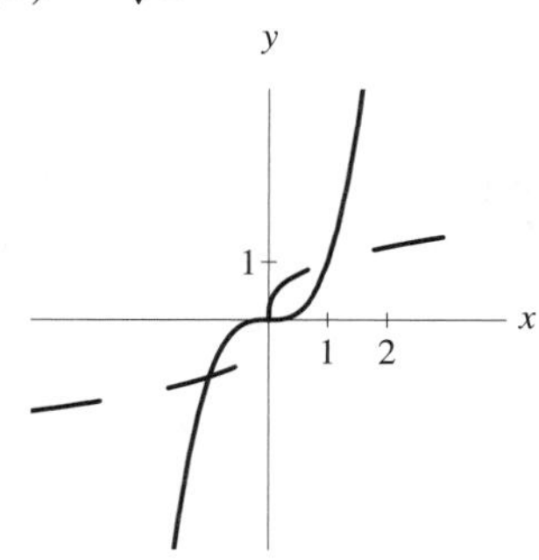

53. $f^{-1}(x) = (x+3)^2$ for $x \geq -3$

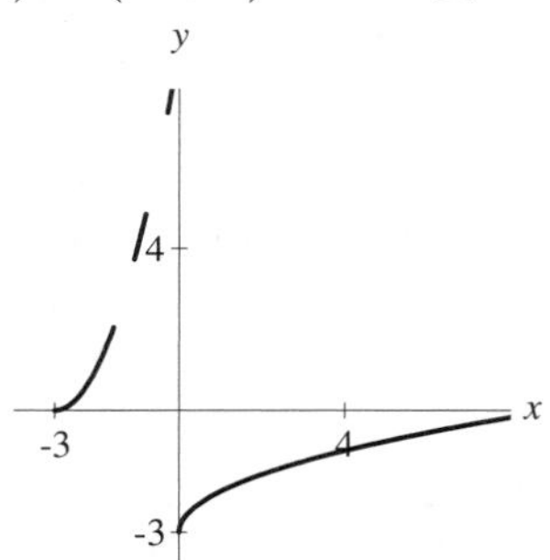

55. The inverse is the composition of subtracting 2 and dividing by 5, i.e., $f^{-1}(x) = \dfrac{x-2}{5}$

57. The inverse is the composition of adding 88 and dividing by 2, i.e.,

$$f^{-1}(x) = \frac{x+88}{2} = \frac{1}{2}x + 44$$

59. The inverse is the composition of subtracting 7 and dividing by -3, i.e.,

$$f^{-1}(x) = \frac{x-7}{-3} = -\frac{1}{3}x + \frac{7}{3}$$

61. The inverse is the composition of subtracting 4 and dividing by -3, i.e.,

$$f^{-1}(x) = \frac{x-4}{-3} = -\frac{1}{3}x + \frac{4}{3}$$

63. The inverse is the composition of adding 9 and multiplying by 2, i.e.,

$$f^{-1}(x) = 2(x+9) = 2x+18$$

65. The inverse is the composition of subtracting 3 and taking the multiplicative inverse, i.e.,

$$f^{-1}(x) = \frac{1}{x-3}$$

67. The inverse is the composition of adding 9 and raising an expression to the third power, i.e.,

$$f^{-1}(x) = (x+9)^3.$$

69. The inverse is the composition of adding 7, dividing by 2, and taking the cube root, i.e.,

$$f^{-1}(x) = \sqrt[3]{\frac{x+7}{2}}.$$

71. $A = d^2/2$

73. Let x be the number of ice cream bars sold. Then $W(x) = (0.20)(0.40x+200) = 0.08x+40$.

75. $C = 1.08P$ expresses the total cost as a function of the purchase price; and $P = C/1.08$ is the purchase price as a function of the total cost.

77.

$$r = \sqrt{5.625 \times 10^{-5} - \frac{V}{500}}$$

where $0 \le V \le 0.028125$

79. When $V = \$18,000$, the depreciation rate is

$$1 - \left(\frac{18,000}{50,000}\right)^{1/5} \approx 0.1848$$

or the depreciation rate is 18.48%.

Solving for V, we obtain

$$\begin{aligned} \left(\frac{V}{50,000}\right)^{1/5} &= 1-r \\ \frac{V}{50,000} &= (1-r)^5 \\ V &= 50,000(1-r)^5. \end{aligned}$$

81. Since $g^{-1}(x) = \dfrac{x+5}{3}$ and $f^{-1}(x) = \dfrac{x-1}{2}$, we have

$$g^{-1} \circ f^{-1}(x) = \frac{\frac{x-1}{2}+5}{3} = \frac{x+9}{6}.$$

Likewise, since $(f \circ g)(x) = 6x-9$, we get

$$(f \circ g)^{-1}(x) = \frac{x+9}{6}.$$

Hence, $(f \circ g)^{-1} = g^{-1} \circ f^{-1}$.

Chapter P Review Exercises

1. The distance is $\sqrt{(-3-2)^2 + (5-(-6))^2} = \sqrt{(-5)^2 + 11^2} = \sqrt{25+121} = \sqrt{146}$.

The midpoint is $\left(\dfrac{-3+2}{2}, \dfrac{5-6}{2}\right) = \left(-\dfrac{1}{2}, -\dfrac{1}{2}\right)$.

3. The distance is $\sqrt{\left(\pi - \dfrac{\pi}{2}\right)^2 + (1-1)^2} = \sqrt{\left(\dfrac{\pi}{2}\right)^2 + 0} = \dfrac{\pi}{2}$. The midpoint is $\left(\dfrac{\pi/2+\pi}{2}, \dfrac{1+1}{2}\right) = \left(\dfrac{3\pi}{4}, 1\right)$.

5. Circle with center at $(0,0)$ and radius 3

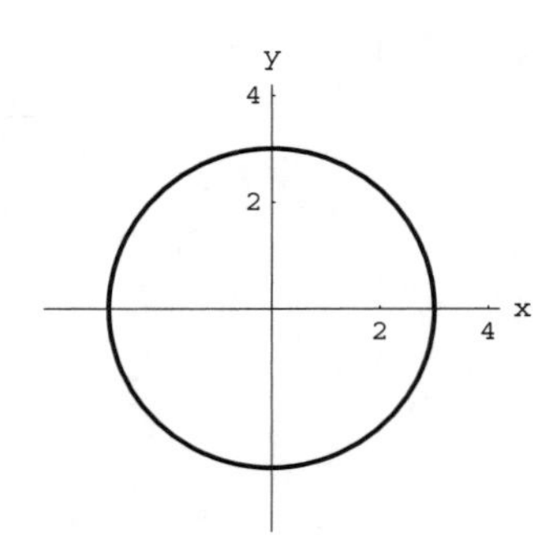

7. Circle with center at $(0, 1)$ and radius 1

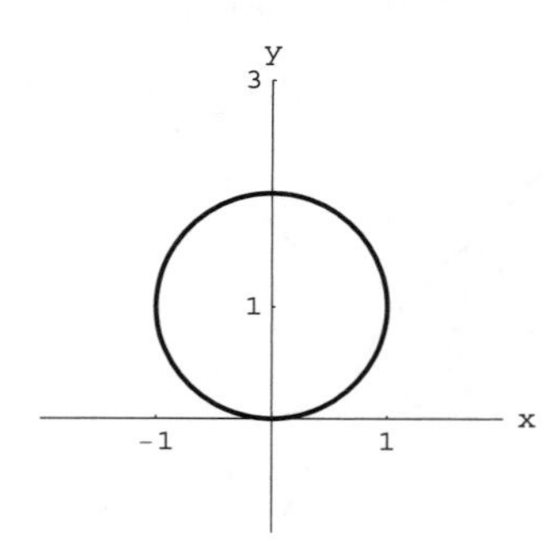

9. The line through the points $(10, 0)$ and $(0, -4)$.

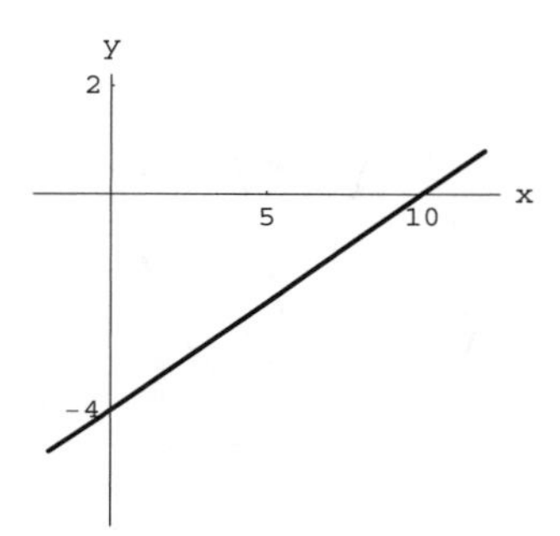

11. The vertical line through the point $(5, 0)$.

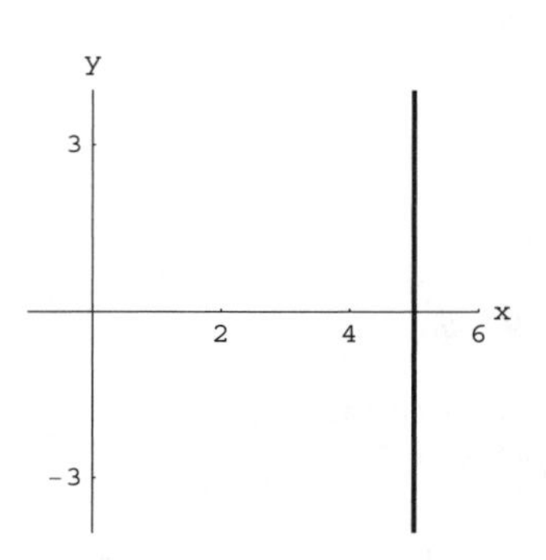

13. Domain $(-\infty, \infty)$ and range $(-\infty, \infty)$

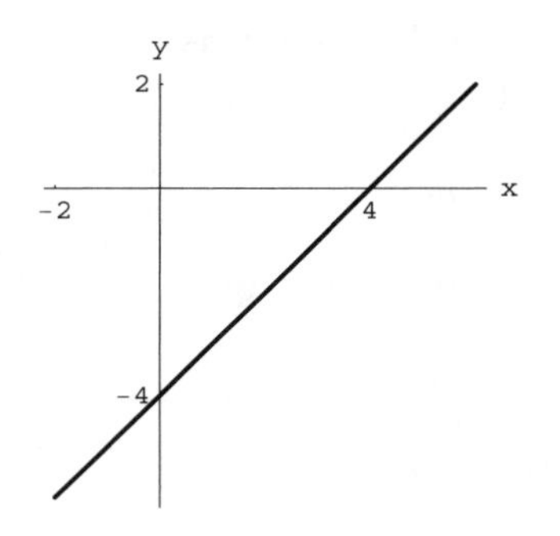

15. Domain $(-\infty, \infty)$ and range $\{4\}$

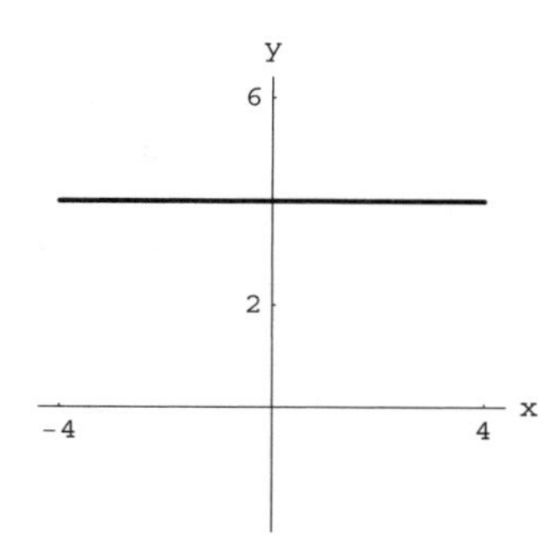

17. Domain $(-\infty, \infty)$ and range $[-3, \infty)$

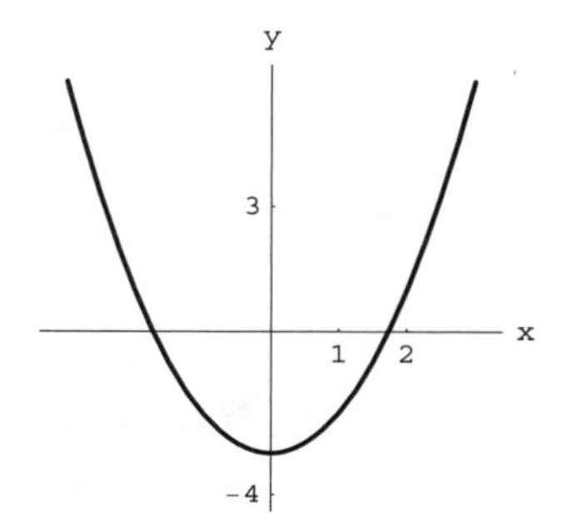

19. Domain $(-\infty, \infty)$ and range $(-\infty, \infty)$

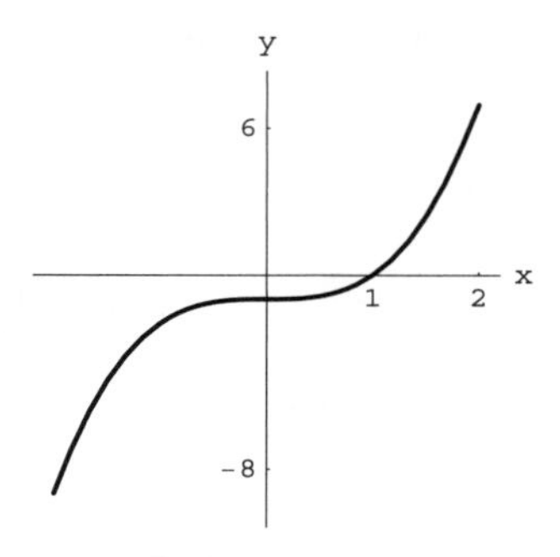

21. Domain $[1, \infty)$ and range $[0, \infty)$

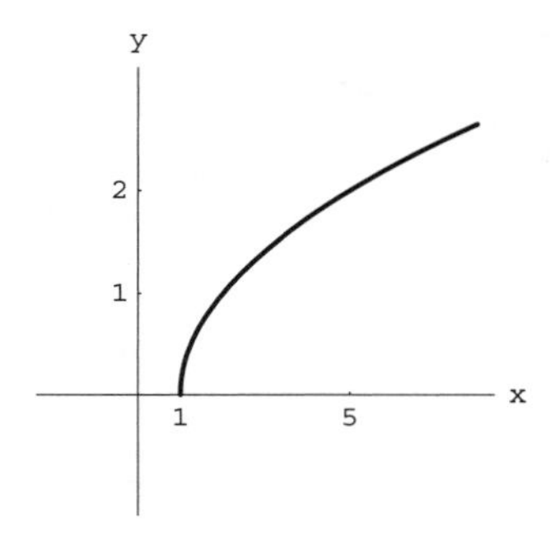

23. Domain $(-\infty, \infty)$ and range $[-4, \infty)$

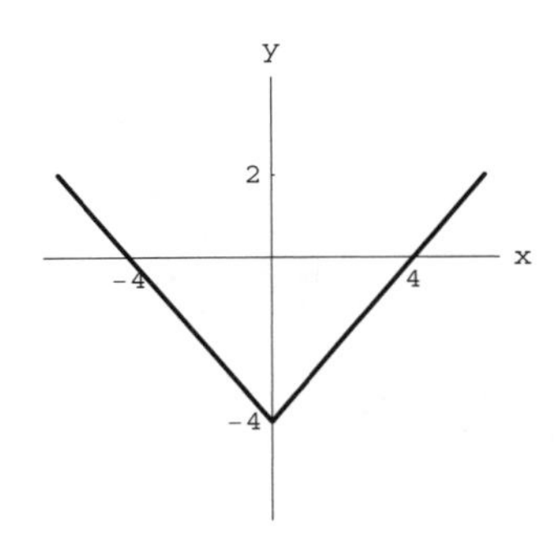

25. $9 + 3 = 12$

27. $24 - 7 = 17$

29. $g(12) = 17$

31. $f(-3) = 12$

33.

$$\begin{aligned} f(g(x)) &= f(2x-7) \\ &= (2x-7)^2 + 3 \\ &= 4x^2 - 28x + 52 \end{aligned}$$

35. $g\left(\dfrac{x+7}{2}\right) = (x+7) - 7 = x$

37. $g^{-1}(x) = \dfrac{x+7}{2}$

39. $f(x) = \sqrt{x}$, $g(x) = 2\sqrt{x+3}$; left by 3, stretch by 2

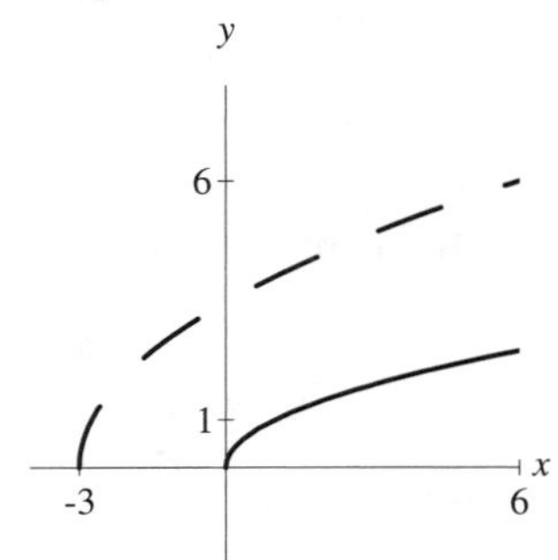

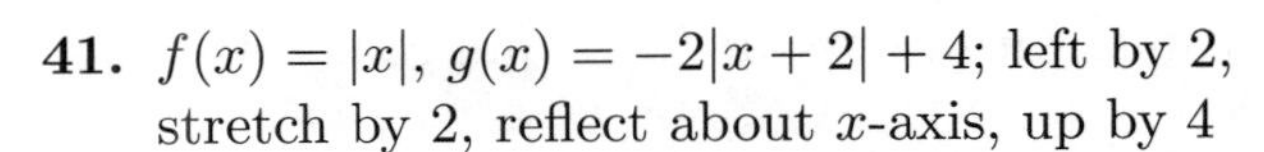
41. $f(x) = |x|$, $g(x) = -2|x+2| + 4$; left by 2, stretch by 2, reflect about x-axis, up by 4

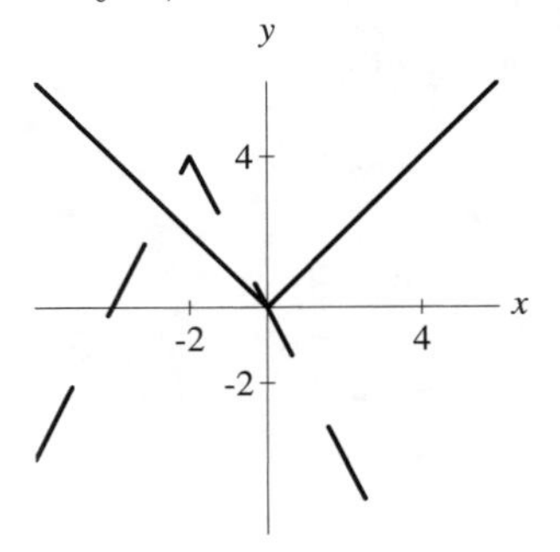

43. $f(x) = x^2$, $g(x) = (x-2)^2 + 1$; right by 2 and up by 1

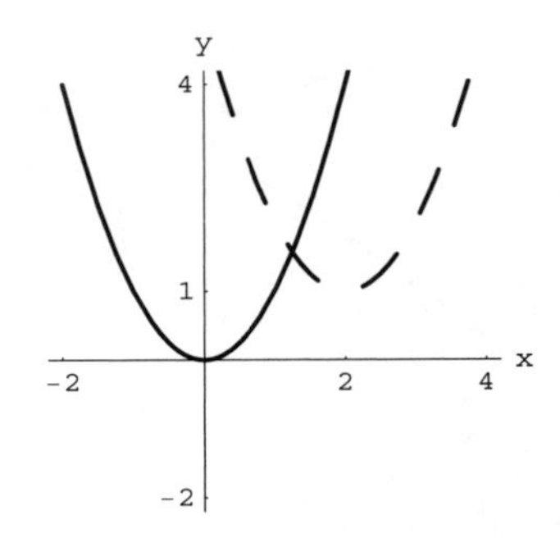

45. $N = h \circ j$

47. $R = h \circ g$

49. $F = f \circ g$

51. $H = f \circ g \circ j$

53. $y = |x| - 3$, domain is $(-\infty, \infty)$, range is $[-3, \infty)$

55. $y = -2|x| + 4$, domain is $(-\infty, \infty)$, range is $(-\infty, 4]$

57. $y = |x+2| + 1$, domain is $(-\infty, \infty)$, range is $[1, \infty)$

59. Symmetry: y-axis

61. Symmetric about the origin

63. Neither symmetry

65. Symmetric about the y-axis

67. inverse functions,
$f(x) = \sqrt{x+3}, g(x) = x^2 - 3$ for $x \geq 0$

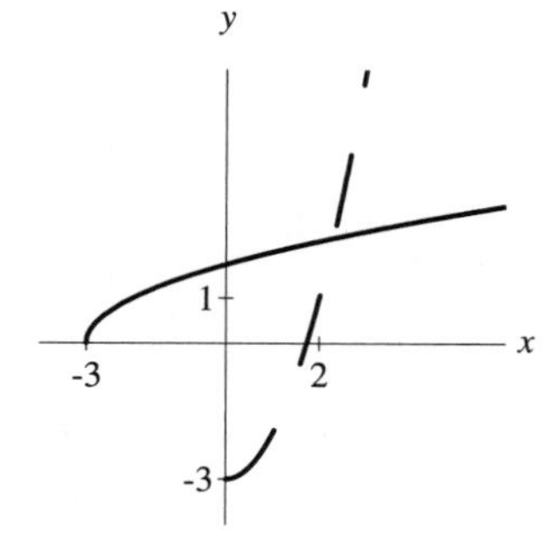

69. inverse functions,

$f(x) = 2x - 4, g(x) = \frac{1}{2}x + 2$

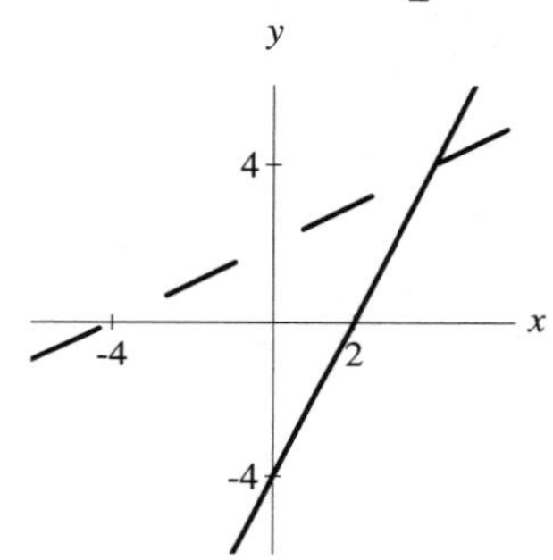

71. $f^{-1}(x) = \frac{x}{5}$

73. $f^{-1}(x) = \frac{x+21}{3}$ or $f^{-1}(x) = \frac{1}{3}x + 7$

75. $f^{-1}(x) = \sqrt[3]{x+1}$

77. $f^{-1}(x) = \frac{1}{x} + 3$

79. $f^{-1}(x) = \sqrt{x}$

81. $\sqrt{(\sqrt{5})^2 + 2^2} = \sqrt{9} = 3$

83. From $3x - 4(0) = 9$, we get that the x-intercept is $(3, 0)$. From $3(0) - 4y = 9$, we find that the y-intercept is $\left(0, -\frac{9}{4}\right)$.

85. $(x+3)^2 + (y-5)^2 = 3$

87. Since $C = 2\pi r$, we get $r = \frac{C}{2\pi}$.

Chapter P Test

1. Domain $(-\infty, \infty)$, range $[0, \infty)$

2. Domain $[9, \infty)$, range $[0, \infty)$

3. Domain $(-\infty, \infty)$, range $(-\infty, 2]$

4. The graph of $y = 2x - 3$ includes the points $(3/2, 0), (0, -3)$

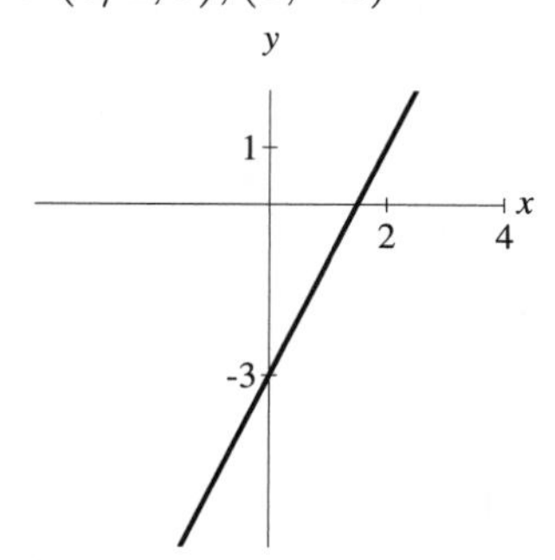

5. The graph of $f(x) = \sqrt{x-5}$ goes through $(5, 0)$, $(9, 2)$.

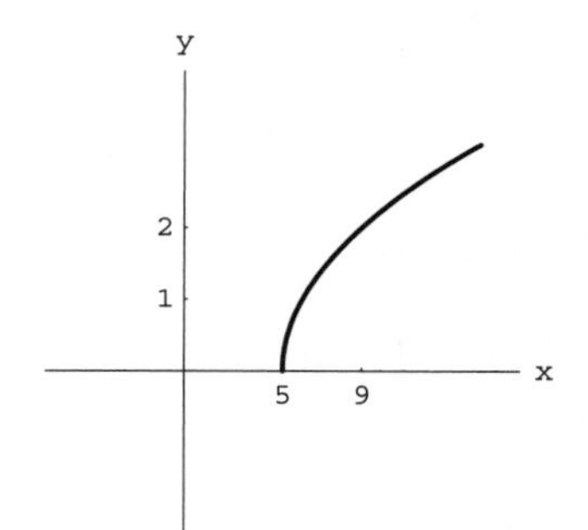

6. The graph of $y = 2|x| - 4$ includes the points $(0, -4), (\pm 3, 2)$

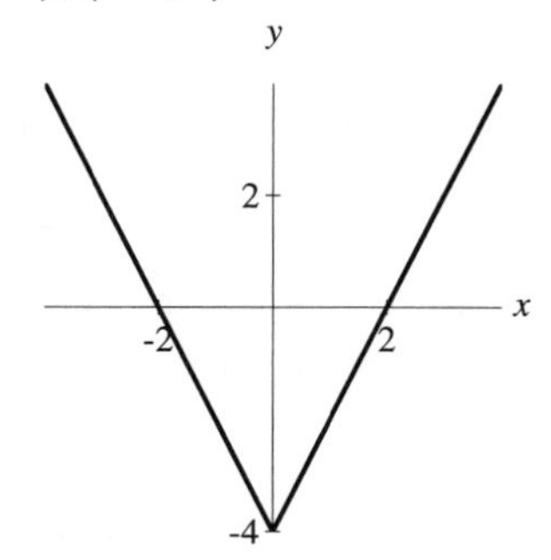

7. The graph of $y = -(x-2)^2 + 5$ is a parabola with vertex $(2, 5)$

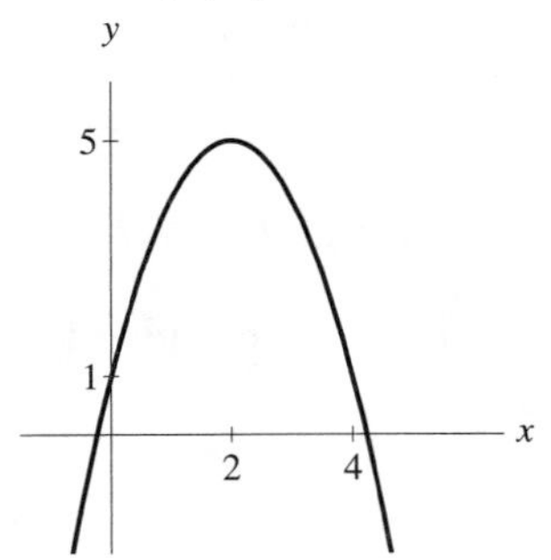

8. $\sqrt{9} = 3$

9. $3(2) - 1 = 5$

10. $f(5) = \sqrt{7}$

11. $g^{-1}(x) = \dfrac{x+1}{3}$ or $g^{-1}(x) = \dfrac{1}{3}x + \dfrac{1}{3}$

12. Using the answer from Exercise 11, we find $g^{-1}(20) = \dfrac{20+1}{3} = 7$.

13. $\left(\dfrac{3+(-3)}{2}, \dfrac{5+6}{2}\right) = \left(0, \dfrac{11}{2}\right)$

14. $\sqrt{(-2-3)^2 + (4-(-1)^2} = \sqrt{25+25} = 5\sqrt{2}$.

15. $(x+4)^2 + (y-1)^2 = 7$

16. Center $(4, -1)$, radius $\sqrt{6}$

17. Solving $3a - 9 = 6$, we obtain $a = 5$.

18. Symmetric about the y-axis since $f(-x) = f(x)$

19. Interchange x and y then solve for y.

$$\begin{aligned} x &= 3 + \sqrt[3]{y-2} \\ x - 3 &= \sqrt[3]{y-2} \\ (x-3)^3 &= y - 2 \\ (x-3)^3 + 2 &= y \end{aligned}$$

The inverse is $g^{-1}(x) = (x-3)^3 + 2$.

20. Since $A = \left(\dfrac{p}{4}\right)^2$, we get $A = \dfrac{p^2}{16}$.

For Thought

1. True **2.** True

3. True, since one complete revolution is $360°$.

4. False, the number of degrees in the intercepted arc of a circle is the degree measure.

5. False, the degree measure is negative if the rotation is clockwise.

6. True, the terminal side of $540°$ lies in the negative x-axis.

7. True

8. False, since $-365°$ lies in quadrant IV while $5°$ lies in quadrant I.

9. True, since $25°60' + 6' = 26°6'$.

10. False, since $25 + 20/60 + 40/3600 = 25.3444....$

1.1 Exercises

1. Substitute $k = 1, 2, -1, -2$ into $60° + k \cdot 360°$. Coterminal angles are $420°, 780°, -300°, -660°$. There are other coterminal angles.

3. Substitute $k = 1, 2, -1, -2$ into $30° + k \cdot 360°$. Coterminal angles are $390°, 750°, -330°, -690°$. There are other coterminal angles.

5. Substitute $k = 1, 2, -1, -2$ into $225° + k \cdot 360°$. Coterminal angles are $585°, 945°, -135°, -495°$. There are other coterminal angles.

7. Substitute $k = 1, 2, -1, -2$ into $-45° + k \cdot 360°$. Coterminal angles are $315°, 675°, -405°, -765°$. There are other coterminal angles.

9. Substitute $k = 1, 2, -1, -2$ into $-90° + k \cdot 360°$. Coterminal angles are $270°, 630°, -450°, -810°$. There are other coterminal angles.

11. Substitute $k = 1, 2, -1, -2$ into $-210° + k \cdot 360°$. Coterminal angles are $150°, 510°, -570°, -930°$. There are other coterminal angles.

13. Yes, since $40° - (-320°) = 360°$

15. No, since $4° - (-364°) = 368° \neq k \cdot 360°$ for any integer k.

17. Yes, since $1235° - 155° = 3 \cdot 360°$.

19. Yes, since $22° - (-1058)° = 3 \cdot 360°$.

21. No, since $312.4° - (-227.6°) = 540° \neq k \cdot 360°$ for any integer k.

23. Quadrant I

25. $-125°$ lies in Quadrant III since $-125° + 360° = 235°$ and $180° < 235° < 270°$

27. -740 lies in Quadrant IV since $-740° + 2 \cdot 360° = -20°$ and $-20°$ lies in Quadrant IV.

29. $933°$ lies in Quadrant III since $933° - 2 \cdot 360° = 213°$ and $213°$ lies in Quadrant III.

31. $-310°$, since $50° - 360° = -310°$

33. $-220°$, since $140° - 360° = -220°$

35. $-90°$, since $270° - 360° = -90°$

37. $40°$, since $400° - 360° = 40°$

39. $20°$, since $-340° + 360° = 20°$

41. $340°$, since $-1100° + 4 \cdot 360° = 340°$

43. $180.54°$, since $900.54° - 2 \cdot 360° = 180.54°$

45. c **47.** e

49. h **51.** g

53. $13° + \dfrac{12}{60}^{\circ} = 13.2°$

55. $-8° - \dfrac{30}{60}^{\circ} - \dfrac{18}{3600}^{\circ} = -8.505°$

57. $28° + \dfrac{5}{60}^{\circ} + \dfrac{9}{3600}^{\circ} \approx 28.0858°$

59. $155° + \dfrac{34}{60}^{\circ} + \dfrac{52}{3600}^{\circ} \approx 155.5811°$

61. $75.5° = 75°30'$ since $0.5(60) = 30$

63. $39.4° = 39°24'$ since $0.4(60) = 24$

65. $-17.33° = -17°19'48''$ since $0.33(60) = 19.8$ and $0.8(60) = 48$

67. $18.123° \approx 18°7'23''$ since $0.123(60) = 7.38$ and $0.38(60) \approx 23$

69. $24°15' + 33°51' = 57°66' = 58°6'$

71. $55°11' - 23°37' = 54°71' - 23°37' = 31°34'$

73. $16°23'41'' + 44°43'39'' = 60°66'80'' = 60°67'20'' = 61°7'20''$

75. $90° - 7°44'35'' = 89°59'60'' - 7°44'35'' = 82°15'25''$

77. $66°43'6'' - 5°51'53'' = 65°102'66'' - 5°51'53'' = 60°51'13''$

79. $2(32°36'37'') = 64°72'74'' = 64°73'14'' = 65°13'14''$

81. $3(15°53'42'') = 45°159'126'' = 45°161'6'' = 47°41'6''$

83. $(43°13'8'')/2 = (42°73'8'')/2 = (42°72'68'')/2 = 21°36'34''$

85. $(13°10'9'')/3 = (12°70'9'')/3 = (12°69'69'')/3 = 4°23'23''$

87. $\alpha = 180° - 88°40' - 37°52' = 180° - 126°32' = 179°60' - 126°32' = 53°28'$

89. $\alpha = 180° - 90° - 48°9'6'' = 89°59'60'' - 48°9'6'' = 41°50'54''$

91. $\alpha = 180° - 140°19'16'' = 179°59'60'' - 140°19'16'' = 39°40'44''$

93. $\alpha = 90° - 75°5'6'' = 89°59'60'' - 75°5'6'' = 14°54'54''$

95. Since $0.17647(60) = 10.5882$ and $0.5882(60) = 35.292$, we find $21.17647° \approx 21°10'35.3''$.

97. Since $73°37' \approx 73.6167°$, $49°41' \approx 49.6833°$, and $56°42' = 56.7000°$, the sum of the numbers in decimal format is $180°$. Also,

$$73°37' + 49°41' + 56°42' = 178°120' = 180°.$$

99. We find $108°24'16'' \approx 108.4044°$, $68°40'40'' \approx 68.6778°$, $84°42'51'' \approx 84.7142°$, and $98°12'13'' \approx 98.2036°$. The sum of these four numbers in decimal format is $360°$. Also, $108°24'16'' + 68°40'40'' + 84°42'51'' + 98°12'13'' = 360°$.

For Thought

1. False, in a negative angle the rotation is clockwise.

2. False, the radius is 1.

3. True, since the circumeference is $2\pi r$ where r is the radius.

4. True, since $s = \alpha r$ **5.** True

6. False, one must multiply by $\frac{\pi}{180}$.

7. True **8.** False, rather $45° = \frac{\pi}{4}$ rad.

9. True

10. True, rather the length of arc is $s = \alpha \cdot r = \frac{\pi}{4} \cdot 4 = 1$.

1.2 Exercises

1. $30° = \frac{\pi}{6}$, $45° = \frac{\pi}{4}$, $60° = \frac{\pi}{3}$, $90° = \frac{\pi}{2}$, $120° = \frac{2\pi}{3}$, $135° = \frac{3\pi}{4}$, $150° = \frac{5\pi}{6}$, $180° = \pi$, $210° = \frac{7\pi}{6}$, $225° = \frac{5\pi}{4}$, $240° = \frac{4\pi}{3}$, $270° = \frac{3\pi}{2}$, $300° = \frac{5\pi}{3}$, $315° = \frac{7\pi}{4}$, $330° = \frac{11\pi}{6}$, $360° = 2\pi$

3. $45 \cdot \frac{\pi}{180} = \frac{\pi}{4}$

5. $90 \cdot \frac{\pi}{180} = \frac{\pi}{2}$

7. $120 \cdot \frac{\pi}{180} = \frac{2\pi}{3}$

9. $150 \cdot \frac{\pi}{180} = \frac{5\pi}{6}$

11. $18 \cdot \frac{\pi}{180} = \frac{\pi}{10}$

13. $\frac{\pi}{3} \cdot \frac{180}{\pi} = 60°$

15. $\frac{5\pi}{12} \cdot \frac{180}{\pi} = 75°$

17. $\frac{3\pi}{4} \cdot \frac{180}{\pi} = 135°$

19. $-6\pi \cdot \frac{180}{\pi} = -1080°$

21. $2.39 \cdot \frac{180}{\pi} \approx 136.937°$

23. $-0.128 \cdot \frac{180}{\pi} \approx -7.334°$

25. $37.4\left(\frac{\pi}{180}\right) \approx 0.653$

27. $\left(-13 - \frac{47}{60}\right) \cdot \frac{\pi}{180} \approx -0.241$

29. $\left(53 + \frac{37}{60} + \frac{6}{3600}\right) \cdot \frac{\pi}{180} \approx 0.936$

31. Substitute $k = 1, 2, -1, -2$ into $\frac{\pi}{3} + k \cdot 2\pi$, coterminal angles are $\frac{7\pi}{3}, \frac{13\pi}{3}, -\frac{5\pi}{3}, -\frac{11\pi}{3}$. There are other coterminal angles.

33. Substitute $k = 1, 2, -1, -2$ into $\frac{\pi}{2} + k \cdot 2\pi$, coterminal angles are $\frac{5\pi}{2}, \frac{9\pi}{2}, -\frac{3\pi}{2}, -\frac{7\pi}{2}$. There are other coterminal angles.

35. Substitute $k = 1, 2, -1, -2$ into $\frac{2\pi}{3} + k \cdot 2\pi$, coterminal angles are $\frac{8\pi}{3}, \frac{14\pi}{3}, -\frac{4\pi}{3}, -\frac{10\pi}{3}$. There are other coterminal angles.

37. Substitute $k = 1, 2, -1, -2$ into $1.2 + k \cdot 2\pi$, coterminal angles are about 7.5, 13.8, -5.1, -11.4. There are other coterminal angles.

39. Quadrant I

41. Quadrant III

43. $-\frac{13\pi}{8}$ lies in Quadrant I since $-\frac{13\pi}{8} + 2\pi = \frac{3\pi}{8}$

45. $\frac{17\pi}{3}$ lies in Quadrant IV since $\frac{17\pi}{3} - 4\pi = \frac{5\pi}{3}$

47. 3 lies in Quadrant II since $\frac{\pi}{2} < 3 < \pi$

49. -7.3 lies in Quadrant IV since $-7.3 + 4\pi \approx 5.3$ and 5.3 lies in Qudrant IV

51. g **53.** b

55. h **57.** d

59. π, since $3\pi - 2\pi = \pi$ and π is the smallest positive coterminal angle

61. $\frac{3\pi}{2}$, since $-\frac{\pi}{2} + 2\pi = \frac{3\pi}{2}$

63. $\frac{9\pi}{2} - 4\pi = \frac{\pi}{2}$

65. $-\frac{5\pi}{3} + 2\pi = \frac{\pi}{3}$

67. $-\frac{13\pi}{3} + 6\pi = \frac{5\pi}{3}$

69. $8.32 - 2\pi \approx 2.04$

71. $\frac{4\pi}{4} - \frac{\pi}{4} = \frac{3\pi}{4}$

73. $\frac{3\pi}{6} + \frac{2\pi}{6} = \frac{5\pi}{6}$

75. $\frac{2\pi}{4} + \frac{\pi}{4} = \frac{3\pi}{4}$

77. $\frac{\pi}{3} - \frac{3\pi}{3} = -\frac{2\pi}{3}$

79. $s = 12 \cdot \frac{\pi}{4} = 3\pi \approx 9.4$ ft

81. $s = 4000 \cdot \frac{3\pi}{180} \approx 209.4$ miles

83. $s = (26.1)(1.3) \approx 33.9$ m

85. radius is $r = \frac{s}{\alpha} = \frac{1}{1} = 1$ mile.

87. radius is $r = \frac{s}{\alpha} = \frac{10}{\pi} \approx 3.2$ km

89. radius is $r = \frac{s}{\alpha} = \frac{500}{\pi/6} \approx 954.9$ ft

91. $A = \frac{\alpha r^2}{2} = \frac{(\pi/6)6^2}{2} = 3\pi$

93. $A = \frac{\alpha r^2}{2} = \frac{(\pi/3)12^2}{2} = 24\pi$

95. Distance from Peshtigo to the North Pole is

$$s = r\alpha = 3950\left(45 \cdot \frac{\pi}{180}\right) \approx 3102 \text{ miles.}$$

97. Central angle is $\alpha = \frac{2000}{3950} \approx 0.506329$ radians

$\approx 0.506329 \cdot \frac{180}{\pi} \approx 29.0^\circ$

99. Since $7^\circ \approx 0.12217305$, the radius of the earth according to Eratosthenes is

$$r = \frac{s}{\alpha} \approx \frac{800}{0.12217305} \approx 6548.089 \text{ km.}$$

Thus, using Eratosthenes' radius, the circumference is

$$2\pi r \approx 41,143 \text{ km.}$$

Using $r = 6378$ km, circumference is 40,074 km.

101. Note, the fraction of the area of the circle of radius r intercepted by the central angle $\frac{\pi}{6}$ is $\frac{1}{12} \cdot \pi r^2$. So, the area watered in one hour is

$$\frac{1}{12} \cdot \pi 150^2 \approx 5890 \text{ ft}^2.$$

103. Note, the radius of the pizza is 8 in. Then the area of each of the six slices is

$$\frac{1}{6} \cdot \pi 8^2 \approx 33.5 \text{ in.}^2$$

105. **a)** Given angle α (in degrees) as in the problem, the radius r of the cone must satisfy $2\pi r = 8\pi - 4\alpha\frac{\pi}{180}$; note, 8π in. is the circumference of a circle with radius 4 in. Then $r = 4 - \frac{\alpha}{90}$. Note, $h = \sqrt{16 - r^2}$ by the Pythagorean theorem. Since the volume $V(\alpha)$ of the cone is $\frac{\pi}{3}r^2 h$, $V(\alpha) = \frac{\pi}{3}\left(4 - \frac{\alpha}{90}\right)^2\sqrt{16 - \left(4 - \frac{\alpha}{90}\right)^2}$. This reduces to

$$V(\alpha) = \frac{\pi(360 - \alpha)^2\sqrt{720\alpha - \alpha^2}}{2,187,000}$$

If $\alpha = 30^\circ$, then $V(30^\circ) \approx 22.5$ inches^3.

b) As shown in part a), the volume of the cone obtained by an overlapping angle α is

$$V(\alpha) = \frac{\pi(360 - \alpha)^2\sqrt{720\alpha - \alpha^2}}{2,187,000}$$

c) The volume of the cone is maximized when $\alpha \approx 66.06^\circ$.

d) The maximum volume is approximately $V(66.06^\circ) \approx 25.8$ cubic inches.

For Thought

1. False, $\frac{240 \text{ rev}}{\text{min}} = \frac{(240)(6) \text{ rev}}{\text{hr}}$.

2. True

3. False, since $\frac{4 \text{ rev}}{\text{sec}} \cdot \frac{2\pi \text{ ft}}{1 \text{ rev}} = \frac{8\pi \text{ ft}}{\text{sec}}$.

4. False, since $\frac{5\pi \text{ rad}}{1 \text{ hr}} \cdot \frac{60 \text{ min}}{1 \text{ hr}} = \frac{300\pi \text{ rad} \cdot \text{min}}{\text{hr}^2}$.

5. False, it is angular velocity.

6. False, it is linear velocity.

7. False, since 40 inches/second is linear velocity.

8. True, since 1 rev/sec is equivalent to $\omega = 2\pi$ radians/sec, we get that the linear velocity is $v = r\omega = 1 \cdot 2\pi = 2\pi$ ft/sec.

9. True

10. False, Miami has a faster linear velocity than Boston. Note, Miami's distance from the axis of the earth is farther than that of Boston's.

1.3 Exercises

1. $\frac{300 \text{ rad}}{60 \text{ min}} = 5 \text{ rad/min}$

3. $\frac{4(2\pi) \text{ rad}}{\text{sec}} = 8\pi \text{ rad/sec}$

5. $\frac{55(6\pi) \text{ ft}}{\text{min}} \approx 1036.7 \text{ ft/min}$

7. $\frac{10(60)\text{ rev}}{2\pi\text{ hr}} \approx 95.5\text{ rev/hr}$

9. $\frac{30\text{ rev}}{\text{min}} \cdot \frac{2\pi\text{ rad}}{\text{rev}} = 60\pi \approx 188.5\text{ rad/min}$

11. $\frac{120\text{ rev}}{\text{hr}} \cdot \frac{1\text{ hr}}{60\text{ min}} = 2\text{ rev/min}$

13. $\frac{180\text{ rev}}{\text{sec}} \cdot \frac{3600\text{ sec}}{1\text{ hr}} \cdot \frac{2\pi\text{ rad}}{1\text{ rev}} \approx$ $4,071,504.1\text{ rad/hr}$

15. $\frac{30\text{ mi}}{\text{hr}} \cdot \frac{1\text{ hr}}{3600\text{ sec}} \cdot \frac{5280\text{ ft}}{1\text{ mi}} \approx$ 44 ft/sec

17. $\frac{500\text{ rev}}{\text{sec}} \cdot \frac{2\pi\text{ rad}}{1\text{ rev}} = 1000\pi \approx 3141.6\text{ rad/sec}$

19. $\frac{433.2\text{ rev}}{\text{min}} \cdot \frac{1\text{ min}}{60\text{ sec}} \cdot \frac{2\pi\text{ rad}}{1\text{ rev}} \approx$ 45.4 rad/sec

21. $\frac{50,000\text{ rev}}{\text{day}} \cdot \frac{1\text{ day}}{3600(24)\text{ sec}} \cdot \frac{2\pi\text{ rad}}{1\text{ rev}} \approx$ 3.6 rad/sec

23. Convert rev/min to rad/hr:

$$\frac{3450\text{ rev}}{\text{min}} \cdot \frac{60\text{ min}}{1\text{ hr}} \cdot \frac{2\pi\text{ rad}}{1\text{ rev}} = \frac{120\pi(3450)\text{ rad}}{1\text{ hr}}.$$

Since arc length is $s = r\alpha$, linear velocity is

$$v = \left((3\text{ in.}) \cdot \frac{1\text{ mi}}{5280(12)\text{ in.}}\right) \cdot \frac{120\pi(3450)\text{ rad}}{1\text{ hr}}$$

≈ 61.6 mph.

25. Convert rev/min to rad/hr:

$$\frac{3450\text{ rev}}{\text{min}} \cdot \frac{60\text{ min}}{1\text{ hr}} \cdot \frac{2\pi\text{ rad}}{1\text{ rev}} = \frac{120\pi(3450)\text{ rad}}{1\text{ hr}}.$$

Since arc length is $s = r\alpha$, linear velocity is

$$v = \left((5\text{ in.}) \cdot \frac{1\text{ mi}}{5280(12)\text{ in.}}\right) \cdot \frac{120\pi(3450)\text{ rad}}{1\text{ hr}}$$

≈ 102.6 mph.

27. Note,

$$\frac{3450\text{ rev}}{\text{min}} \cdot \frac{60\text{ min}}{1\text{ hr}} \cdot \frac{2\pi\text{ rad}}{1\text{ rev}} = \frac{120\pi(3450)\text{ rad}}{1\text{ hr}}$$

and arc length is $s = r\alpha$.
The linear velocity is

$$v = \left((7\text{ in.}) \cdot \frac{1\text{ mi}}{5280(12)\text{ in.}}\right) \cdot \frac{120\pi(3450)\text{ rad}}{1\text{ hr}}$$

≈ 143.7 mph

29. The angular velocity is

$$w = 45\,(2\pi) = 90\pi \approx 282.7\text{ rad/min}.$$

Sinc arc length $s = r\alpha$, linear velocity is

$$v = 6.25(90\pi) \approx 918.9\text{ in./min}.$$

31. Linear velocity is

$$v = (125\text{ft}) \cdot \frac{2\text{ rev}}{\text{hr}} \cdot \frac{2\pi\text{ rad}}{1\text{ rev}} \cdot \frac{1\text{ hr}}{3600\text{ sec}} \approx$$

0.4 ft/sec.

33. The radius of the bit is $r = 0.5$ in. $=$ $0.5 \cdot \frac{1}{12 \cdot 5280}$ mi. ≈ 0.0000079 mi.
Since the angle made in one hour is $\alpha = 45,000 \cdot 2\pi \cdot 60 = 5,400,000\pi$, we get

$$v = r\alpha \approx (0.0000079) \cdot (5,400,000\pi)$$

≈ 133.9 mph.

35. The radius of the tire in miles is

$$r = \frac{13}{12(5280)} \approx 0.00020517677\text{ mi}.$$

So $w = \frac{\alpha}{t} = \frac{\alpha}{1} = \alpha = \frac{s}{r} \approx$

$$\frac{55}{0.00020517677} \approx 268,061.5\text{ radians/hr}.$$

37. The angular velocity is $w = \frac{\pi}{12}$ rad/hr or about 0.26 rad/hr. Let r be the distance between Peshtigo and the point on the x-axis closest to Peshtigo. Since Peshtigo is on the 45th parallel, that point on the x-axis is r miles from the center of the earth.
By the Pythagorean Theorem, $r^2 + r^2 = 3950^2$ or $r \approx 2,793.0718$ miles.
The linear velocity is

$$v = w \cdot r \approx \frac{\pi}{12} \cdot 2,793.0718 \approx 731.2\text{ mph}.$$

39. **a)** The linear velocity is $v = r\alpha =$ (10 meters)$(3(2\pi)$ rad/min$) =$ 60π meters/minute.

b) Angular velocity is $w = 3(2\pi) =$ 6π rad/min.

c) The arc length between two adjacent seats is $s = r\alpha = 10 \cdot \frac{2\pi}{8} = \frac{5\pi}{2}$ meters.

41. Since the velocity at point A is 10 ft/sec, the linear velocity at B and C are both 10 ft/sec. The angular velocity at B is

$$\omega = \frac{v}{r} = \frac{10 \text{ ft/sec}}{3/12 \text{ ft}} = 40 \text{ rad/sec}.$$

and the angular velocity at C is

$$\omega = \frac{v}{r} = \frac{10 \text{ ft/sec}}{5/12 \text{ ft}} = 24 \text{ rad/sec}$$

43. Since the chain ring which has 52 teeth turns at the rate of 1 rev/sec, the cog with 26 teeth will turn at the rate of 2 rev/sec. Thus, the linear velocity of the bicycle with 13.5-in.-radius wheels is

$$\frac{2\text{rev}}{\text{sec}} \cdot \frac{2\pi(13.5)\text{in.}}{\text{rev}} \cdot \frac{1\text{mile}}{63,360\text{in.}} \cdot \frac{3600\text{sec}}{1\text{hr}} \approx 9.6 \text{ mph}.$$

For Thought

1. True, since $(0,1)$ is on the positive y-axis and $\sin 90° = \frac{y}{r} = \frac{1}{1} = 1.$

2. True, since $(1,0)$ is on the positive x-axis and $\sin 0° = \frac{x}{r} = \frac{0}{1} = 0.$

3. True, since $(1,1)$ is on the terminal side of 45° and $r = \sqrt{2}$, and so

$$\cos 45° = \frac{x}{r} = \frac{1}{\sqrt{2}}.$$

4. True, since $(1,1)$ is on the terminal side of 45° we get

$$\tan 45° = \frac{y}{x} = \frac{1}{1} = 1.$$

5. True, since $(\sqrt{3},1)$ is on the terminal side of 30° and $r = 2$ we find

$$\sin 30° = \frac{y}{r} = \frac{1}{2}.$$

6. True, since $(0,1)$ is on the positive y-axis we get $\cos(\pi/2) = \frac{x}{r} = \frac{0}{1} = 1.$

7. True, since the terminal sides of 390° and 30° are the same.

8. False, since $\sin(-\pi/3) = -\frac{\sqrt{3}}{2}$ and $\sin(\pi/3) = \frac{\sqrt{3}}{2}.$

9. True, since $\csc\alpha = \frac{1}{\sin\alpha} = \frac{1}{1/5} = 5.$

10. True, since $\sec\alpha = \frac{1}{\cos\alpha} = \frac{1}{2/3} = 1.5.$

1.4 Exercises

1. Note $r = \sqrt{1^2+2^2} = \sqrt{5}$. Then

$$\sin\alpha = \frac{y}{r} = \frac{2}{\sqrt{5}} = \frac{2\sqrt{5}}{5},$$
$$\cos\alpha = \frac{x}{r} = \frac{1}{\sqrt{5}} = \frac{\sqrt{5}}{5},$$
$$\tan\alpha = \frac{y}{x} = \frac{2}{1} = 2, \ \csc\alpha = \frac{1}{\sin\alpha} = \frac{\sqrt{5}}{2},$$
$$\sec\alpha = \frac{1}{\cos\alpha} = \frac{\sqrt{5}}{1} = \sqrt{5}, \text{ and}$$
$$\cot\alpha = \frac{1}{\tan\alpha} = \frac{1}{2}.$$

3. Note $r = \sqrt{0^2+1^2} = 1$. Then

$$\sin\alpha = \frac{y}{r} = \frac{1}{1} = 1, \ \cos\alpha = \frac{x}{r} = \frac{0}{1} = 0,$$
$$\tan\alpha = \frac{y}{x} = \frac{1}{0} = \text{undefined},$$
$$\csc\alpha = \frac{1}{\sin\alpha} = \frac{1}{1} = 1,$$
$$\sec\alpha = \frac{1}{\cos\alpha} = -\frac{1}{0} = \text{undefined, and}$$
$$\cot\alpha = \frac{x}{y} = \frac{0}{1} = 0.$$

5. Note $r = \sqrt{1^2+1^2} = \sqrt{2}$. Then

$$\sin\alpha = \frac{y}{r} = \frac{1}{\sqrt{2}} = \frac{\sqrt{2}}{2},$$
$$\cos\alpha = \frac{x}{r} = \frac{1}{\sqrt{2}} = \frac{\sqrt{2}}{2},$$
$$\tan\alpha = \frac{y}{x} = \frac{1}{1} = 1,$$
$$\csc\alpha = \frac{1}{\sin\alpha} = \frac{1}{1/\sqrt{2}} = \sqrt{2},$$
$$\sec\alpha = \frac{1}{\cos\alpha} = \frac{1}{1/\sqrt{2}} = \sqrt{2}, \text{ and}$$
$$\cot\alpha = \frac{1}{\tan\alpha} = \frac{1}{1} = 1.$$

7. Note $r = \sqrt{(-2)^2 + 2^2} = 2\sqrt{2}$. Then

$$\sin\alpha = \frac{y}{r} = \frac{2}{2\sqrt{2}} = \frac{\sqrt{2}}{2},$$

$$\cos\alpha = \frac{x}{r} = \frac{-2}{2\sqrt{2}} = -\frac{\sqrt{2}}{2},$$

$$\tan\alpha = \frac{y}{x} = \frac{2}{-2} = -1,$$

$$\csc\alpha = \frac{1}{\sin\alpha} = \frac{1}{2/(2\sqrt{2})} = \sqrt{2},$$

$$\sec\alpha = \frac{1}{\cos\alpha} = \frac{1}{-2/(2\sqrt{2})} = -\sqrt{2}, \text{ and}$$

$$\cot\alpha = \frac{1}{\tan\alpha} = \frac{1}{-1} = -1.$$

9. Note $r = \sqrt{(-4)^2 + (-6)^2} = 2\sqrt{13}$. Then

$$\sin\alpha = \frac{y}{r} = \frac{-6}{2\sqrt{13}} = -\frac{3\sqrt{13}}{13},$$

$$\cos\alpha = \frac{x}{r} = \frac{-4}{2\sqrt{13}} = \frac{-2\sqrt{13}}{13},$$

$$\tan\alpha = \frac{y}{x} = \frac{-6}{-4} = \frac{3}{2},$$

$$\csc\alpha = \frac{1}{\sin\alpha} = -\frac{13}{3\sqrt{13}} = -\frac{\sqrt{13}}{3},$$

$$\sec\alpha = \frac{1}{\cos\alpha} = -\frac{13}{2\sqrt{13}} = -\frac{\sqrt{13}}{2}, \text{ and}$$

$$\cot\alpha = \frac{1}{\tan\alpha} = \frac{2}{3}.$$

11. 0 **13.** -1 **15.** 0

17. Undefined **19.** -1 **21.** -1

23. $\frac{\sqrt{2}}{2}$ **25.** $\frac{\sqrt{2}}{2}$

27. -1 **29.** $\sqrt{2}$

31. $\frac{1}{2}$ **33.** $\frac{1}{2}$

35. $-\frac{\sqrt{3}}{3}$ **37.** -2

39. 2 **41.** $\sqrt{3}$

43. $\frac{\cos(\pi/3)}{\sin(\pi/3)} = \frac{1/2}{\sqrt{3}/2} = \frac{\sqrt{3}}{3}$

45. $\frac{\sin(7\pi/4)}{\cos(7\pi/4)} = \frac{-\sqrt{2}/2}{\sqrt{2}/2} = -1$

47. $\sin\left(\frac{\pi}{3} + \frac{\pi}{6}\right) = \sin\left(\frac{\pi}{2}\right) = 1$

49. $\frac{1-\cos(5\pi/6)}{\sin(5\pi/6)} = \frac{1-(-\sqrt{3}/2)}{1/2} \cdot \frac{2}{2} = 2 + \sqrt{3}$

51. $\frac{\sqrt{2}}{2} + \frac{\sqrt{2}}{2} = \sqrt{2}$

53. $\cos(45^\circ)\cos(60^\circ) - \sin(45^\circ)\sin(60^\circ) =$

$$\frac{\sqrt{2}}{2} \cdot \frac{1}{2} - \frac{\sqrt{2}}{2} \cdot \frac{\sqrt{3}}{2} = \frac{\sqrt{2}-\sqrt{6}}{4}$$

55. $\frac{1-\cos(\pi/3)}{2} = \frac{1-1/2}{2} = \frac{1}{4}.$

57. $2\cos(210^\circ) = 2 \cdot \frac{-\sqrt{3}}{2} = -\sqrt{3}.$

59. 0.6820

61. -0.6366

63. 0.0105

65. $\frac{1}{\sin(23^\circ 48')} \approx 2.4780$

67. $\frac{1}{\cos(-48^\circ 3' 12'')} \approx 1.4960$

69. $\frac{1}{\tan(\pi/9)} \approx 2.7475$

71. 0.8578

73. 0.2679

75. 0.9894

77. 2.9992

79. $\sin(2 \cdot \pi/4) = \sin(\pi/2) = 1$

81. $\cos(2 \cdot \pi/6) = \cos(\pi/3) = \frac{1}{2}$

83. $\sin((3\pi/2)/2) = \sin(3\pi/4) = \frac{\sqrt{2}}{2}$

85.

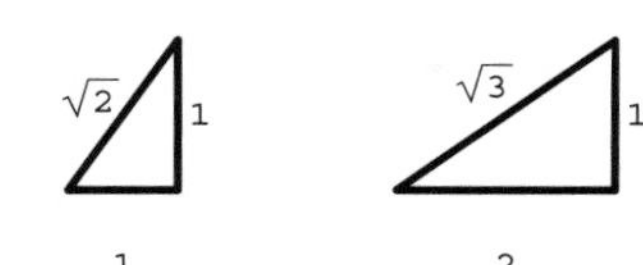

87. $\csc\alpha = \dfrac{1}{\sin\alpha} = \dfrac{1}{3/4} = \dfrac{4}{3}$

89. $\cos\alpha = \dfrac{1}{\sec\alpha} = \dfrac{1}{10/3} = \dfrac{3}{10}$

91. **a)** II, since $y > 0$ and $x < 0$ in Quadrant II

b) IV, since $y < 0$ and $x > 0$ in Quadrant IV

c) III, since $y/x > 0$ and $x < 0$ in Quadrant III

d) II, since $y/x < 0$ and $y > 0$ in Quadrant II

For Thought

1. True

2. False, rather $\alpha = \pi/6$ since $\cos(\pi/6) = \dfrac{\sqrt{3}}{2}$.

3. False, since $\sin^{-1}(\sqrt{2}/2) = 45°$.

4. False, since $\cos^{-1}(1/2) = 60°$.

5. False, since $\tan^{-1}(1) = 45°$.

6. False, since $c = \sqrt{2^2 + 4^2} = \sqrt{20}$.

7. True, since $c = \sqrt{3^2 + 4^2} = \sqrt{25} = 5$.

8. False, since $c = \sqrt{4^2 + 8^2} = \sqrt{80}$.

9. True, since $\alpha + \beta = 90°$.

10. False, otherwise $1 = \sin 90° = \dfrac{\text{hyp}}{\text{adj}}$ and we find hyp = adj, which is impossible. The hypotenuse is longer than each of the legs of a right triangle.

1.5 Exercises

1. 45° **3.** 60°

5. 60° **7.** 0°

9. 83.6°

11. 67.6°

13. 26.1°

15. 29.1°

17. Note, the hypotenuse is hyp $= \sqrt{13}$.

Then $\sin\alpha = \dfrac{\text{opp}}{\text{hyp}} = \dfrac{2}{\sqrt{13}} = \dfrac{2\sqrt{13}}{13}$,

$\cos\alpha = \dfrac{\text{adj}}{\text{hyp}} = \dfrac{3}{\sqrt{13}} = \dfrac{3\sqrt{13}}{13}$,

$\tan\alpha = \dfrac{\text{opp}}{\text{adj}} = \dfrac{2}{3}$,

$\csc\alpha = \dfrac{\text{hyp}}{\text{opp}} = \dfrac{\sqrt{13}}{2}$,

$\sec\alpha = \dfrac{\text{hyp}}{\text{adj}} = \dfrac{\sqrt{13}}{3}$, and

$\cot\alpha = \dfrac{\text{adj}}{\text{opp}} = \dfrac{3}{2}$.

19. Note, the hypotenuse is $4\sqrt{5}$. Then $\sin(\alpha) = \sqrt{5}/5, \cos(\alpha) = 2\sqrt{5}/5$, $\tan(\alpha) = 1/2$, $\sin(\beta) = 2\sqrt{5}/5$, $\cos(\beta) = \sqrt{5}/5$, and $\tan(\beta) = 2$.

21. Note, the hypotenuse is $2\sqrt{34}$. $\sin(\alpha) = 3\sqrt{34}/34$, $\cos(\alpha) = 5\sqrt{34}/34$, $\tan(\alpha) = 3/5$, $\sin(\beta) = 5\sqrt{34}/34$, $\cos(\beta) = 3\sqrt{34}/34$, and $\tan(\beta) = 5/3$.

23. Note, the side adjacent to β has length 12. Then $\sin(\alpha) = 4/5$, $\cos(\alpha) = 3/5$, $\tan(\alpha) = 4/3$, $\sin(\beta) = 3/5$, $\cos(\beta) = 4/5$, and $\tan(\beta) = 3/4$.

25. Form the right triangle with $a = 6$, $b = 10$.

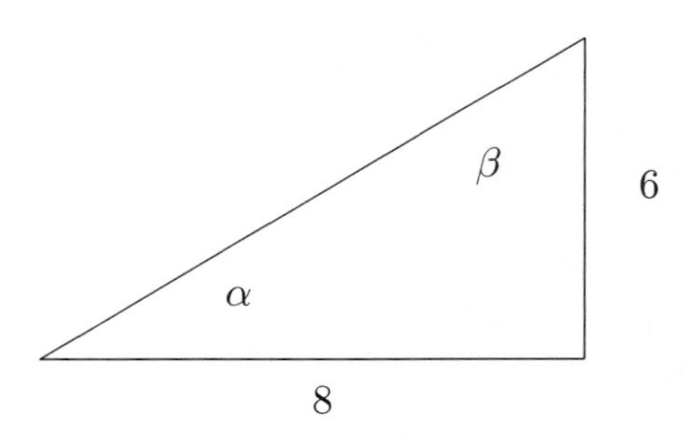

Note, we have

$$c = \sqrt{6^2 + 8^2} = 10$$

and $\tan(\alpha) = 6/8$. Then

$$\alpha = \tan^{-1}(6/8) \approx 36.9°$$

and $\beta \approx 53.1°$.

27. Form the right triangle with $b = 6$, $c = 8.3$.

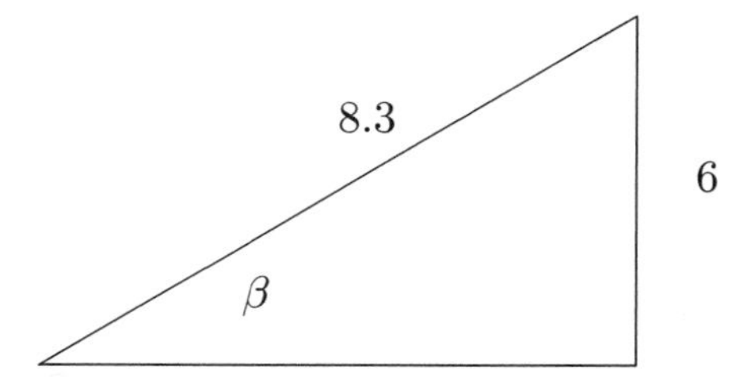

Note, we have

$$a = \sqrt{8.3^2 - 6^2} \approx 5.73$$

and $\sin(\beta) = 6/8.3$. Then $\beta = \sin^{-1}(6/8.3) \approx 46.3°$ and $\alpha \approx 43.7°$.

29. Form the right triangle with $\alpha = 16°$, $c = 20$.

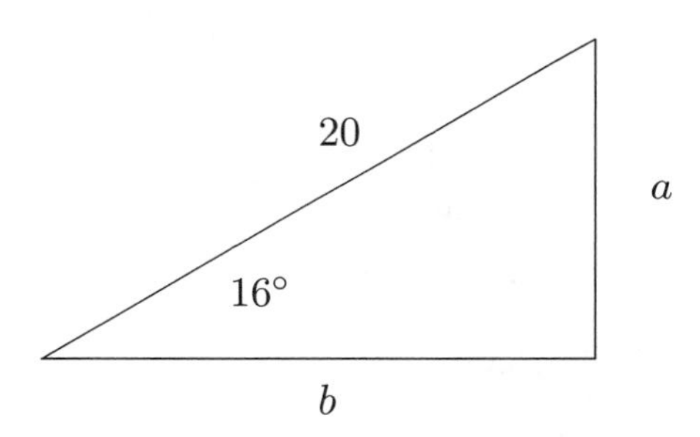

Since $\sin(16°) = a/20$ and $\cos(16°) = b/20$,
$a = 20\sin(16°) \approx 5.5$ and
$b = 20\cos(16°) \approx 19.2$. Also $\beta = 74°$.

31. Form the right triangle with $\alpha = 39°9'$, $a = 9$.

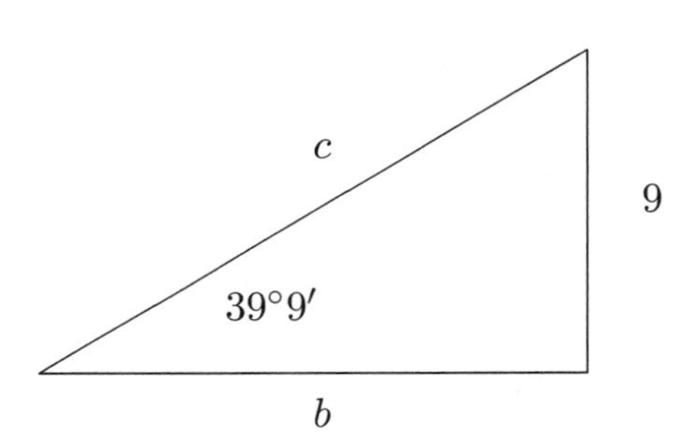

Since $\sin(39°9') = 9/c$ and $\tan(39°9') = 9/b$, then $c = 9/\sin(39°9') \approx 14.3$ and
$b = 9/\tan(39°9') \approx 11.1$. Also $\beta = 50°51'$.

33. Let h be the height of the buliding.

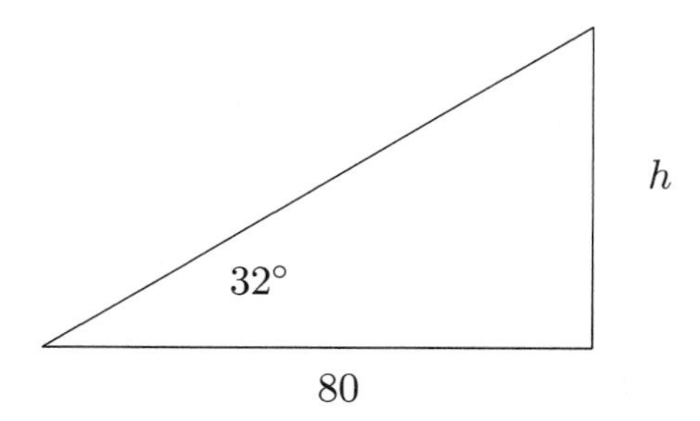

Since $\tan(32°) = h/80$, we obtain
$h = 80 \cdot \tan(32°) \approx 50$ ft.

35. Let x be the distance between Muriel and the road at the time she encountered the swamp.

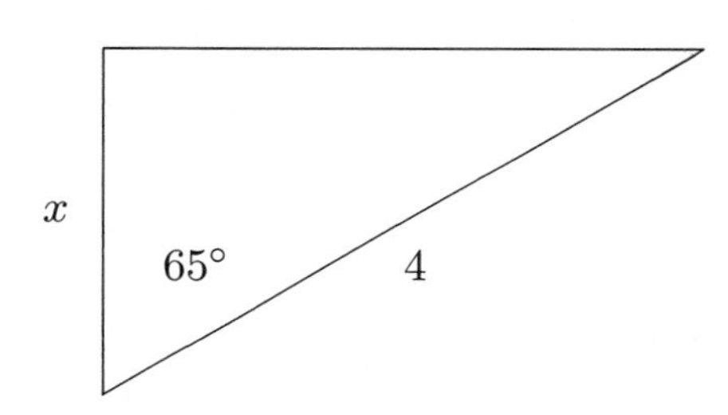

Since $\cos(65°) = x/4$, we obtain
$x = 4 \cdot \cos(65°) \approx 1.7$ miles.

37. Let x be the distance between the car and a point on the highway directly below the overpass.

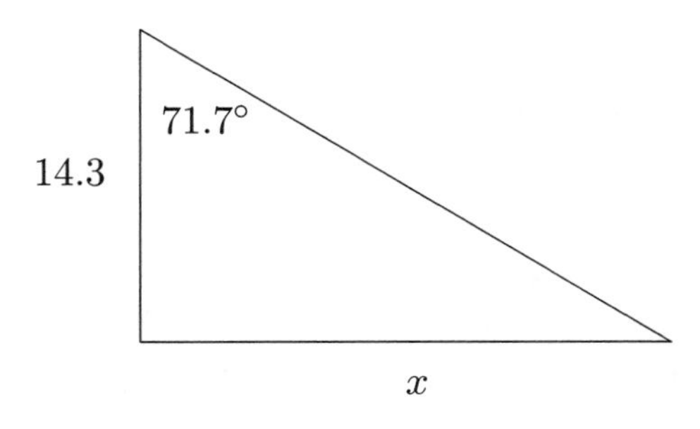

Since $\tan(71.7°) = x/14.3$, we obtain
$x = 14.3 \cdot \tan(71.7°) \approx 43.2$ miles.

39. Let h be the height as in the picture below.

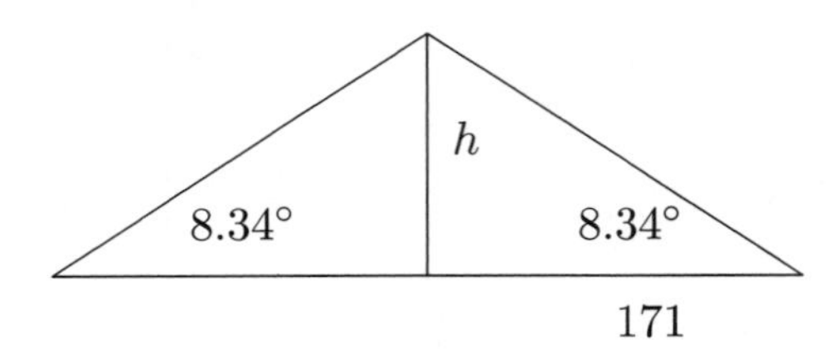

Since $\tan(8.34°) = h/171$, we obtain $h = 171 \cdot \tan(8.34°) \approx 25.1$ ft.

41. Let α be the angle the guy wire makes with the ground.

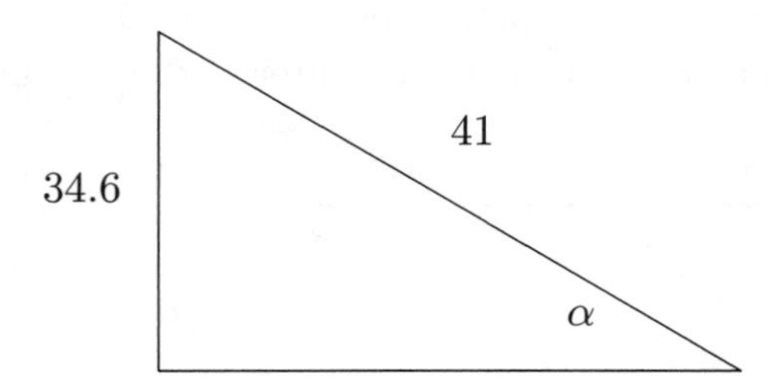

From the Pythagorean Theorem, the distance of the point to the base of the antenna is

$$\sqrt{41^2 - 34.6^2} \approx 22 \text{ meters.}$$

Also, $\alpha = \sin^{-1}(34.6/41) \approx 57.6°$.

43. Note, 1.75 sec. = 1.75/3600 hour.

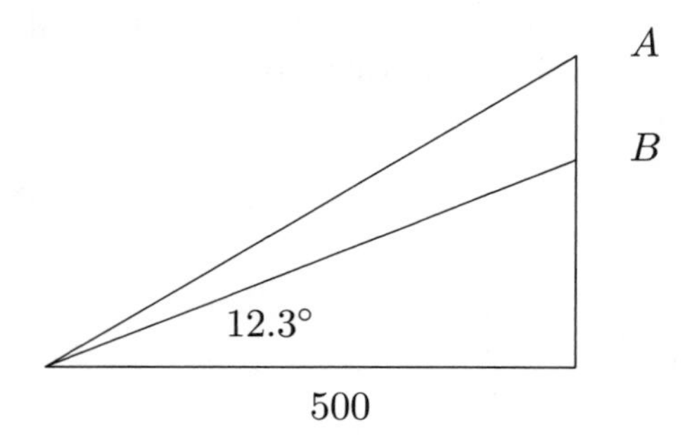

The distance in miles between A and B is

$$\frac{500\,(\tan(15.4°) - \tan(12.3°))}{5280} \approx 0.0054366$$

The speed is

$$\frac{0.0054366}{(1.75/3600)} \approx 11.2 \text{ mph}$$

and the car is not speeding.

45. Let h be the height.

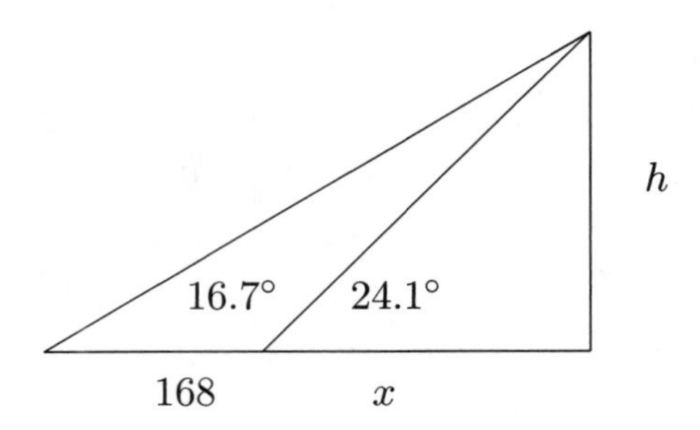

Note, $\tan 24.1° = \dfrac{h}{x}$ and $\tan 16.7° = \dfrac{h}{168 + x}$. Solve for h in the second equation and substitute $x = \dfrac{h}{\tan 24.1°}$.

$$h = \tan(16.7°) \cdot \left(168 + \frac{h}{\tan 24.1°}\right)$$

$$h - \frac{h\tan(16.7°)}{\tan 24.1°} = \tan(16.7°) \cdot 168$$

$$h = \frac{168 \cdot \tan(16.7°)}{1 - \tan(16.7°)/\tan(24.1°)}$$

$$h \approx 153.1\text{meters}$$

The height is 153.1 meters.

47. Let y be the closest distance the boat can come to the lighthouse LH.

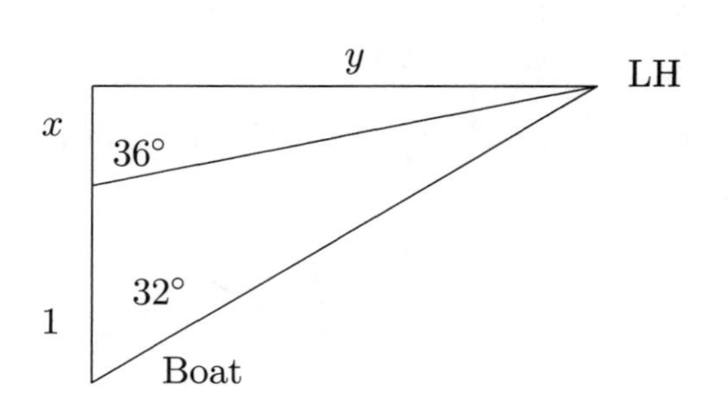

Since $\tan(36°) = y/x$ and $\tan(32°) = y/(1 + x)$, we obtain

$$\tan(32°) = \frac{y}{1 + y/\tan(36°)}$$

$$\tan(32°) + \frac{\tan(32°)y}{\tan(36°)} = y$$

$$\tan(32^\circ) = y\left(1 - \frac{\tan(32^\circ)}{\tan(36^\circ)}\right)$$

$$y = \frac{\tan(32^\circ)}{1 - \tan(32^\circ)/\tan(36^\circ)}$$

$$y \approx 4.5 \text{ km.}$$

The closest the boat will come to the lighthouse is 4.5 km.

49. Let x be the number of miles in one parsec.

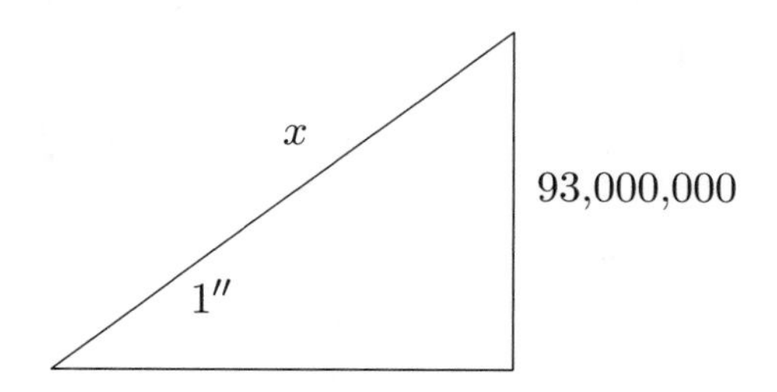

Since $\sin(1'') = \dfrac{93,000,000}{x}$, we obtain

$$x = \frac{93,000,000}{\sin(1/3600^\circ)} \approx 1.9 \times 10^{13} \text{ miles.}$$

Light travels one parsec in 3.26 years since

$$\frac{x}{193,000(63,240)} \approx 3.26 \text{ years.}$$

51. In the triangle below CE stands for the center of the earth, and LS is a point on the surface of the earth lying in the line of sight of the satellite.

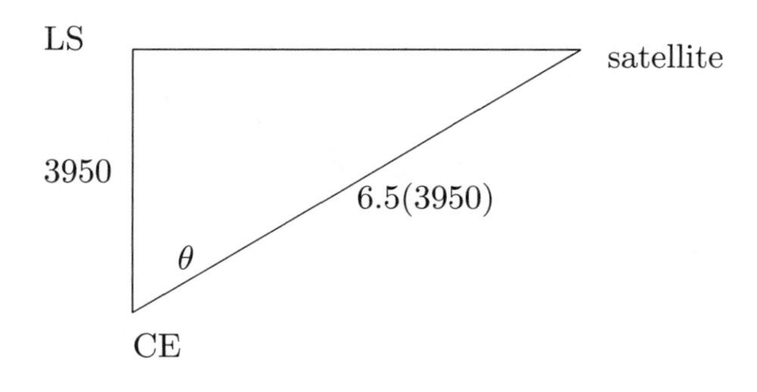

Since $\cos(\theta) = \dfrac{3950}{6.5(3950)} = 1/6.5$, we get $\theta = \cos^{-1}(1/6.5) \approx 1.41634$ radians. But 2θ is the widest angle formed by a sender and receiver of a signal with vertex CE. The maximum distance is the arclength subtended by 2θ, i.e., $s = r \cdot 2\theta = 3950 \cdot 2 \cdot 1.41634 \approx 11,189$ miles.

53. First, consider the figure below.

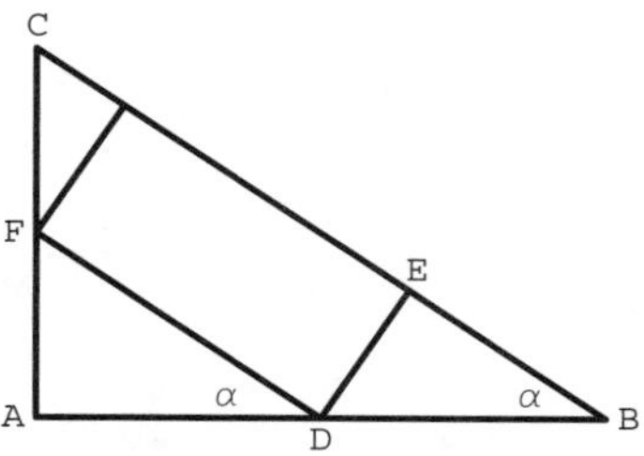

Suppose $AC = s$, $AB = \dfrac{3s}{2}$, $BD = h$, $DE = w$, and $DF = L$. Note, $\tan\alpha = \dfrac{AC}{AB} = \dfrac{2}{3}$. Since $\sin\alpha = \dfrac{2}{\sqrt{13}}$ and $\sin\alpha = \dfrac{DE}{BD}$, we obtain $w = \dfrac{2h}{\sqrt{13}}$. Since $\cos\alpha = \dfrac{AD}{DF}$, we get

$$\cos\alpha = \frac{\frac{3s}{2} - h}{L}.$$

Solving for L. we find $L = \dfrac{\frac{3s}{2} - h}{\cos\alpha}$ and since $\cos\alpha = \dfrac{3}{\sqrt{13}}$, we get

$$L = \frac{\sqrt{13}}{3}\left(\frac{3s}{2} - \frac{w\sqrt{13}}{2}\right).$$

So, the area of the parking lot is

$$wL = \frac{w\sqrt{13}}{3}\left(\frac{3s}{2} - \frac{w\sqrt{13}}{2}\right).$$

Since this area represents a quadratic function of w, one can find the vertex of its graph and conclude that the maximum area of the rectangle is obtained if one chooses $w = \dfrac{3s}{2\sqrt{13}}$.

Correspondingly, we obtain $L = \dfrac{s\sqrt{13}}{4}$.

Finally, given $s = 100$ feet, the dimensions of the house with maximum area are

$w = \dfrac{3(100)}{2\sqrt{13}} \approx 41.60$ ft and $L = \dfrac{100\sqrt{13}}{4} \approx 90.14$ ft.

55. Consider the right triangle formed by the hook, the center of the circle, and a point on the circle where the chain is tangent to the circle.

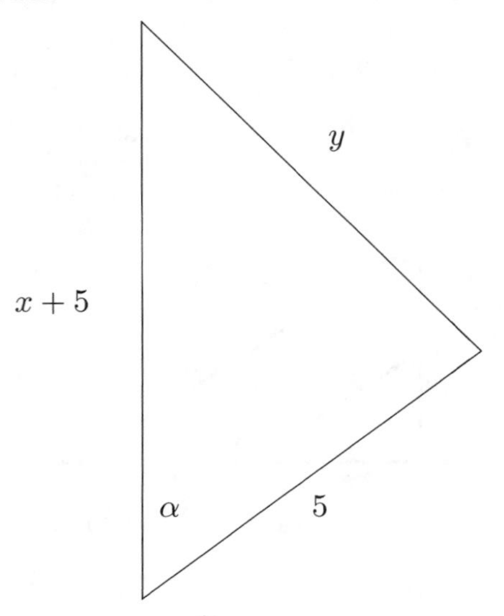

Then $\tan\alpha = \dfrac{y}{5}$ or $y = 5\tan\alpha$. Since the chain is 40 ft long and the angle $2\pi - 2\alpha$ intercepts an arc around the pipe where the chain wraps around the circle, we obtain

$$2y + 5(2\pi - 2\alpha) = 40.$$

By substitution, we get

$$10\tan\alpha + 10\pi - 10\alpha = 40.$$

With a graphing calculator, we obtain $\alpha \approx 1.09835$ radians. From the figure above, we get $\cos\alpha = \dfrac{5}{5+x}$. Solving for x, we obtain

$$x = \frac{5 - 5\cos\alpha}{\cos\alpha} \approx 5.987 \text{ ft}.$$

57. In the figure, r is the radius of the circle.

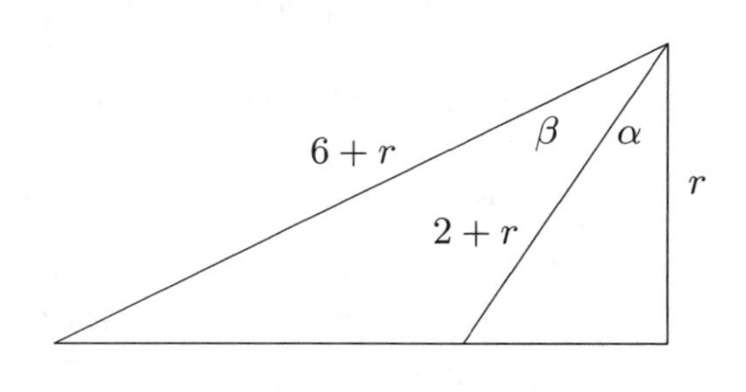

Note,

$$\alpha = \cos^{-1}\left(\frac{r}{2+r}\right) \text{ and } \beta = \cos^{-1}\left(\frac{r}{6+r}\right)$$

which we assume are in degree measure. With the aid of a graphing calculator, we find that the solutions to

$$\cos^{-1}\left(\frac{r}{6+r}\right) - \cos^{-1}\left(\frac{r}{2+r}\right) - 18 = 0$$

are $r \approx 3.626$ ft and $r \approx 9.126$ ft.

When we use $19°$, with a graphing calculator we see that

$$\cos^{-1}\left(\frac{r}{6+r}\right) - \cos^{-1}\left(\frac{r}{2+r}\right) - 19 = 0$$

has no solution.

59. In the figure below,

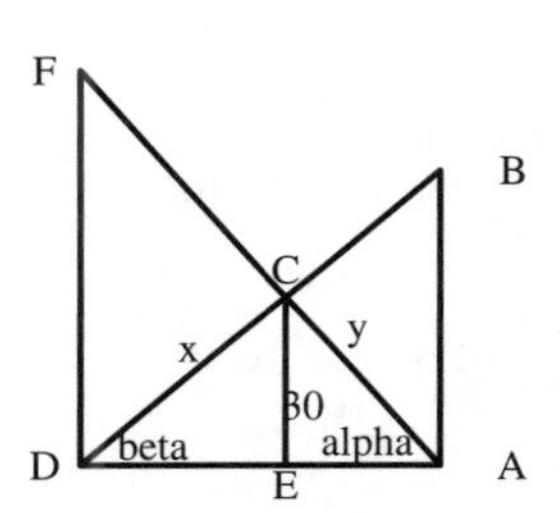

let $AF = 120$, $AC = y$, $BD = 90$, and $CD = x$. Also, let α be the angle formed by AE and AC, and let β be the angle formed by CD and DE.

Choose the point Q in DF so that DF is perpendicular to CQ. In $\triangle CQF$, we have

$$\sin(\pi/2 - \alpha) = \frac{QC}{120 - y}.$$

In $\triangle CQD$, we get

$$\sin(\pi/2 - \beta) = \frac{QC}{x}.$$

Therefore,

$$\begin{aligned} \sin(\pi/2 - \alpha)(120 - y) &= x\sin(\pi/2 - \beta) \\ \frac{x}{\sin(\pi/2 - \alpha)} &= \frac{120 - y}{\sin(\pi/2 - \beta)} \\ \frac{x}{\cos\alpha} &= \frac{120 - y}{\cos\beta}. \end{aligned}$$

Similarly, $\dfrac{90-x}{\cos\alpha} = \dfrac{y}{\cos\beta}$. Thus,

$$\frac{x}{120-y} = \frac{\cos\alpha}{\cos\beta} = \frac{90-x}{y}.$$

Consequently, $4x + 3y = 360$ or equivalently

$$y = \frac{360-4x}{3}.$$

From the right triangles △ ACE and △ CDE, one finds $\sin\alpha = \dfrac{30}{y}$ and $\sin\beta = \dfrac{30}{x}$.

Consequently,

$$\cos\alpha = \frac{\sqrt{y^2-30^2}}{y} \quad \text{and} \quad \cos\beta = \frac{\sqrt{x^2-30^2}}{x}.$$

Moreover, using triangles △ ABD and △ ADF, one obtains

$$\cos\beta = \frac{AD}{90}$$

and

$$\cos\alpha = \frac{AD}{120}.$$

Combining all of these, one derives

$$\begin{aligned}
90\cos\beta &= 120\cos\alpha \\
\frac{3}{4} &= \frac{\cos\alpha}{\cos\beta} \\
\frac{3}{4} &= \frac{\frac{\sqrt{y^2-30^2}}{y}}{\frac{\sqrt{x^2-30^2}}{x}} \\
\frac{3}{4} &= \frac{\sqrt{y^2-30^2}}{\sqrt{x^2-30^2}} \cdot \frac{x}{y} \\
\frac{3}{4} &= \frac{\sqrt{\left(\frac{360-4x}{3}\right)^2 - 30^2}}{\sqrt{x^2-30^2}} \cdot \frac{x}{\frac{360-4x}{3}} \\
\frac{3}{4} &= \frac{3x}{360-4x} \cdot \frac{\sqrt{\left(\frac{360-4x}{3}\right)^2 - 30^2}}{\sqrt{x^2-30^2}} \\
\frac{1}{16} &= \frac{x^2}{(360-4x)^2} \cdot \frac{\left(\frac{360-4x}{3}\right)^2 - 30^2}{x^2-30^2}
\end{aligned}$$

$$(360-4x)^2(x^2-30^2) =$$

$$= 16x^2\left[\left(\frac{360-4x}{3}\right)^2 - 30^2\right]$$

Using a graphing calculator, for $0 < x < 90$, one finds $x \approx 60.4$. Working backwards, one derives

$$\sin\beta = \frac{30}{60.4} \quad \text{or} \quad \beta \approx 29.8^\circ$$

and

$$\cos(29.8) = \frac{AD}{90}.$$

Hence, the width of the property is

$$AD = 78.1 \text{ feet}.$$

For Thought

1. True, since $\sin^2\alpha + \cos^2\alpha = 1$ for any real number α.

2. True

3. False, since α is in Quadrant IV.

4. True

5. False, rather $\sin\alpha = -\dfrac{1}{2}$.

6. True **7.** True

8. False, since the reference angle is $\dfrac{\pi}{3}$.

9. True, since $\cos 120^\circ = -\dfrac{1}{2} = -\cos 60^\circ$.

10. True, since $\sin(7\pi/6) = -\dfrac{1}{2} = -\sin(\pi/6)$.

1.6 Exercises

1. $\sin\alpha = \pm\sqrt{1-\cos^2\alpha} = \pm\sqrt{1-1^2} = 0$

3. Use the Fundamental Identity.

$$\begin{aligned}
\left(\frac{5}{13}\right)^2 + \cos^2(\alpha) &= 1 \\
\frac{25}{169} + \cos^2(\alpha) &= 1 \\
\cos^2(\alpha) &= \frac{144}{169} \\
\cos(\alpha) &= \pm\frac{12}{13}
\end{aligned}$$

Since α is in quadrant II, $\cos(\alpha) = -12/13$.

5. Use the Fundamental Identity.

$$\begin{aligned}\left(\frac{3}{5}\right)^2 + \sin^2(\alpha) &= 1\\ \frac{9}{25} + \sin^2(\alpha) &= 1\\ \sin^2(\alpha) &= \frac{16}{25}\\ \sin(\alpha) &= \pm\frac{4}{5}\end{aligned}$$

Since α is in quadrant IV, $\sin(\alpha) = -4/5$.

7. Use the Fundamental Identity.

$$\begin{aligned}\left(\frac{1}{3}\right)^2 + \cos^2(\alpha) &= 1\\ \frac{1}{9} + \cos^2(\alpha) &= 1\\ \cos^2(\alpha) &= \frac{8}{9}\\ \cos(\alpha) &= \pm\frac{2\sqrt{2}}{3}\end{aligned}$$

Since $\cos(\alpha) > 0$, $\cos(\alpha) = \dfrac{2\sqrt{2}}{3}$.

9. $30°, \pi/6$

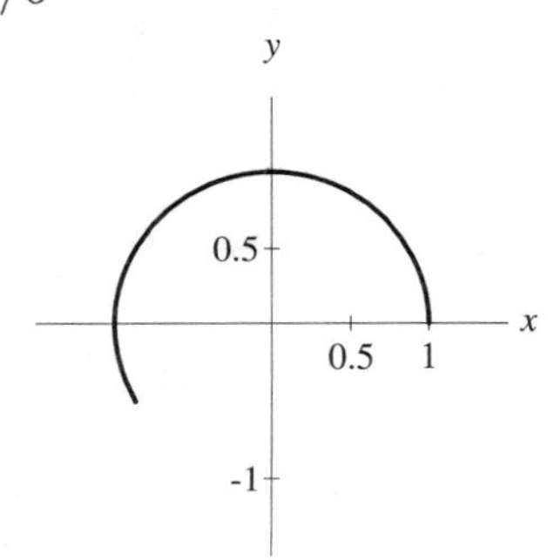

11. $60°, \pi/3$

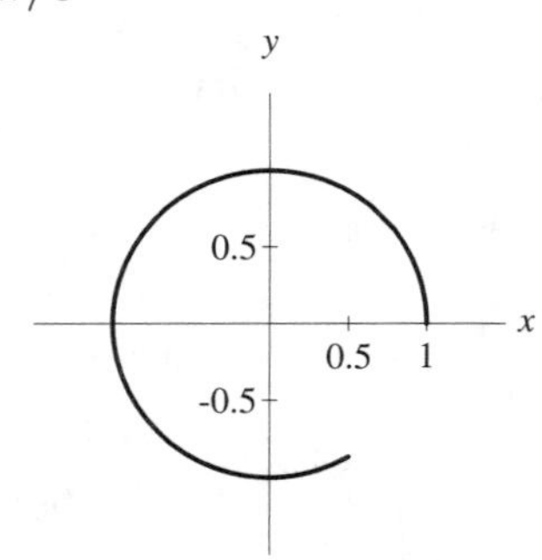

13. $60°, \pi/3$

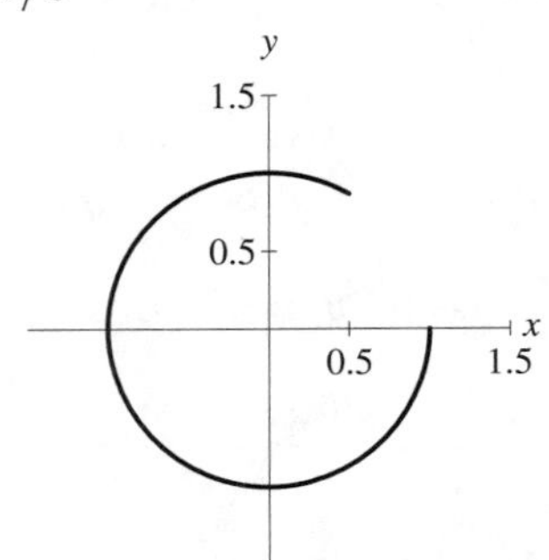

15. $30°, \pi/6$

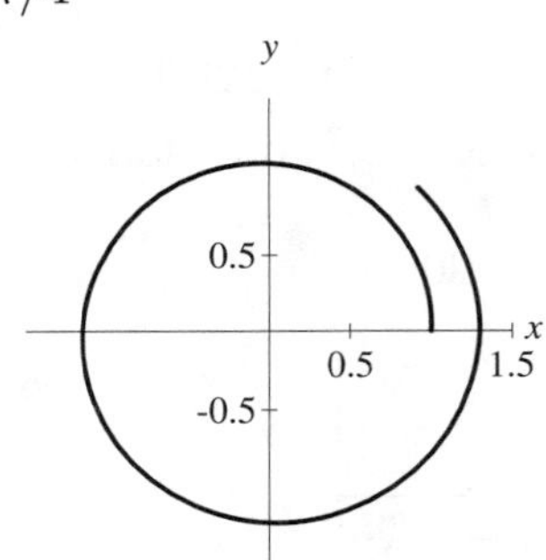

17. $45°, \pi/4$

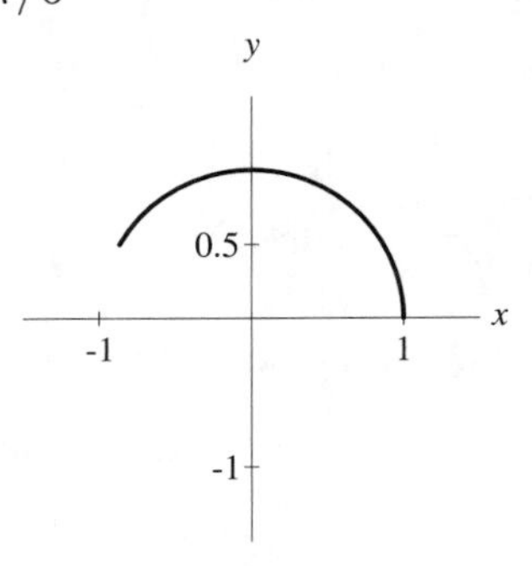

19. $45°, \pi/4$

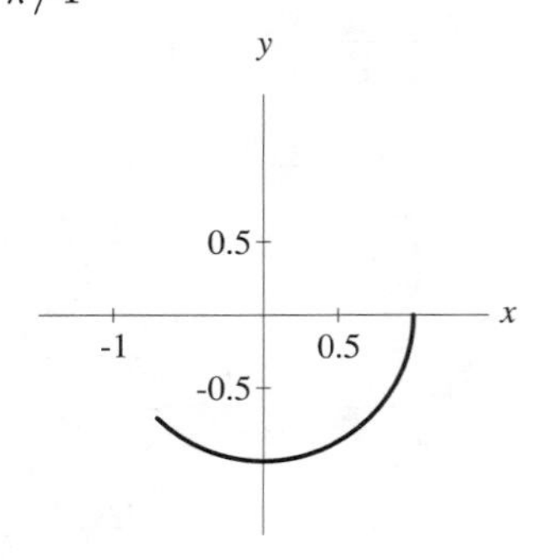

21. $\sin(135°) = \sin(45°) = \dfrac{\sqrt{2}}{2}$

23. $\cos\left(\dfrac{5\pi}{3}\right) = \cos\left(\dfrac{\pi}{3}\right) = \dfrac{1}{2}$

25. $\sin\left(\dfrac{7\pi}{4}\right) = -\sin\left(\dfrac{\pi}{4}\right) = -\dfrac{\sqrt{2}}{2}$

27. $\cos\left(-\dfrac{17\pi}{6}\right) = -\cos\left(\dfrac{\pi}{6}\right) = -\dfrac{\sqrt{3}}{2}$

29. $\sin(-45°) = -\sin(45°) = -\dfrac{\sqrt{2}}{2}$

31. $\cos(-240°) = -\cos(60°) = -\dfrac{1}{2}$

33. The reference angle of $3\pi/4$ is $\pi/4$.

Then $\sin(3\pi/4) = \sin(\pi/4) = \dfrac{\sqrt{2}}{2}$,

$\cos(3\pi/4) = -\cos(\pi/4) = -\dfrac{\sqrt{2}}{2}$,
$\tan(3\pi/4) = -\tan(\pi/4) = -1$,
$\csc(3\pi/4) = \csc(\pi/4) = \sqrt{2}$,
$\sec(3\pi/4) = -\sec(\pi/4) = -\sqrt{2}$, and
$\cot(3\pi/4) = -\cot(\pi/4) = -1$.

35. The reference angle of $4\pi/3$ is $\pi/3$.

Then $\sin(4\pi/3) = -\sin(\pi/3) = -\dfrac{\sqrt{3}}{2}$,

$\cos(4\pi/3) = -\cos(\pi/3) = -\dfrac{1}{2}$,
$\tan(4\pi/3) = \tan(\pi/3) = \sqrt{3}$,
$\csc(4\pi/3) = -\csc(\pi/3) = -\dfrac{2\sqrt{3}}{3}$,
$\sec(4\pi/3) = -\sec(\pi/3) = -2$, and
$\cot(4\pi/3) = \cot(\pi/3) = \dfrac{\sqrt{3}}{3}$.

37. The reference angle of $300°$ is $60°$.

Then $\sin(300°) = -\sin(60°) = -\dfrac{\sqrt{3}}{2}$,

$\cos(300°) = \cos(60°) = \dfrac{1}{2}$,
$\tan(300°) = -\tan(60°) = -\sqrt{3}$,
$\csc(300°) = -\csc(60°) = -\dfrac{2\sqrt{3}}{3}$,
$\sec(300°) = \sec(60°) = 2$, and
$\cot(300°) = -\cot(60°) = -\dfrac{\sqrt{3}}{3}$.

39. The reference angle of $-135°$ is $45°$.

Then $\sin(-135°) = -\sin(45°) = -\dfrac{\sqrt{2}}{2}$,

$\cos(-135°) = -\cos(45°) = -\dfrac{\sqrt{2}}{2}$,
$\tan(-135°) = \tan(45°) = 1$,
$\csc(-135°) = -\csc(45°) = -\sqrt{2}$,
$\sec(-135°) = -\sec(45°) = -\sqrt{2}$, and
$\cot(-135°) = \cot(45°) = 1$.

41. False, since $\sin 210° = -\sin 30°$.

43. True, since $\cos 330° = \dfrac{\sqrt{3}}{2} = \cos 30°$.

45. False, since $\sin 179° = \sin 1°$.

47. True, for the reference angle is $\pi/7$, $6\pi/7$ is in Quadrant II, and cosine is negative in Quadrant II.

49. False, since $\sin(23\pi/24) = \sin(\pi/24)$.

51. True, for the reference angle is $\pi/7$, $13\pi/7$ is in Quadrant IV, and cosine is positive in Quadrant IV.

53. If $h = 18$, then

$$\begin{aligned} T &= 18\sin\left(\frac{\pi}{12}(6)\right) + 102 \\ &= 18\sin\left(\frac{\pi}{2}\right) + 102 = 18 + 102 \\ &= 120°\text{F}. \end{aligned}$$

If $h = 6$, then

$$\begin{aligned} T &= 18\sin\left(\frac{\pi}{12}(-6)\right) + 102 \\ &= 18\sin\left(-\frac{\pi}{2}\right) + 102 = -18 + 102 \\ &= 84°\text{F}. \end{aligned}$$

55. Note, $x(t) = 4\sin(t) + 3\cos(t)$.

a) The initial position is

$$x(0) = 3\cos 0 = 3.$$

b) If $t = 5\pi/4$, the position is

$$\begin{aligned} x(5\pi/4) &= 4\sin(5\pi/4) + 3\cos(5\pi/4) \\ &= -2\sqrt{2} - \frac{3\sqrt{2}}{2} \\ &= -\frac{7\sqrt{2}}{2}. \end{aligned}$$

57. The angle between the tips of two adjacent teeth is $\frac{2\pi}{22} = \frac{\pi}{11}$. The actual distance is

$$c = 6\sqrt{2 - 2\cos(\pi/11)} \approx 1.708 \text{ in.}$$

The length of the arc is

$$s = 6 \cdot \frac{\pi}{11} \approx 1.714 \text{ in.}$$

59. Solving for v_o, one finds

$$\begin{aligned} 367 &= \frac{v_o^2}{32}\sin 86^\circ \\ \sqrt{\frac{32(367)}{\sin 86^\circ}} \text{ ft/sec} &= v_o \\ \sqrt{\frac{32(367)}{\sin 86^\circ}\frac{3600}{5280}} \text{ mph} &= v_o \\ 74 \text{ mph} &\approx v_o. \end{aligned}$$

Chapter 1 Review Exercises

1. $388^\circ - 360^\circ = 28^\circ$

3. $-153^\circ 14' 27'' + 359^\circ 59' 60'' = 206^\circ 45' 33''$

5. 180°

7. $13\pi/5 - 2\pi = 3\pi/5 = 3 \cdot 36^\circ = 108^\circ$

9. $5\pi/3 = 5 \cdot 60^\circ = 300^\circ$

11. 270° **13.** $11\pi/6$ **15.** $-5\pi/3$

17.

θ deg	0	30	45	60	90	120	135	150	180
θ rad	0	$\frac{\pi}{6}$	$\frac{\pi}{4}$	$\frac{\pi}{3}$	$\frac{\pi}{2}$	$\frac{2\pi}{3}$	$\frac{3\pi}{4}$	$\frac{5\pi}{6}$	π
$\sin\theta$	0	$\frac{1}{2}$	$\frac{\sqrt{2}}{2}$	$\frac{\sqrt{3}}{2}$	1	$\frac{\sqrt{3}}{2}$	$\frac{\sqrt{2}}{2}$	$\frac{1}{2}$	0
$\cos\theta$	1	$\frac{\sqrt{3}}{2}$	$\frac{\sqrt{2}}{2}$	$\frac{1}{2}$	0	$-\frac{1}{2}$	$-\frac{\sqrt{2}}{2}$	$-\frac{\sqrt{3}}{2}$	-1

19. $-\sqrt{2}/2$ **21.** $\sqrt{3}$ **23.** $-2\sqrt{3}/3$

25. 0 **27.** 0

29. -1 **31.** $\cot(60^\circ) = \sqrt{3}/3$ **33.** $-\sqrt{2}/2$

35. -2 **37.** $-\sqrt{3}/3$

39. 0.6947 **41.** -0.0923 **43.** 0.1869

45. $\dfrac{1}{\cos(105^\circ 4')} \approx -3.8470$

47. $\dfrac{1}{\sin(\pi/9)} \approx 2.9238$

49. $\dfrac{1}{\tan(33^\circ 44')} \approx 1.4975$

51. 45° **53.** 0°

55. 30° **57.** 30°

59. Note, the the length of the hypotenuse is 13. Then $\sin(\alpha) = \text{opp}/\text{hyp} = 5/13$,
$\cos(\alpha) = \text{adj}/\text{hyp} = 12/13$,
$\tan(\alpha) = \text{opp}/\text{adj} = 5/12$,
$\csc(\alpha) = \text{hyp}/\text{opp} = 13/5$,
$\sec(\alpha) = \text{adj}/\text{hyp} = 13/12$, and
$\cot(\alpha) = \text{adj}/\text{opp} = 12/5$.

61. Form the right triangle with $a = 2$, $b = 3$.

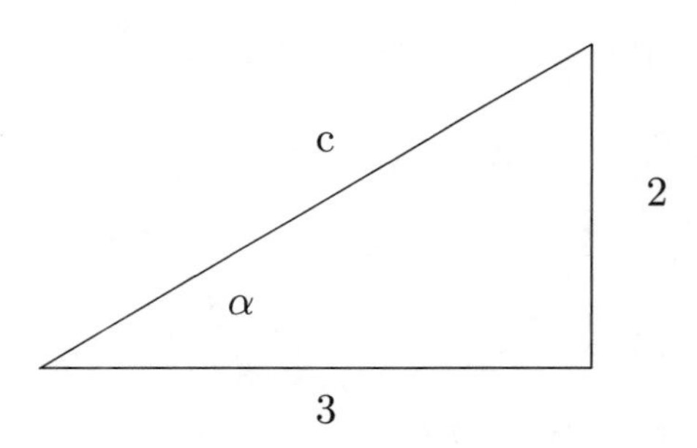

Note that

$$c = \sqrt{2^2 + 3^2} = \sqrt{13}$$

and $\tan(\alpha) = 2/3$. Then

$$\alpha = \tan^{-1}(2/3) \approx 33.7^\circ$$

and $\beta \approx 56.3^\circ$.

63. Form the right triangle with $a = 3.2$ and $\alpha = 21.3°$.

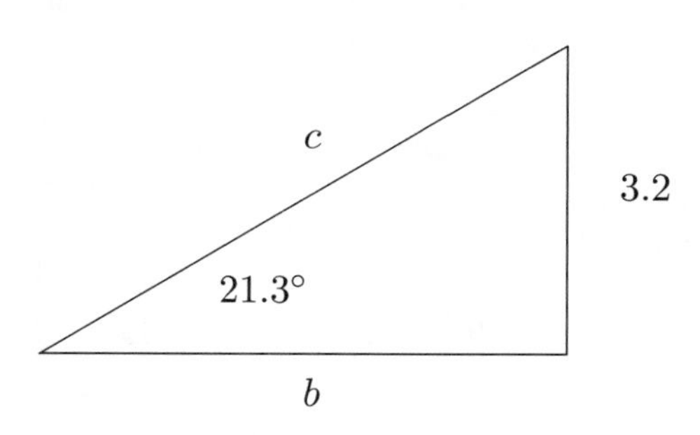

Since $\sin 21.3° = \dfrac{3.2}{c}$ and

$\tan 21.3° = \dfrac{3.2}{b}$, $c = \dfrac{3.2}{\sin 21.3°} \approx 8.8$

and $b = \dfrac{3.2}{\tan 21.3°} \approx 8.2$

Also, $\beta = 90° - 21.3° = 68.7°$

65. $\sin(\alpha) = -\sqrt{1 - \left(\frac{1}{5}\right)^2} = -\sqrt{\frac{24}{25}} = \frac{-2\sqrt{6}}{5}$

67. In one hour, the nozzle revolves through an angle of $\frac{2\pi}{8}$. The linear velocity is

$v = r \cdot \alpha = 120 \cdot \frac{2\pi}{8} \approx 94.2$ ft/hr.

69. The height of the man is

$s = r \cdot \alpha = 1000(0.4) \cdot \frac{\pi}{180} \approx 6.9813$ ft.

71. Form the right triangle below.

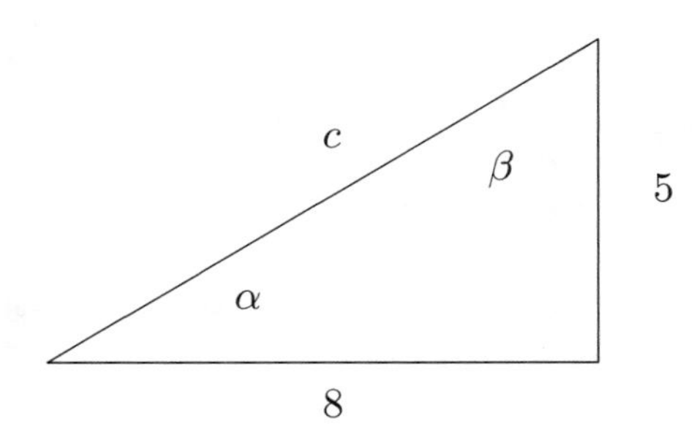

By the Pythagorean Theorem, we get

$$c = \sqrt{8^2 + 5^2} = \sqrt{89} \text{ ft.}$$

Note, $\alpha = \tan^{-1}\left(\frac{5}{8}\right) \approx 32.0°$ and

$\beta = 90° - \alpha \approx 58.0°$.

73. Form the right triangle below.

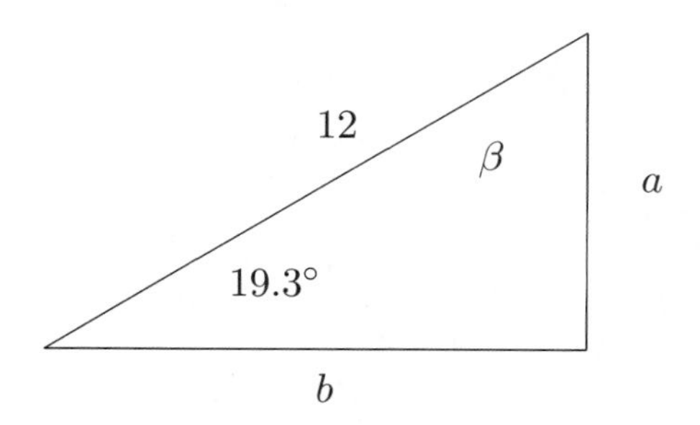

Note, $\beta = 90° - 19.3° = 70.7°$.
Also, $a = 12\sin(19.3°) \approx 4.0$ ft and
$b = 12\cos(19.3°) \approx 11.3$ ft.

75. Let s be the height of the shorter building and let $a + b$ the height of the taller building.

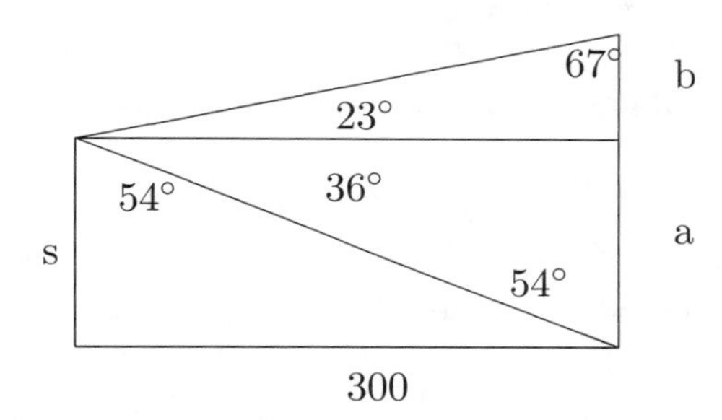

Since $\tan 54° = \dfrac{300}{s}$, get

$$s = \frac{300}{\tan 54°} \approx 218 \text{ ft.}$$

Similarly, since $\tan 36° = \dfrac{a}{300}$ and

$\tan 23° = \dfrac{b}{300}$, the height of the taller building is

$$a + b = 300\tan 36° + 300\tan 23° \approx 345 \text{ ft.}$$

77. Let h be the height of the tower.

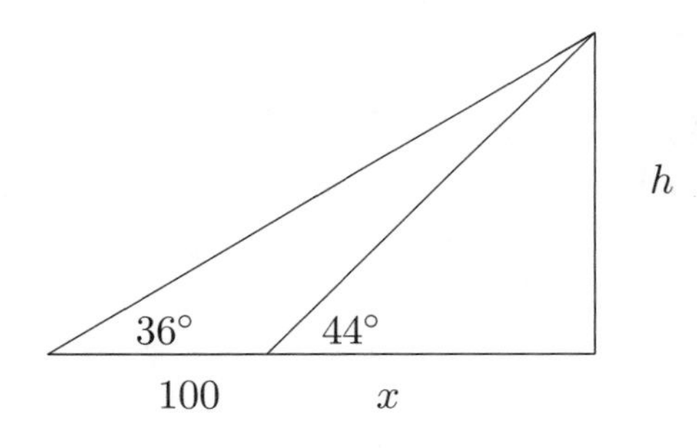

Since

$$h = (100 + x)\tan 36^\circ$$

and

$$x = \frac{h}{\tan 44^\circ}$$

we find

$$\begin{aligned} h &= \left(100 + \frac{h}{\tan 44^\circ}\right)\tan 36^\circ \\ h &= 100\tan 36^\circ + h\frac{\tan 36^\circ}{\tan 44^\circ} \\ h &= \frac{100\tan 36^\circ}{1 - \frac{\tan 36^\circ}{\tan 44^\circ}} \\ h &\approx 293 \text{ ft} \end{aligned}$$

Chapter 1 Test

1. Since 60° is the reference angle, we get

$$\cos 420^\circ = \cos(60^\circ) = 1/2.$$

2. Since 30° is the reference angle, we get

$$\sin(-390^\circ) = -\sin(30^\circ) = -\frac{1}{2}.$$

3. $\frac{\sqrt{2}}{2}$ **4.** $\frac{1}{2}$ **5.** $\frac{\sqrt{3}}{3}$ **6.** $\sqrt{3}$

7. Undefined, since $\dfrac{1}{\cos(\pi/2)} = \dfrac{1}{0}$

8. $\dfrac{1}{\sin(-\pi/2)} = \dfrac{1}{-1} = -1$

9. Undefined, since $(-1, 0)$ lies on the terminal side of angle -3π and

$$\cot(-3\pi) = \frac{x}{y} = \frac{-1}{0}.$$

10. Since $(-1, -1)$ lies on the terminal side of angle 225°, we get

$$\cot(225^\circ) = \frac{x}{y} = \frac{-1}{-1} = 1.$$

11. $\frac{\pi}{4}$ or 45° **12.** $\frac{\pi}{6}$ or 30°

13. Since $46^\circ 24' 6'' \approx 0.8098619$, the arclength is $s = r\alpha = 35.62(0.8098619) \approx 28.85$ meters.

14. $2.34 \cdot \dfrac{180^\circ}{\pi} \approx 134.07^\circ$

15. Coterminal since

$$2200^\circ - 40^\circ = 2160^\circ = 6 \cdot 360^\circ.$$

16. $\cos(\alpha) = -\sqrt{1 - \left(\frac{1}{4}\right)^2} = -\dfrac{\sqrt{15}}{4}$

17. $\omega = 103 \cdot 2\pi \approx 647.2$ radians/minute

18. In one minute, the wheel turns through an arclength of $13(103 \cdot 2\pi)$ inches. Multiplying this by $\dfrac{60}{12 \cdot 5280}$ results in the speed in mph which is 7.97 mph.

19. Since we have

$$r = \sqrt{x^2 + y^2} = \sqrt{5^2 + (-2)^2} = \sqrt{29}$$

we find $\sin\alpha = \dfrac{y}{r} = \dfrac{-2}{\sqrt{29}} = \dfrac{-2\sqrt{29}}{29}$,

$$\cos\alpha = \frac{x}{r} = \frac{5}{\sqrt{29}} = \frac{5\sqrt{29}}{29},$$

$$\tan\alpha = \frac{y}{x} = \frac{-2}{5},$$

$$\csc\alpha = \frac{r}{y} = \frac{\sqrt{29}}{-2} = -\frac{\sqrt{29}}{2},$$

$$\sec\alpha = \frac{r}{x} = \frac{\sqrt{29}}{5}, \text{ and}$$

$$\cot\alpha = \frac{x}{y} = \frac{5}{-2} = -\frac{5}{2}.$$

20. Consider the right triangle below.

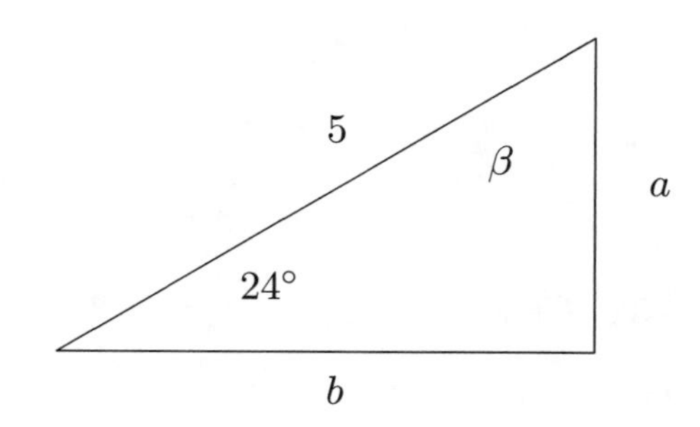

Note,

$$\beta = 90° - 24° = 66°.$$

Then

$$a = 5\sin(24°) \approx 2.0 \text{ ft}$$

and

$$b = 5\cos(24°) \approx 4.6 \text{ ft}.$$

21. Let h be the height of the head.

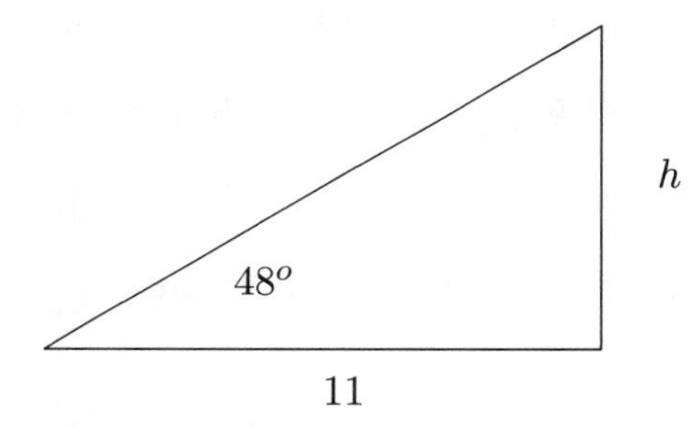

Since

$$\tan(48^o) = \frac{h}{11}$$

we find

$$h = 11\tan 48^o \approx 12.2 \text{ m}.$$

22. Let h be the height of the building.

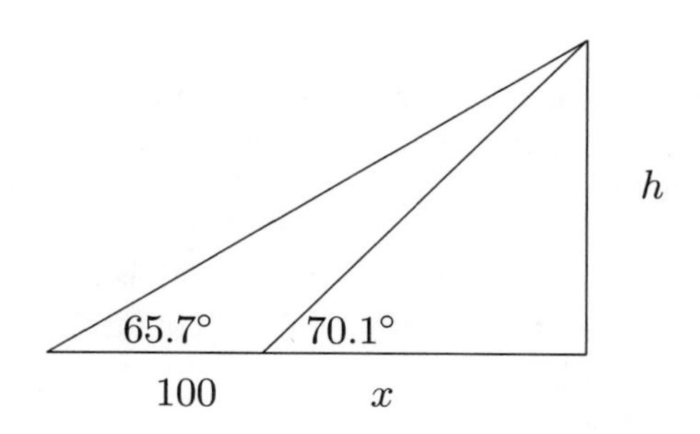

Since $\tan 70.1^o = \dfrac{h}{x}$ and $\tan 65.7° = \dfrac{h}{100+x}$, we obtain

$$\tan(65.7°) = \frac{h}{100 + h/\tan(70.1°)}$$

$$100 \cdot \tan(65.7°) + h \cdot \frac{\tan(65.7^o)}{\tan(70.1°)} = h$$

$$100 \cdot \tan(65.7°) = h\left(1 - \frac{\tan(65.7°)}{\tan(70.1°)}\right)$$

$$h = \frac{100 \cdot \tan(65.7°)}{1 - \tan(65.7°)/\tan(70.1°)}$$

$$h \approx 1117 \text{ ft}.$$

For Thought

1. True

2. False, since the range of $y = 4\sin(x)$ is $[-4, 4]$, we get that the range of $y = 4\sin(x) + 3$ is $[-4+3, 4+3]$ or $[-1, 7]$.

3. True, since the range of $y = \cos(x)$ is $[-1, 1]$, we find that the range of $y = \cos(x) - 5$ is $[-1-5, 1-5]$ or $[-6, -4]$.

4. False, the phase shift is $-\pi/6$.

5. False, the graph of $y = \sin(x + \pi/6)$ lies $\pi/6$ to the *left* of the graph of $y = \sin(x)$.

6. True, since $\cos(5\pi/6 - \pi/3) = \cos(\pi/2) = 0$ and $\cos(11\pi/6 - \pi/3) = \cos(3\pi/2) = 0$.

7. False, for if $x = \pi/2$ we find $\sin(\pi/2) = 1 \neq \cos(\pi/2 + \pi/2) = \cos(\pi) = -1$.

8. False, the minimum value is -3.

9. True, since the maximum value of $y = -2\cos(x)$ is 2, we get that the maximum value of $y = -2\cos(x) + 4$ is $2 + 4$ or 6.

10. True, since $(-\pi/6 + \pi/3, 0) = (\pi/6, 0)$.

2.1 Exercises

1. 0

3. $\dfrac{\sin(\pi/3)}{\cos(\pi/3)} = \dfrac{\sqrt{3}/2}{1/2} = \sqrt{3}$.

5. 1/2

7. $\dfrac{1}{\cos(\pi/3)} = \dfrac{1}{1/2} = 2$.

9. 0

11. $(0 + \pi/4, 0) = (\pi/4, 0)$

13. $(\pi/2 + \pi/4, 3) = (3\pi/4, 3)$

15. $(-\pi/2 + \pi/4, -1) = (-\pi/4, -1)$

17. $(\pi + \pi/4, 0) = (5\pi/4, 0)$

19. $(\pi/3 - \pi/3, 0) = (0, 0)$

21. $(\pi - \pi/3, 1) = (2\pi/3, 1)$

23. $(\pi/2 - \pi/3, -1) = (\pi/6, -1)$

25. $(-\pi - \pi/3, 1) = (-4\pi/3, 1)$

27. $(\pi + \pi/6, -1 + 2) = (7\pi/6, 1)$

29. $(\pi/2 + \pi/6, 0 + 2) = (2\pi/3, 2)$

31. $(-3\pi/2 + \pi/6, 1 + 2) = (-4\pi/3, 3)$

33. $(2\pi + \pi/6, -4 + 2) = (13\pi/6, -2)$

35. $\left(\dfrac{\pi + 2\pi}{2}, 0\right) = \left(\dfrac{3\pi}{2}, 0\right)$

37. $\left(\dfrac{0 + \pi/4}{2}, 2\right) = \left(\dfrac{\pi}{8}, 2\right)$

39. $\left(\dfrac{\pi/6 + \pi/2}{2}, 1\right) = \left(\dfrac{\pi}{3}, 1\right)$

41. $\left(\dfrac{\pi/3 + \pi/2}{2}, -4\right) = \left(\dfrac{5\pi}{12}, -4\right)$

43. $P(0, 0)$, $Q(\pi/4, 2)$, $R(\pi/2, 0)$, $S(3\pi/4, -2)$

45. $P(\pi/4, 0)$, $Q(5\pi/8, 2)$, $R(\pi, 0)$, $S(11\pi/8, -2)$

47. $P(0, 2)$, $Q(\pi/12, 3)$, $R(\pi/6, 2)$, $S(\pi/4, 1)$

49. Amplitude 2, period 2π, phase shift 0, range $[-2, 2]$

51. Amplitude 1, period 2π, phase shift $\pi/2$, range $[-1, 1]$

53. Amplitude 2, period 2π, phase shift $-\pi/3$, range $[-2, 2]$

55. Amplitude 1, phase shift 0, range $[-1, 1]$, some points are $(0, 0)$, $\left(\dfrac{\pi}{2}, -1\right)$, $(\pi, 0)$, $\left(\dfrac{3\pi}{2}, 1\right)$, $(2\pi, 0)$

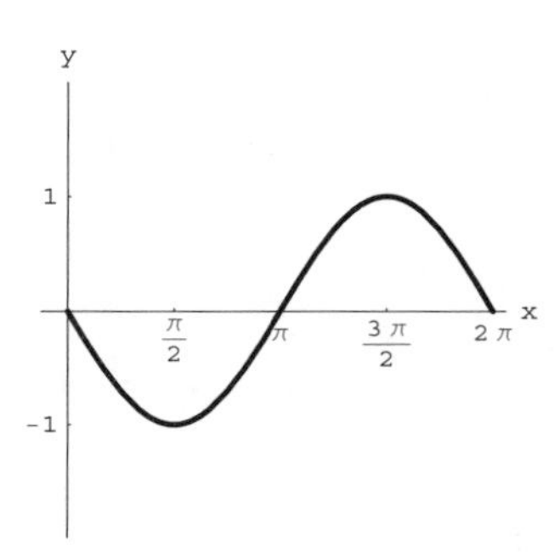

57. Amplitude 3, phase shift 0, range $[-3, 3]$, some points are $(0,0)$, $\left(\frac{\pi}{2}, -3\right)$, $(\pi, 0)$, $\left(\frac{3\pi}{2}, 3\right)$, $(2\pi, 0)$

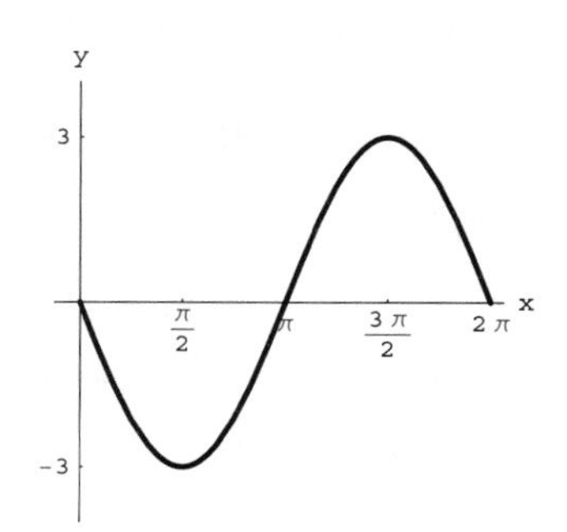

59. Amplitude $1/2$, phase shift 0, range $[-1/2, 1/2]$, some points are $(0, 1/2)$, $(\pi/2, 0)$, $(\pi, -1/2)$ $(3\pi/2, 0)$, $(2\pi, 1/2)$

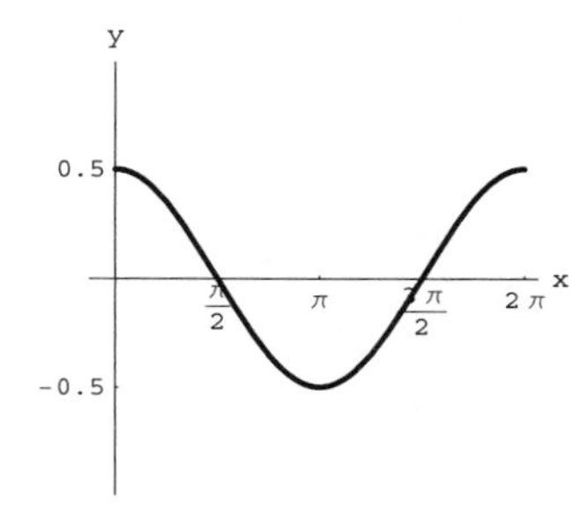

61. Amplitude 1, phase shift $-\pi$, range $[-1, 1]$, some points are $(-\pi, 0)$, $(-\pi/2, 1)$, $(0, 0)$, $\left(\frac{\pi}{2}, -1\right)$, $(\pi, 0)$

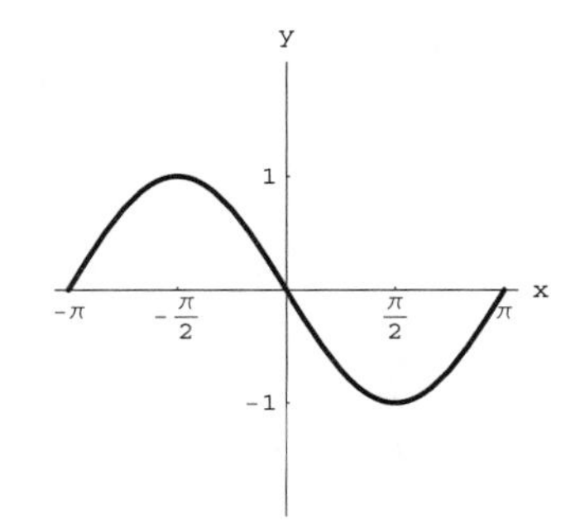

63. Amplitude 1, phase shift $\pi/3$, range $[-1, 1]$, some points are $\left(\frac{\pi}{3}, 1\right)$, $\left(\frac{5\pi}{6}, 0\right)$, $\left(\frac{4\pi}{3}, -1\right)$, $\left(\frac{11\pi}{6}, 0\right)$, $\left(\frac{7\pi}{3}, 1\right)$,

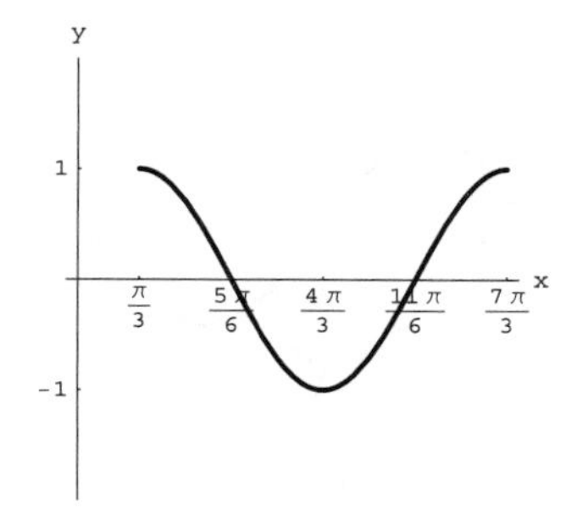

65. Amplitude 1, phase shift 0, range $[1, 3]$, some points are $(0, 3)$, $\left(\frac{\pi}{2}, 2\right)$, $(\pi, 1)$ $\left(\frac{3\pi}{2}, 2\right)$, $(2\pi, 3)$

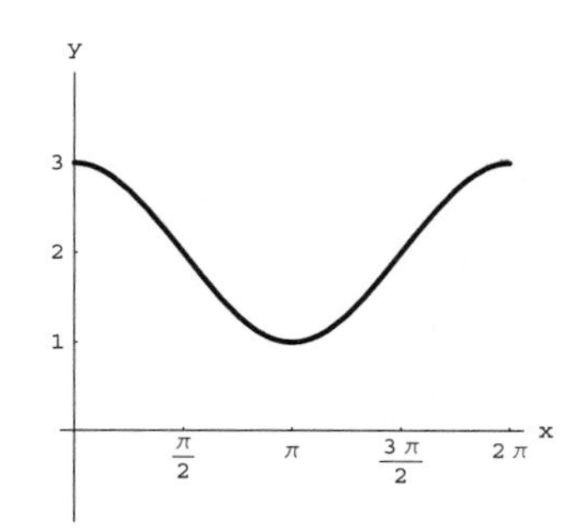

67. Amplitude 1, phase shift 0, range $[-2, 0]$, some points are $(0, -1)$, $\left(\frac{\pi}{2}, -2\right)$, $(\pi, -1)$, $\left(\frac{3\pi}{2}, 0\right)$, $(2\pi, -1)$

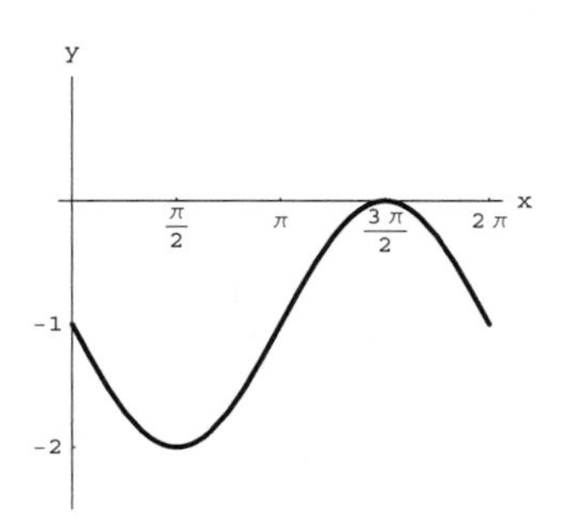

69. Amplitude 1, phase shift $-\pi/4$, range $[1, 3]$, some points are $\left(-\frac{\pi}{4}, 2\right)$, $\left(\frac{\pi}{4}, 3\right)$, $\left(\frac{3\pi}{4}, 2\right)$, $\left(\frac{5\pi}{4}, 1\right)$, $\left(\frac{7\pi}{4}, 2\right)$

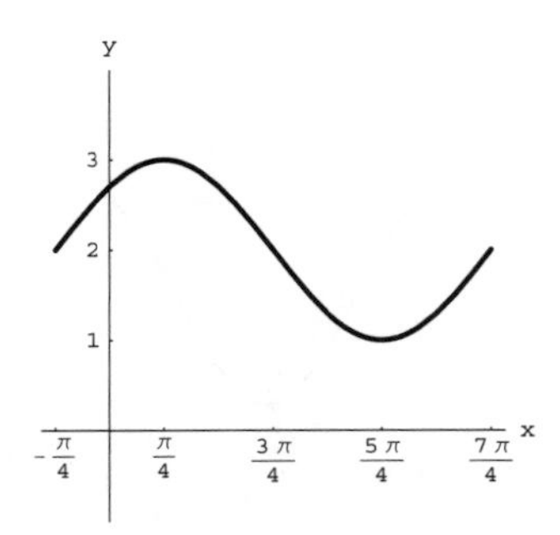

71. Amplitude 2, phase shift $-\pi/6$, range $[-1, 3]$, some points are $\left(-\frac{\pi}{6}, 3\right)$, $\left(\frac{\pi}{3}, 1\right)$, $\left(\frac{5\pi}{6}, -1\right)$, $\left(\frac{4\pi}{3}, 1\right)$, $\left(\frac{11\pi}{6}, 3\right)$

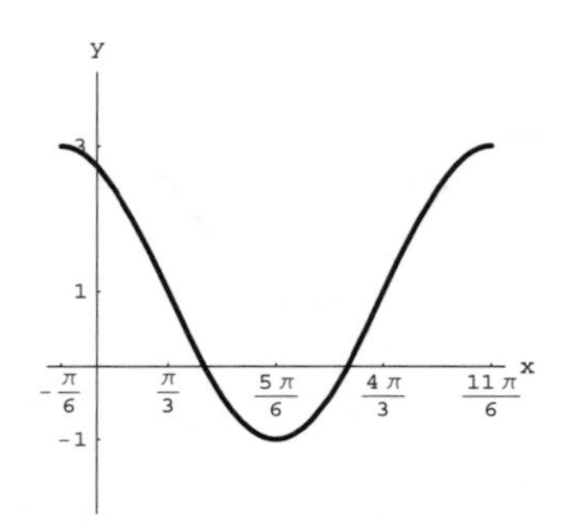

73. Amplitude 2, phase shift $\pi/3$, range $[-1, 3]$, some points are $\left(\frac{\pi}{3}, 1\right)$, $\left(\frac{5\pi}{6}, -1\right)$, $\left(\frac{4\pi}{3}, 1\right)$, $\left(\frac{11\pi}{6}, 3\right)$ $\left(\frac{7\pi}{3}, 1\right)$

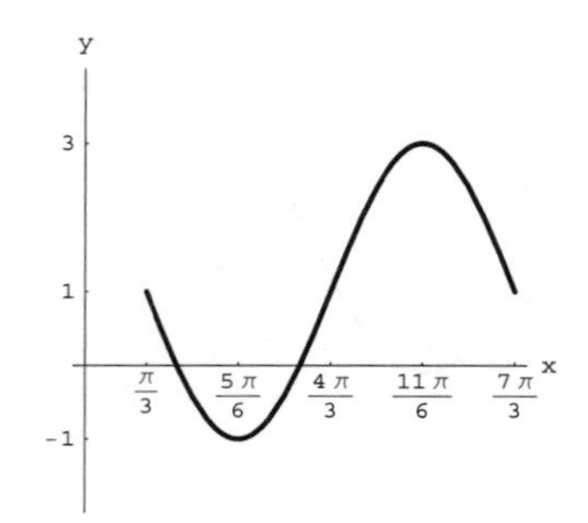

75. Note the amplitude is 2, phase shift is $\pi/2$, and the period is 2π. Then $A = 2$, $C = \pi/2$, and $B = 1$. An equation is $y = 2\sin\left(x - \frac{\pi}{2}\right)$.

77. Note the amplitude is 2, phase shift $-\pi$, the period is 2π, and the vertical shift is 1 unit up. Then $A = 2$, $C = -\pi$, $B = 1$, and $D = 1$.

An equation is $y = 2\sin(x + \pi) + 1$.

79. Note the amplitude is 2, phase shift is $\pi/2$, and the period is 2π. Then $A = 2$, $C = \pi/2$, and $B = 1$. An equation is $y = 2\cos\left(x - \frac{\pi}{2}\right)$.

81. Note, the amplitude is 2, phase shift is π, the period is 2π, and the vertical shift is 1 unit up. Then $A = 2$, $C = \pi$, $B = 1$, and $D = 1$.

An equation is $y = 2\cos(x - \pi) + 1$.

83. $y = \sin\left(x - \frac{\pi}{4}\right)$

85. $y = \sin\left(x + \frac{\pi}{2}\right)$

87. $y = -\cos\left(x - \frac{\pi}{5}\right)$

89. $y = -\cos\left(x - \frac{\pi}{8}\right) + 2$

91. $y = -3\cos\left(x + \frac{\pi}{4}\right) - 5$

93. A determines the stretching, shrinking, or reflection about the x-axis, C is the phase shift, and D is the vertical translation.

For Thought

1. True, since $B = 4$ and the period is $\frac{2\pi}{B} = \frac{\pi}{2}$.

2. False, since $B = 2\pi$ and the period is $\frac{2\pi}{B} = 1$.

3. True, since $B = \pi$ and the period is $\frac{2\pi}{B} = 2$.

4. True, since $B = 0.1\pi$ and the period is $\frac{2\pi}{0.1\pi} =$ 20.

5. False, the phase shift is $-\frac{\pi}{12}$.

6. False, the phase shift is $-\dfrac{\pi}{8}$.

7. True, since the period is $P = 2\pi$ the frequency is $\dfrac{1}{P} = \dfrac{1}{2\pi}$.

8. True, since the period is $P = 2$ the frequency is $\dfrac{1}{P} = \dfrac{1}{2}$.

9. False, rather the graphs of $y = \cos(x)$ and $y = \sin\left(x + \dfrac{\pi}{2}\right)$ are identical.

10. True

2.2 Exercises

1. Amplitude 3, period $\dfrac{2\pi}{4}$ or $\dfrac{\pi}{2}$, and phase shift 0

3. Since $y = -2\cos\left(2\left(x + \dfrac{\pi}{4}\right)\right) - 1$, we get amplitude 2, period $\dfrac{2\pi}{2}$ or π, and phase shift $-\dfrac{\pi}{4}$.

5. Since $y = -2\sin(\pi(x-1))$, we get amplitude 2, period $\dfrac{2\pi}{\pi}$ or 2, and phase shift 1.

7. Period $2\pi/3$, phase shift 0, range $[-1, 1]$, labeled points are $(0,0)$, $\left(\dfrac{\pi}{6}, 1\right)$, $\left(\dfrac{\pi}{3}, 0\right)$, $\left(\dfrac{\pi}{2}, -1\right)$, $\left(\dfrac{2\pi}{3}, 0\right)$

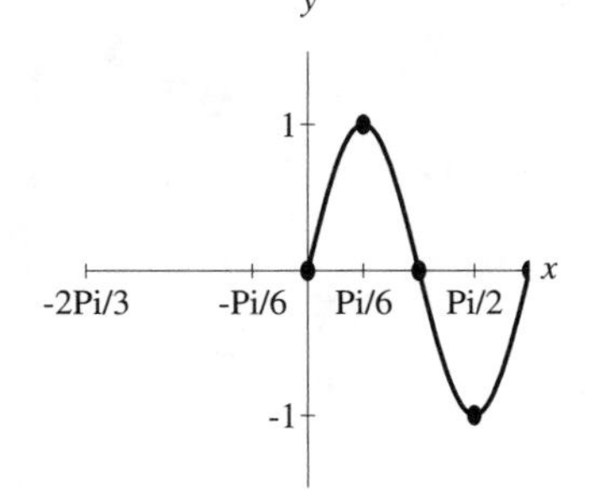

9. Period π, phase shift 0, range $[-1, 1]$, labeled points are $(0,0)$, $\left(\dfrac{\pi}{4}, -1\right)$, $\left(\dfrac{\pi}{2}, 0\right)$, $\left(\dfrac{3\pi}{4}, 1\right)$, $(\pi, 0)$

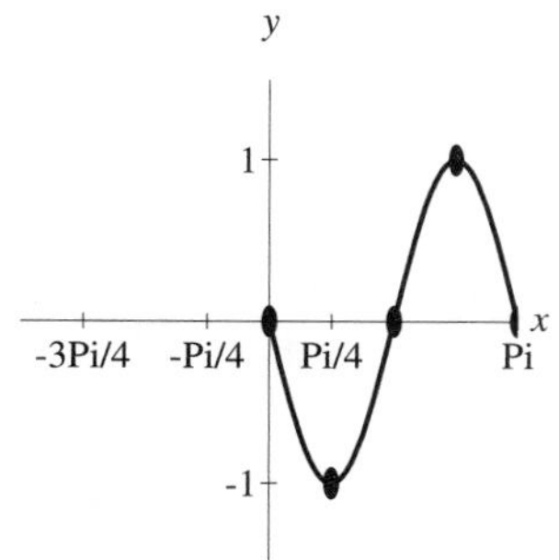

11. Period $\pi/2$, phase shift 0, range $[1, 3]$, labeled points are $(0,3)$, $\left(\dfrac{\pi}{8}, 2\right)$, $\left(\dfrac{\pi}{4}, 1\right)$, $\left(\dfrac{3\pi}{8}, 2\right)$, $\left(\dfrac{\pi}{2}, 3\right)$

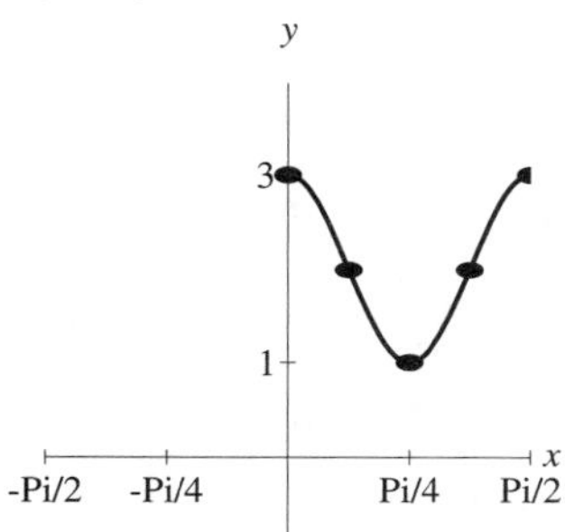

13. Period 8π, phase shift 0, range $[1, 3]$, labeled points are $(0,2)$, $(2\pi, 1)$, $(4\pi, 2)$, $(6\pi, 3)$, $(8\pi, 2)$

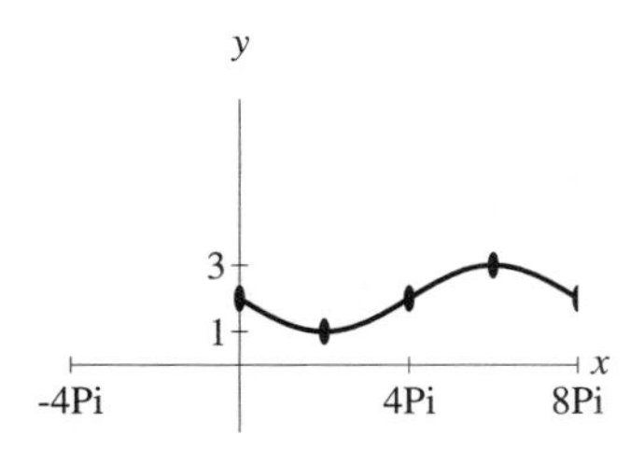

15. Period 6, phase shift 0, range $[-1,1]$, labeled points are $(0,0)$, $(1.5,1)$, $(3,0)$, $(4.5,-1)$, $(6,0)$

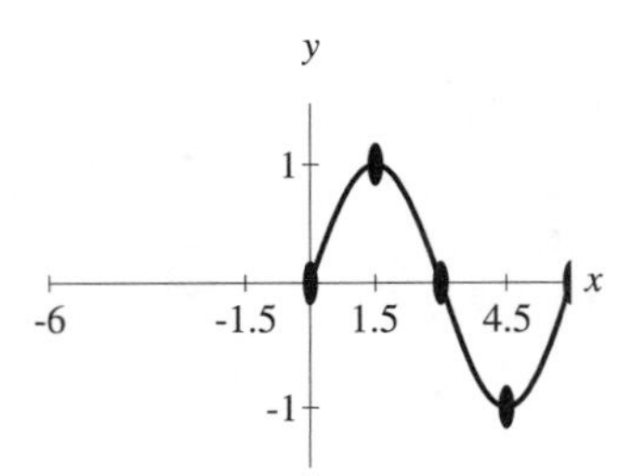

17. Period π, phase shift $\pi/2$, range $[-1,1]$, labeled points are $\left(\frac{\pi}{2},0\right)$, $\left(\frac{3\pi}{4},1\right)$, $(\pi,0)$, $\left(\frac{5\pi}{4},-1\right)$, $\left(\frac{3\pi}{2},0\right)$

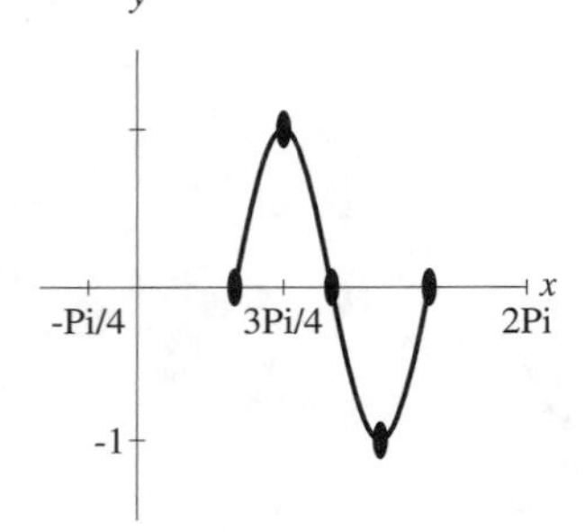

19. Period 4, phase shift -3, range $[-1,1]$, labeled points are $(-3,0)$, $(-2,1)$, $(-1,0)$, $(0,-1)$, $(1,0)$

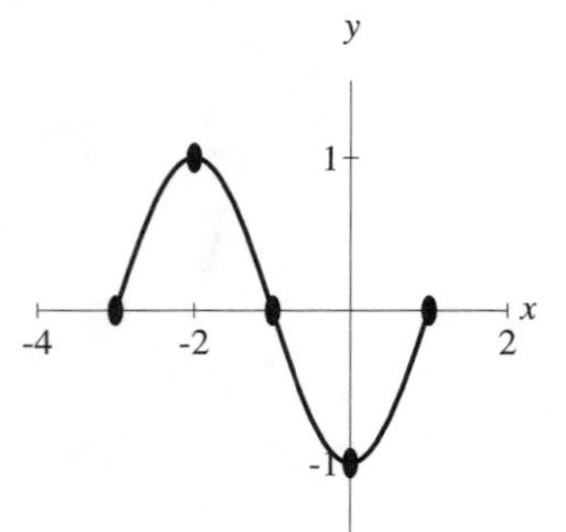

21. Period π, phase shift $-\pi/6$, range $[-1,3]$, labeled points are $\left(-\frac{\pi}{6},3\right)$, $\left(\frac{\pi}{12},1\right)$, $\left(\frac{\pi}{3},-1\right)$, $\left(\frac{7\pi}{12},1\right)$, $\left(\frac{5\pi}{6},3\right)$

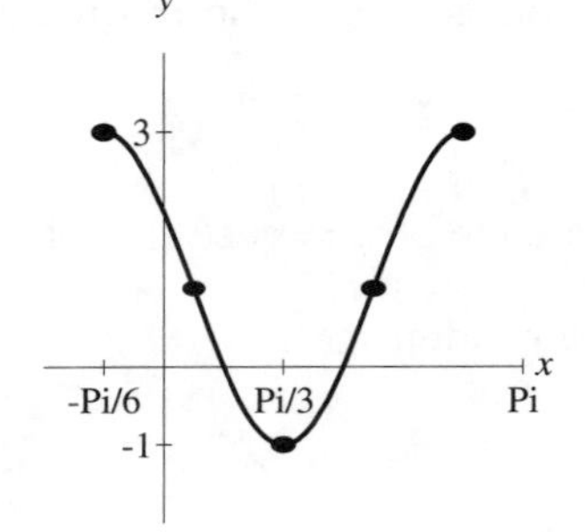

23. Period $\frac{2\pi}{3}$, phase shift $\frac{\pi}{6}$, range $\left[-\frac{3}{2},-\frac{1}{2}\right]$, labeled points are $\left(\frac{\pi}{6},-1\right)$, $\left(\frac{\pi}{3},-\frac{3}{2}\right)$, $\left(\frac{\pi}{2},-1\right)$, $\left(\frac{2\pi}{3},-\frac{1}{2}\right)$, $\left(\frac{5\pi}{6},-1\right)$

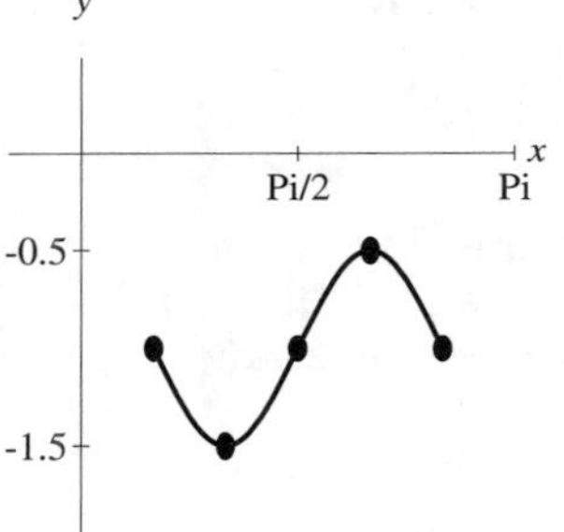

25. Note, $A=2$, period is π and so $B=2$, phase shift is $C=\frac{\pi}{4}$, and $D=0$ or no vertical shift.

Then $y=2\sin\left(2\left(x-\frac{\pi}{4}\right)\right)$.

27. Note, $A=3$, period is $\frac{4\pi}{3}$ and so $B=\frac{3}{2}$, phase shift is $C=-\frac{\pi}{3}$, and $D=3$ since the vertical shift is three units up.

Then $y=3\sin\left(\frac{3}{2}\left(x+\frac{\pi}{3}\right)\right)+3$.

29. $\frac{\pi}{4}-\frac{\pi}{4}=0$

31. $f\left(g\left(\frac{\pi}{4}\right)\right) = f(0) = \sin(0) = 0$

33. $h\left(f\left(g\left(\frac{\pi}{4}\right)\right)\right) = h\left(f\left(0\right)\right) = h\left(\sin\left(0\right)\right) =$ $h(0) = 3 \cdot 0 = 0$

35. $f\left(g\left(x\right)\right) = f\left(x - \frac{\pi}{4}\right) = \sin\left(x - \frac{\pi}{4}\right)$

37. $h\left(f\left(g\left(x\right)\right)\right) = h\left(f\left(x - \frac{\pi}{4}\right)\right) =$ $h\left(\sin\left(x - \frac{\pi}{4}\right)\right) = 3\sin\left(x - \frac{\pi}{4}\right)$

39. 100 cycles/sec since the frequency is the reciprocal of the period

41. Frequency is $\frac{1}{0.025} = 40$ cycles per hour

43. Substitute $v_o = 6$, $\omega = 2$, and $x_o = 0$ into $x(t) = \frac{v_o}{\omega} \cdot \sin(\omega t) + x_o \cdot \cos(\omega t)$.
Then $x(t) = 3\sin(2t)$.
The amplitude is 3 and the period is π.

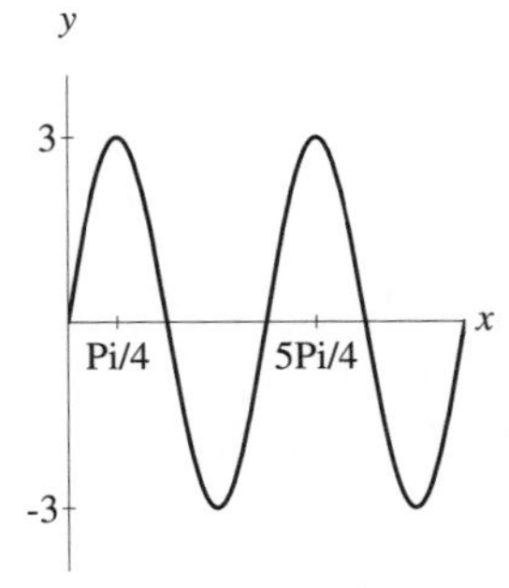

45. 11 years

47. Note, the range of $v = 400\sin(60\pi t) + 900$ is $[-400 + 900, 400 + 900]$ or $[500, 1300]$.
(a) Maximum volume is 1300 cc and mininum volume is 500 cc
(b) The runner takes a breath every 1/30 (which is the period) of a minute. So a runner makes 30 breaths in one minute.

49. Period is 12, amplitude is 15,000, phase-shift is -3, vertical translation is 25,000, a formula for the curve is

$$y = 15,000\sin\left(\frac{\pi}{6}x + \frac{\pi}{2}\right) + 25,000;$$

for April (when $x = 4$), the revenue is

$$15,000\sin\left(\frac{\pi}{6}x + \frac{\pi}{2}\right) + 25,000 \approx \$17,500.$$

51.

a) period is 40, amplitude is 65, an equation for the sine wave is

$$y = 65\sin\left(\frac{\pi}{20}x\right)$$

b) 40 days

c) $65\sin\left(\frac{\pi}{20}(36)\right) \approx -38.2$ meters/second

d) The new planet is between Earth and Rho.

53. Since the period is $20 = \frac{2\pi}{B}$, we get $B = \frac{\pi}{10}$. Also, the amplitude is 1 and the vertical translation is 1. An equation for the swell is

$$y = \sin\left(\frac{\pi}{10}x\right) + 1.$$

55. With a calculator we obtain the sine regression equation is $y = a\sin(bx + c) + d$ where $a = 51.62635869$, $b = 0.1985111953$, $c = 1.685588984$, $d = 50.42963472$, or about

$$y = 51.6\sin(0.20x + 1.69) + 50.43.$$

The period is

$$\frac{2\pi}{b} = \frac{2\pi}{0.1985111953} \approx 31.7 \text{ days.}$$

When $x = 35$, we find

$$a\sin(b(35) + c) + d \approx 87\%$$

of the moon is illuminated on February 4,2010.

Shown below is a graph of the regression equation and the data points.

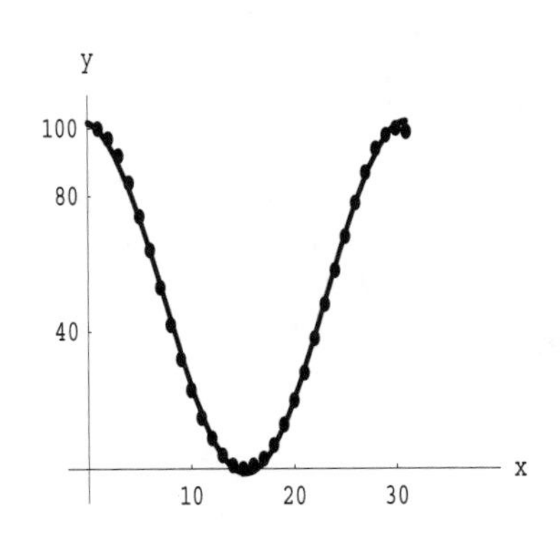

For Thought

1. False, since $\sec(\pi/4) = \dfrac{1}{\cos(\pi/4)}$.

2. True, since $\csc(x) = \dfrac{1}{\sin(x)}$.

3. True, since $\csc(\pi/2) = \dfrac{1}{\sin(\pi/2)} = \dfrac{1}{1} = 1$.

4. False, since $\sec(\pi/2) = \dfrac{1}{\cos(\pi/2)} = \dfrac{1}{0}$ is undefined.

5. True, since $B = 2$ and $\dfrac{2\pi}{B} = \dfrac{2\pi}{2} = \pi$.

6. True, since $B = \pi$ and $\dfrac{2\pi}{B} = \dfrac{2\pi}{\pi} = 2$.

7. False, rather the graphs of $y = 2\csc x$ and $y = \dfrac{2}{\sin x}$ are identical.

8. True, since the maximum and minimum of $0.5\csc x$ are ± 0.5.

9. True, since $\dfrac{\pi}{2} + k\pi$ are the zeros of $y = \cos x$ we get that the asymptotes of $y = \sec(2x)$ are $2x = \dfrac{\pi}{2} + k\pi$ or $x = \dfrac{\pi}{4} + \dfrac{k\pi}{2}$. If $k = \pm 1$, we get the asymptotes $x = \pm\dfrac{\pi}{4}$.

10. True, since if we substitute $x = 0$ in $\dfrac{1}{\csc(4x)}$ we get $\dfrac{1}{\csc(0)} = \dfrac{1}{0}$ which is undefined.

2.3 Exercises

1. $\dfrac{1}{\cos(\pi/3)} = \dfrac{1}{1/2} = 2$

3. $\dfrac{1}{\sin(-\pi/4)} = \dfrac{1}{-1/\sqrt{2}} = -\sqrt{2}$

5. Undefined, since $\dfrac{1}{\cos(\pi/2)} = \dfrac{1}{0}$

7. Undefined, since $\dfrac{1}{\sin(\pi)} = \dfrac{1}{0}$

9. $\dfrac{1}{\cos 1.56} \approx 92.6$

11. $\dfrac{1}{\sin 0.01} \approx 100.0$

13. $\dfrac{1}{\sin 3.14} \approx 627.9$

15. $\dfrac{1}{\cos 4.71} \approx -418.6$

17. Since $B = 2$, the period is $\dfrac{2\pi}{B} = \dfrac{2\pi}{2}$ or π

19. Since $B = \dfrac{3}{2}$, the period is $\dfrac{2\pi}{B} = \dfrac{2\pi}{3/2}$ or $\dfrac{4\pi}{3}$

21. Since $B = \pi$, the period is $\dfrac{2\pi}{B} = \dfrac{2\pi}{\pi}$ or 2

23. $(-\infty, -2] \cup [2, \infty)$

25. $(-\infty, -1/2] \cup [1/2, \infty)$

27. Since the range of $y = \sec(\pi x - 3\pi)$ is

$$(-\infty, -1] \cup [1, \infty),$$

the range of $y = \sec(\pi x - 3\pi) - 1$ is

$$(-\infty, -1 - 1] \cup [1 - 1, \infty)$$

or equivalently

$$(-\infty, -2] \cup [0, \infty).$$

29. period 2π, asymptotes $x = \dfrac{\pi}{2} + k\pi$, range $(-\infty, -2] \cup [2, \infty)$

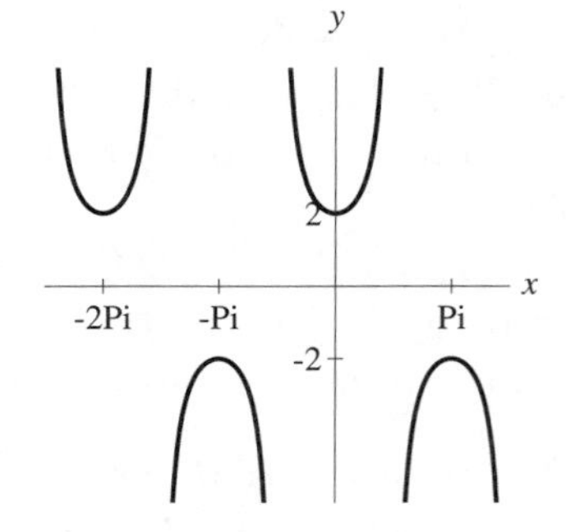

31. period $2\pi/3$, asymptotes $3x = \frac{\pi}{2} + k\pi$ or

$x = \frac{\pi}{6} + \frac{k\pi}{3}$, range $(-\infty, -1] \cup [1, \infty)$

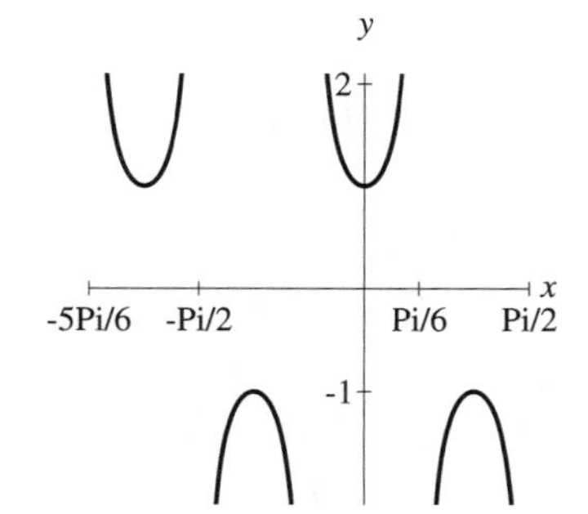

33. period $\frac{2\pi}{1} = 2\pi$,asymptotes $x + \frac{\pi}{4} = \frac{\pi}{2} + k\pi$

or $x = \frac{\pi}{4} + k\pi$, range $(-\infty, -1] \cup [1, \infty)$

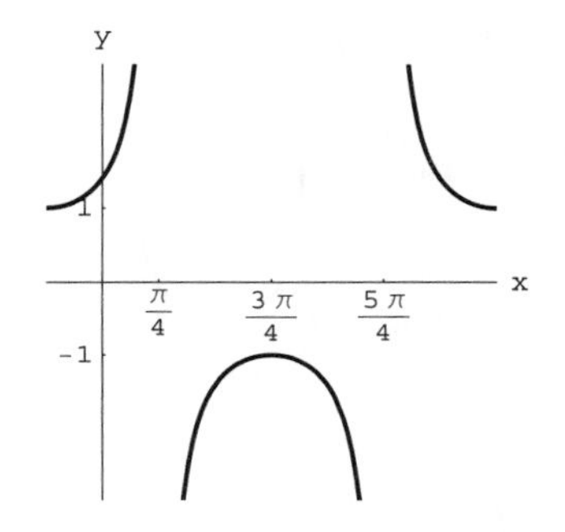

35. period $\frac{2\pi}{1/2} = 4\pi$, asymptotes $\frac{x}{2} = \frac{\pi}{2} + k\pi$ or

$x = \pi + 2k\pi$, range $(-\infty, -1] \cup [1, \infty)$

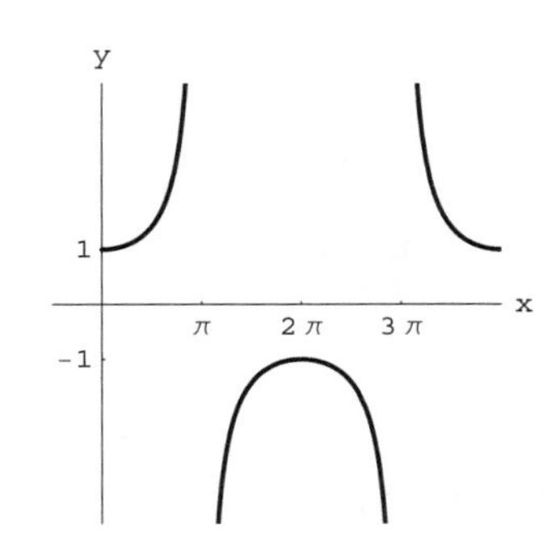

37. period $\frac{2\pi}{\pi} = 2$, asymptotes $\pi x = \frac{\pi}{2} + k\pi$ or

$x = \frac{1}{2} + k$, range $(-\infty, -2] \cup [2, \infty)$

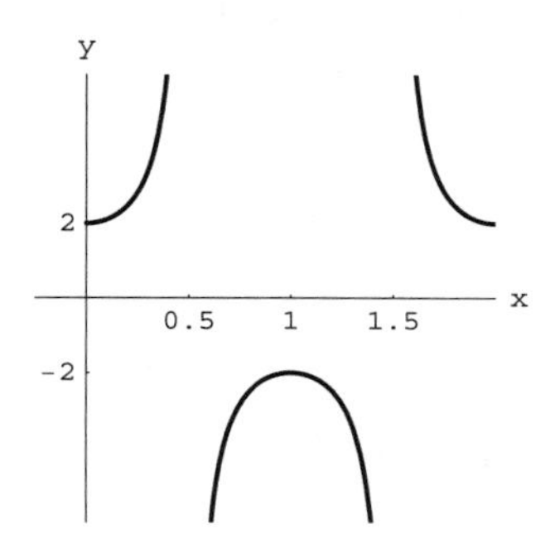

39. period $\frac{2\pi}{2} = \pi$, asymptotes $2x = \frac{\pi}{2} + k\pi$ or

$x = \frac{\pi}{4} + \frac{k\pi}{2}$, and since the range of
$y = 2\sec(2x)$ is $(-\infty, -2] \cup [2, \infty)$ then
the range of $y = 2 + 2\sec(2x)$ is
$(-\infty, -2 + 2] \cup [2 + 2, \infty)$ or $(-\infty, 0] \cup [4, \infty)$.

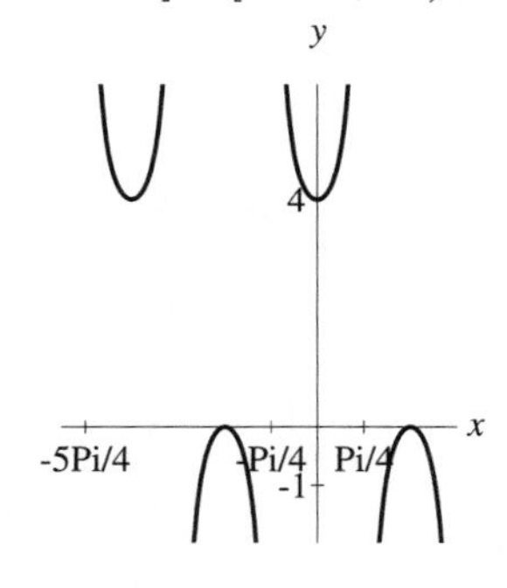

41. period 2π, asymptotes $x = k\pi$,

range $(-\infty, -2] \cup [2, \infty)$

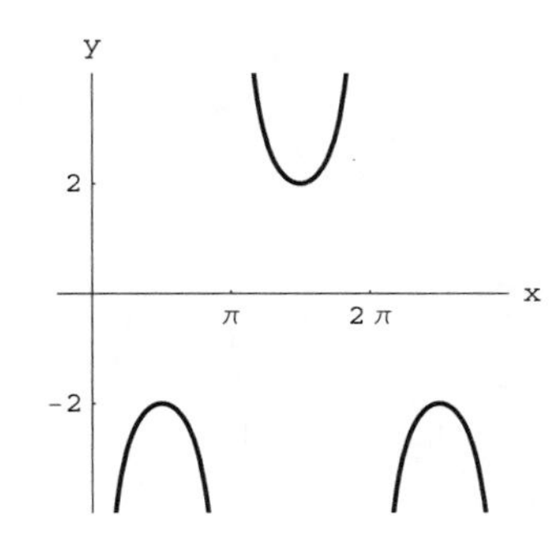

43. period 2π ,asymptotes $x+\dfrac{\pi}{2}=k\pi$ or $x=-\dfrac{\pi}{2}+k\pi$ or $x=\dfrac{\pi}{2}+k\pi$, range $(-\infty,-1]\cup[1,\infty)$

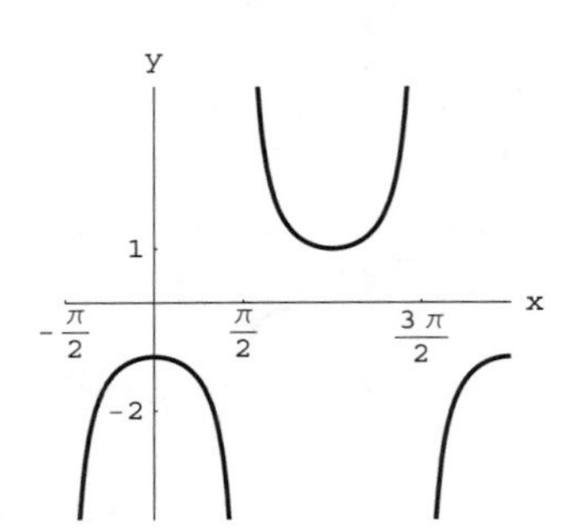

45. period $\dfrac{2\pi}{2}$ or π, asymptotes $2x-\dfrac{\pi}{2}=k\pi$ or $x=\dfrac{\pi}{4}+\dfrac{k\pi}{2}$, range $(-\infty,-1]\cup[1,\infty)$

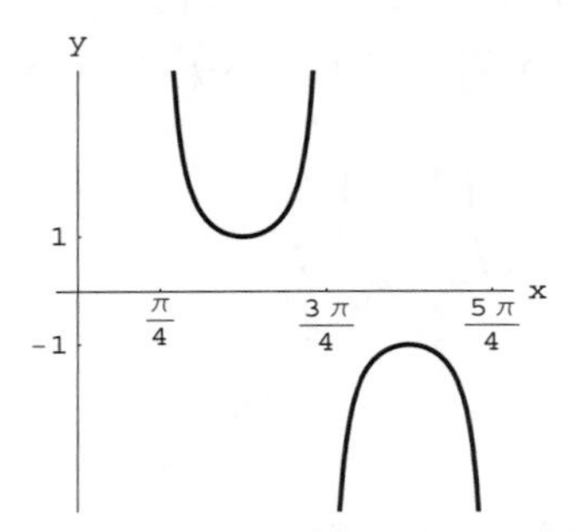

47. period $\dfrac{2\pi}{\pi/2}$ or 4,asymptotes $\dfrac{\pi x}{2}-\dfrac{\pi}{2}=k\pi$ or $x=1+2k$, range $(-\infty,-1]\cup[1,\infty)$

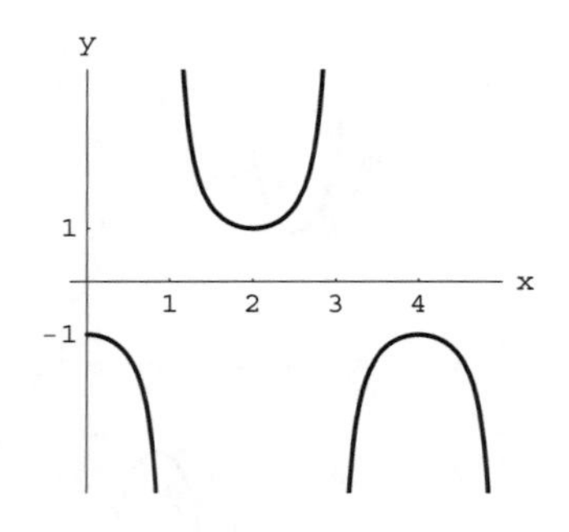

49. period $\dfrac{2\pi}{2}$ or π, asymptotes $2x-\dfrac{\pi}{2}=k\pi$ or $x=\dfrac{\pi}{4}+\dfrac{k\pi}{2}$, range $(-\infty,-1]\cup[1,\infty)$

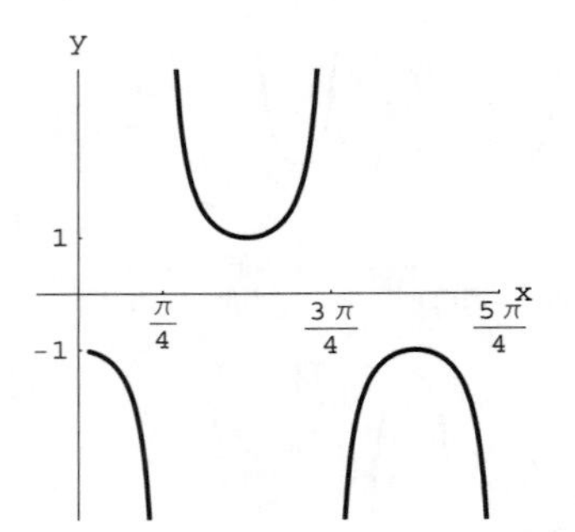

51. period $\dfrac{2\pi}{\pi/2}$ or 4, asymptotes $\dfrac{\pi x}{2}+\dfrac{\pi}{2}=k\pi$ or $x=-1+2k$ or $x=1+2k$, range $(-\infty,-1]\cup[1,\infty)$

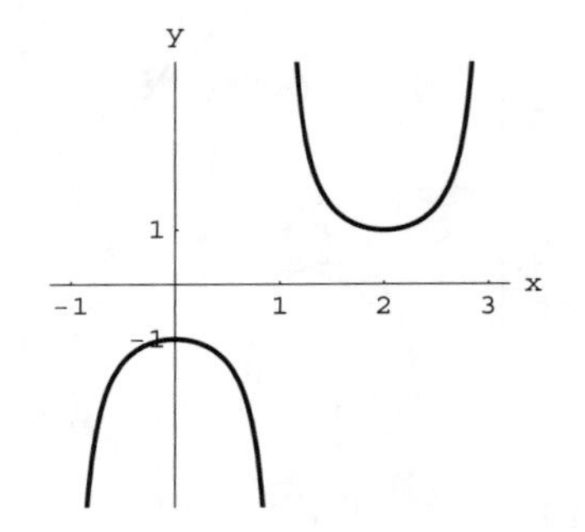

53. $y=\sec\left(x-\dfrac{\pi}{2}\right)+1$

55. $y=-\csc(x+1)+4$

57. Since the zeros of $y=\cos x$ are $x=\dfrac{\pi}{2}+k\pi$, the vertical asymptotes of $y=\sec x$ are $x=\dfrac{\pi}{2}+k\pi$.

59. Note, the zeros of $y=\sin x$ are $x=k\pi$. To find the asymptotes of $y=\csc(2x)$, let $2x=k\pi$. The asymptotes are $x=\dfrac{k\pi}{2}$.

61. Note, the zeros of $y=\cos x$ are $x=\dfrac{\pi}{2}+k\pi$. To find the asymptotes of $y=\sec\left(x-\dfrac{\pi}{2}\right)$, let $x-\dfrac{\pi}{2}=\dfrac{\pi}{2}+k\pi$. Solving for x, we get $x=\pi+k\pi$ or equivalently $x=k\pi$. The asymptotes are $x=k\pi$.

63 Note, the zeros of $y = \sin x$ are $x = k\pi$. To find the asymptotes of $y = \csc(2x - \pi)$, let $2x - \pi = k\pi$. Solving for x, we get $x = \frac{\pi}{2} + \frac{k\pi}{2}$ or equivalently $x = \frac{k\pi}{2}$. The asymptotes are $x = \frac{k\pi}{2}$.

65 Note, the zeros of $y = \sin x$ are $x = k\pi$. To find the asymptotes of $y = \frac{1}{2}\csc(2x) + 4$, let $2x = k\pi$. Solving for x, we get that the asymptotes are $x = \frac{k\pi}{2}$.

67. Note, the zeros of $y = \cos x$ are $x = \frac{\pi}{2} + k\pi$. To find the asymptotes of $y = \sec(\pi x + \pi)$, let $\pi x + \pi = \frac{\pi}{2} + k\pi$. Solving for x, we get $x = -\frac{1}{2} + k$ or equivalently $x = \frac{1}{2} + k$. The asymptotes are $x = \frac{1}{2} + k$.

69. Since the range of $y = A\sec(B(x - C))$ is $(-\infty, |A|] \cup [|A|, \infty)$, the range of $y = A\sec(B(x - C)) + D$ is

$$(-\infty, |A| + D] \cup [|A| + D, \infty).$$

For Thought

1. True, since $\tan x = \frac{\sin x}{\cos x}$.

2. True, since $\cot x = \frac{1}{\tan x}$ provided $\tan x \neq 0$.

3. False, since $\cot\left(\frac{\pi}{2}\right) = 0$ and $\tan\left(\frac{\pi}{2}\right)$ is undefined.

4. True, since $\frac{\sin 0}{\cos 0} = \frac{0}{1} = 0$.

5. False, since $\frac{\sin(\pi/2)}{\cos(\pi/2)} = \frac{1}{0}$ is undefined.

6. False, since $\frac{\sin(5\pi/2)}{\cos(5\pi/2)} = \frac{1}{0}$ is undefined.

7. True

8. False, the range of $y = \cot x$ is $(-\infty, \infty)$.

9. True, since $\tan\left(3 \cdot \left(\pm\frac{\pi}{6}\right)\right) = \tan\left(\pm\frac{\pi}{2}\right) = \frac{\pm 1}{0}$ is undefined.

10. True, since $\cot\left(4 \cdot \left(\pm\frac{\pi}{4}\right)\right) = \cot(\pm\pi) = \frac{\pm 1}{0}$ is undefined.

2.4 Exercises

1. $\frac{\sin(\pi/3)}{\cos(\pi/3)} = \frac{\sqrt{3}/2}{1/2} = \frac{\sqrt{3}}{2} \cdot \frac{2}{1} = \sqrt{3}$

3. Undefined, since $\frac{\sin(\pi/2)}{\cos(\pi/2)} = \frac{1}{0}$

5. $\frac{\sin(\pi)}{\cos(\pi)} = \frac{0}{-1} = 0$

7. $\frac{\cos(\pi/4)}{\sin(\pi/4)} = \frac{\sqrt{2}/2}{\sqrt{2}/2} = 1$

9. Undefined, since $\frac{\cos(0)}{\sin(0)} = \frac{1}{0}$

11. $\frac{\cos(\pi/2)}{\sin(\pi/2)} = \frac{0}{1} = 0$

13. 92.6

15. -108.6

17. $\frac{1}{\tan 0.002} \approx 500.0$

19. $\frac{1}{\tan(-0.002)} \approx -500.0$

21. Since $B = 8$, the period is $\frac{\pi}{B} = \frac{\pi}{8}$.

23. Since $B = \pi$, the period is $\frac{\pi}{B} = \frac{\pi}{\pi} = 1$.

25. Since $B = \frac{\pi}{3}$, the period is $\frac{\pi}{B} = \frac{\pi}{\pi/3} = 3$.

27. Since $B = 3$, the period is $\frac{\pi}{B} = \frac{\pi}{3}$.

29. $y = \tan(3x)$ has period $\dfrac{\pi}{3}$, and if $3x = \dfrac{\pi}{2} + k\pi$ then the asymptotes are $x = \dfrac{\pi}{6} + \dfrac{k\pi}{3}$ for any integer k.

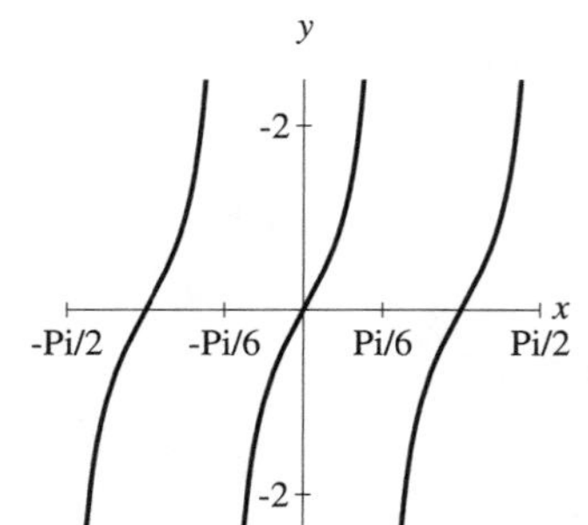

31. $y = \tan(\pi x)$ has period 1, and if $\pi x = \dfrac{\pi}{2} + k\pi$ then the asymptotes are $x = \dfrac{1}{2} + k$ for any integer k.

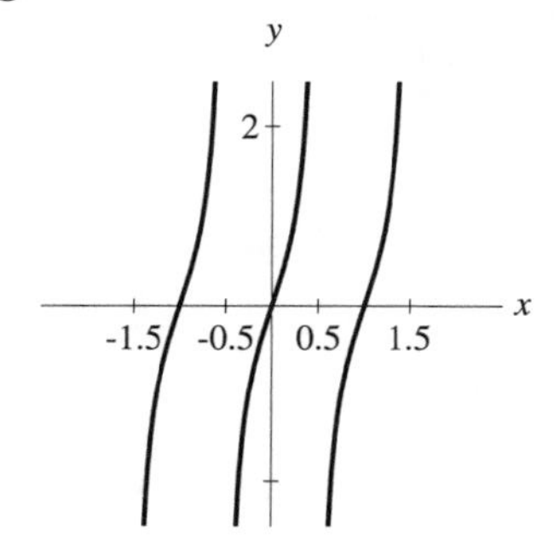

33. $y = -2\tan(x) + 1$ has period π, and the asymptotes are $x = \dfrac{\pi}{2} + k\pi$ where k is an integer

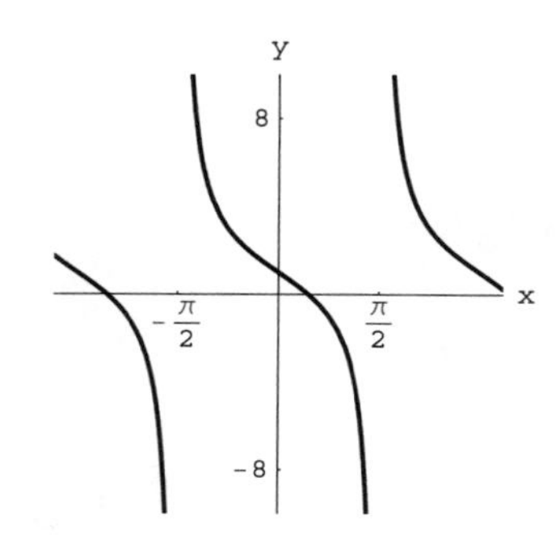

35. $y = -\tan(x - \pi/2)$ has period π, and if $x - \pi/2 = \dfrac{\pi}{2} + k\pi$ then the asymptotes are $x = \pi + k\pi$ or $x = k\pi$ for any integer k

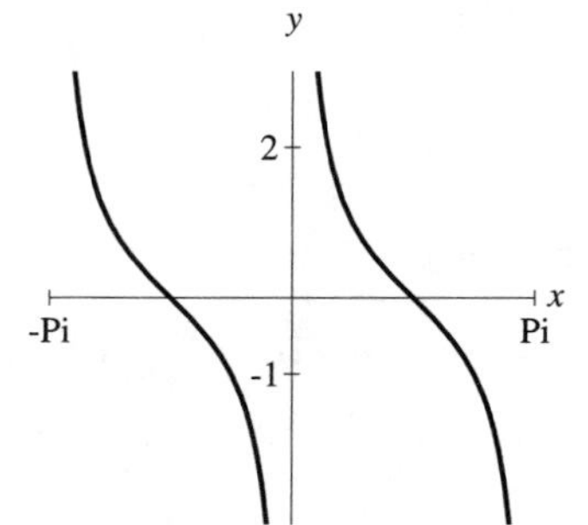

37. $y = \tan\left(\dfrac{\pi}{2}x - \dfrac{\pi}{2}\right)$ has period $\dfrac{\pi}{\pi/2}$ or 2, and if $\dfrac{\pi}{2}x - \dfrac{\pi}{2} = \dfrac{\pi}{2} + k\pi$ then the asymptotes are $x = 2 + 2k$ or $x = 2k$ for any integer k

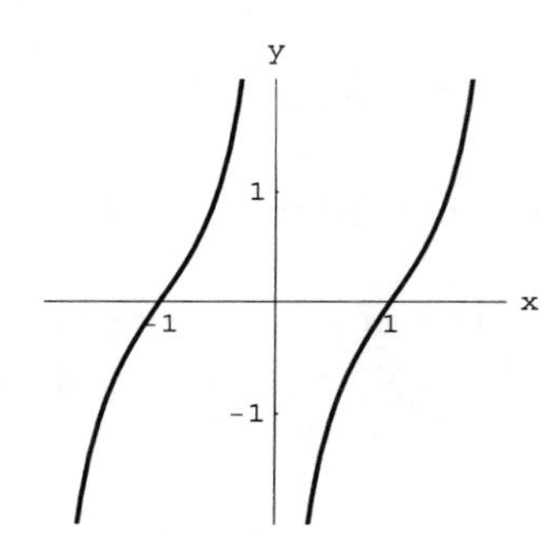

39. $y = \cot(x + \pi/4)$ has period π, and if $x + \dfrac{\pi}{4} = k\pi$ then the asymptotes are $x = -\dfrac{\pi}{4} + k\pi$ or $x = \dfrac{3\pi}{4} + k\pi$ for any integer k

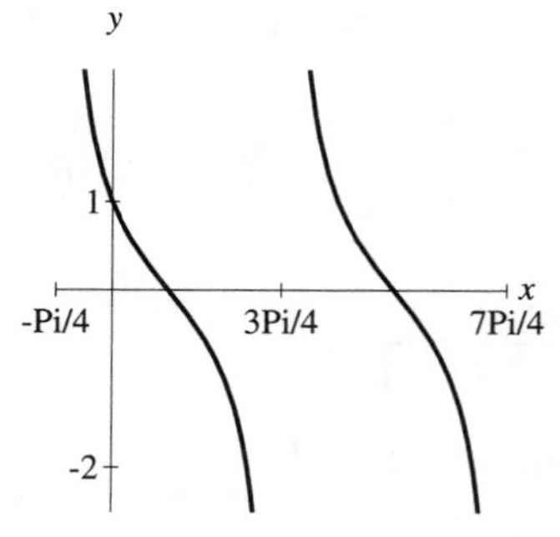

41. $y = \cot(x/2)$ has period 2π, and if $\frac{x}{2} = k\pi$ then the asymptotes are $x = 2k\pi$ for any integer k

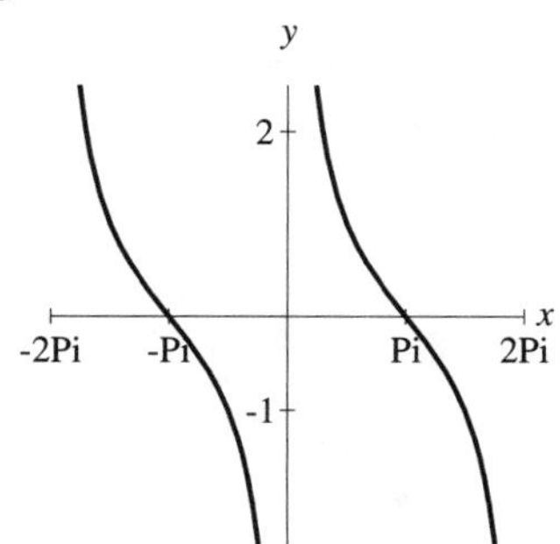

43. $y = -\cot(x + \pi/2)$ has period π, and if $x + \frac{\pi}{2} = k\pi$ then the asymptotes are $x = -\frac{\pi}{2} + k\pi$ or $x = \frac{\pi}{2} + k\pi$ for any integer k

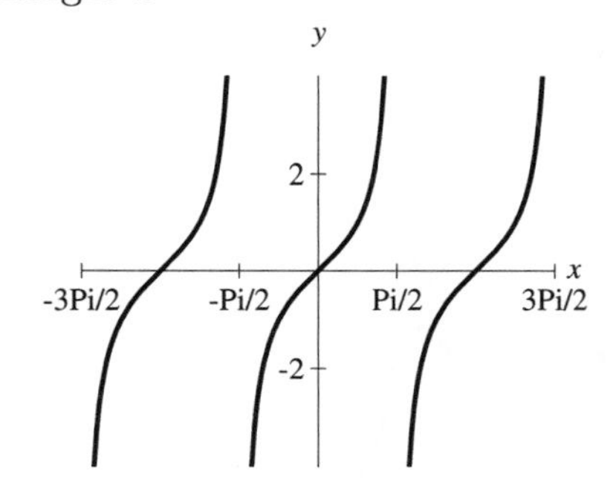

45. $y = \cot(2x - \pi/2) - 1$ has period $\frac{\pi}{2}$, and if $2x - \frac{\pi}{2} = k\pi$ then the asymptotes are $x = \frac{\pi}{4} + \frac{k\pi}{2}$ for any integer k

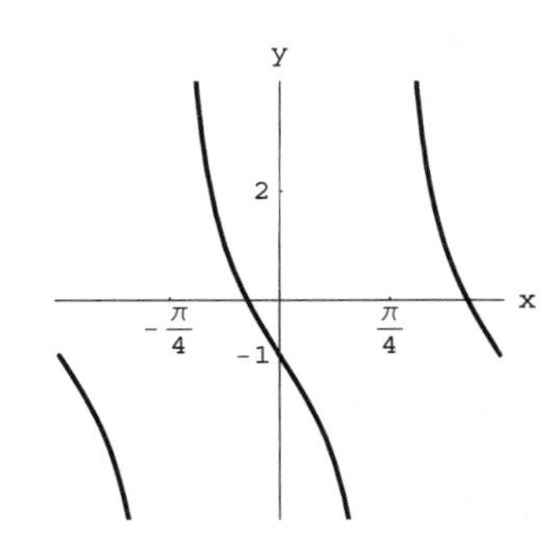

47. $y = 3\tan\left(x - \frac{\pi}{4}\right) + 2$

49. $y = -\cot\left(x + \frac{\pi}{2}\right) + 1$

51. Note, the period is $\frac{\pi}{2}$. So $\frac{\pi}{B} = \frac{\pi}{2}$ and $B = 2$. The phase shift is $\frac{\pi}{4}$ and $A = 1$ since $\left(\frac{3\pi}{8}, 1\right)$ is a point on the graph.

An equation is $y = \tan\left(2\left(x - \frac{\pi}{4}\right)\right)$.

53. Note, the period is 2. So $\frac{\pi}{B} = 2$ and $B = \frac{\pi}{2}$. Since the graph is reflected about the x-axis and $\left(\frac{1}{2}, -1\right)$ is a point on the graph, we find $A = -1$. An equation is $y = -\tan\left(\frac{\pi}{2}x\right)$.

55. $f(g(-3)) = f(0) = \tan(0) = 0$

57. Undefined, since $\tan(\pi/2)$ is undefined and $g(h(f(\pi/2))) = g(h(\tan(\pi/2)))$

59. $f(g(h(x))) = f(g(2x)) = f(2x + 3) = \tan(2x + 3)$

61. $g(h(f(x))) = g(h(\tan(x))) = g(2\tan(x)) = 2\tan(x) + 3$

63. Note $m = \tan(\pi/4) = 1$. Since the line passes through $(2, 3)$, we get $y - 3 = 1 \cdot (x - 2)$. Solving for y, we obtain $y = x + 1$.

65. Note $m = \tan(\pi/3) = \sqrt{3}$. Since the line passes through $(3, -1)$, we get

$$y + 1 = \sqrt{3}(x - 3).$$

Solving for y, we obtain $y = \sqrt{3}x - 3\sqrt{3} - 1$.

67.

a) Period is about 2.3 years

b) It looks like the graph of a tangent function.

For Thought

1. False, since the graph of $y = x + \sin x$ does not duplicate itself.

2. True, since the range of $y = \sin x$ is $[-1, 1]$ it follows that $y = x + \sin x$ oscillates about $y = x$.

3. True, since if x is small then the value of y in $y = \dfrac{1}{x}$ is a large number.

4. True, since $y = 0$ is the horizontal asymptote of $y = \dfrac{1}{x}$ and the x-axis is the graph of $y = 0$.

5. True, since the range of $y = \sin x$ is $[-1, 1]$ it follows that $y = \dfrac{1}{x} + \sin x$ oscillates about $y = \dfrac{1}{x}$.

6. True, since $\sin x = 1$ whenever $x = \dfrac{\pi}{2} + k2\pi$ and $\sin x = -1$ whenever $x = \dfrac{3\pi}{2} + k2\pi$, and since $\dfrac{1}{x}$ is approximately zero when x is a large numbers, then $\dfrac{1}{x} + \sin x = 0$ has many solutions in x for 0 is between 1 and -1.

7. False, since $\cos(\pi/6)+\cos(2\cdot\pi/6) = \dfrac{\sqrt{3}+1}{2} > 1$ then 1 is not the maximum of $y = \cos(x) + \cos(2x)$.

8. False, since $\sin(x) + \cos(x) = \sqrt{2}\sin\left(x + \dfrac{\pi}{4}\right)$ then the maximum value of $\sin(x) + \cos(x)$ is $\sqrt{2}$, and not 2.

9. True, since on the interval $[0, 2\pi]$ we find that $\sin(x) = 0$ for $x = 0, \pi, 2\pi$.

10. True, since $B = \pi$ and the period is $\dfrac{2\pi}{B} = \dfrac{2\pi}{\pi}$ or 2.

2.5 Exercises

1. For each x-coordinate, the y-coordinate of $y = x + \cos x$ is the sum of the y-coordinates of $y_1 = x$ and $y_2 = \cos x$. Note, the graph below oscillates about $y_1 = x$.

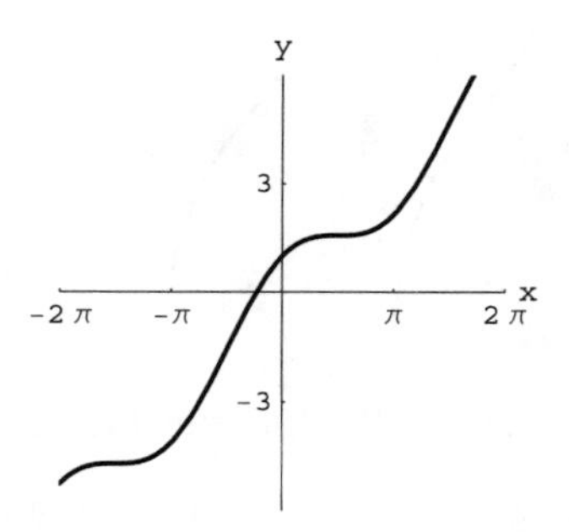

3. For each x-coordinate, the y-coordinate of $y = \dfrac{1}{x} - \sin x$ is obtained by subtracting the y-coordinate of $y_2 = \sin x$ from the y-coordinate of $y_1 = \dfrac{1}{x}$.

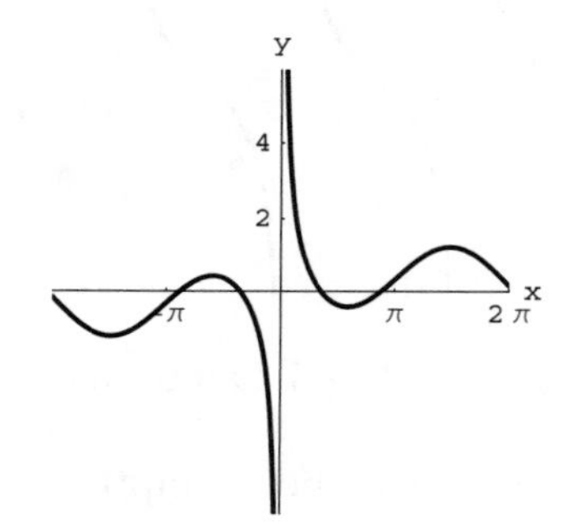

5. For each x-coordinate, the y-coordinate of $y = \dfrac{1}{2}x + \sin x$ is the sum of the y-coordinates of $y_1 = \dfrac{1}{2}x$ and $y_2 = \sin x$.

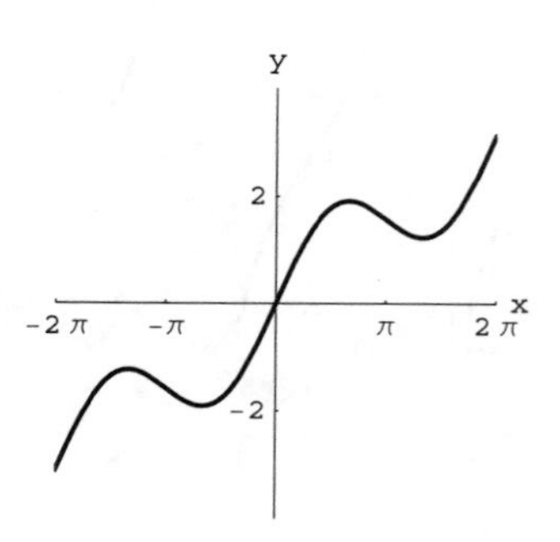

7. For each x-coordinate, the y-coordinate of $y = x^2 + \sin x$ is the sum of the y-coordinates of $y_1 = x^2$ and $y_2 = \sin x$. Note, $y_2 = \sin x$ oscillates about $y_1 = x^2$.

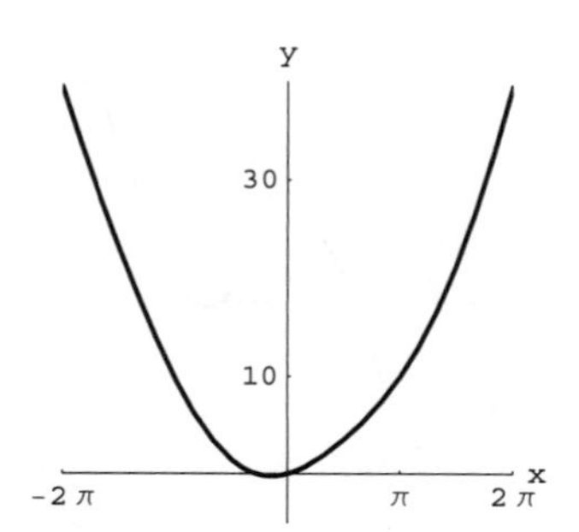

9. For each x-coordinate, the y-coordinate of $y = \sqrt{x} + \cos x$ is the sum of the y-coordinates of $y_1 = \sqrt{x}$ and $y_2 = \cos x$.

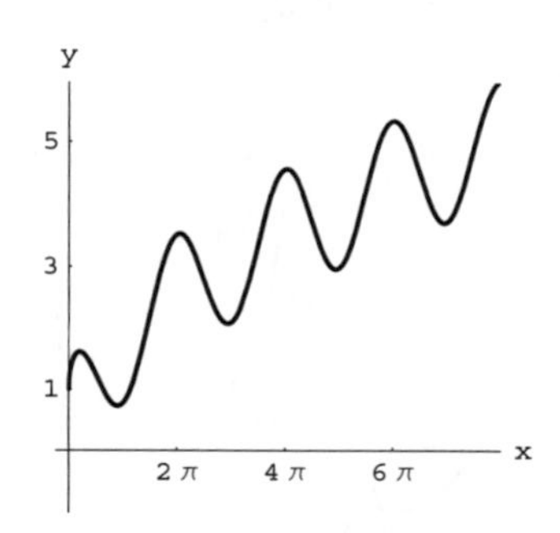

11. For each x-coordinate, the y-coordinate of $y = |x| + 2\sin x$ is the sum of the y-coordinates of $y_1 = |x|$ and $y_2 = 2\sin x$. Note, the graph below oscillates about $y_1 = |x|$.

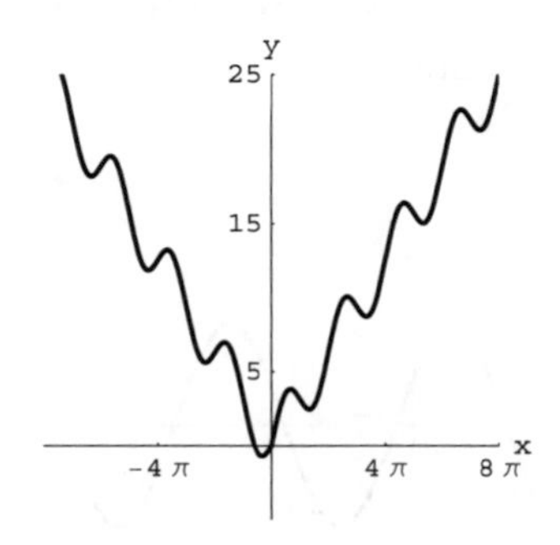

13. For each x-coordinate, the y-coordinate of $y = \cos(x) + 2\sin(x)$ is obtained by adding the y-coordinates of $y_1 = \cos x$ and $y_2 = 2\sin x$. Note, $y = \cos(x) + 2\sin(x)$ is a periodic function since it is the sum of two periodic functions.

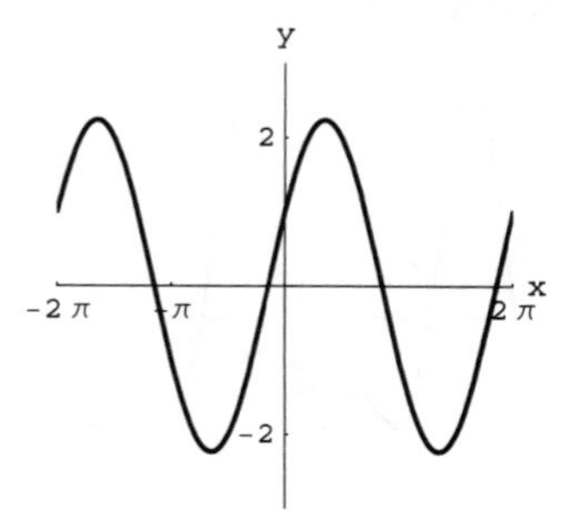

15. For each x-coordinate, the y-coordinate of $y = \sin(x) - \cos(x)$ is obtained by subtracting the y-coordinate of $y_2 = \cos x$ from the y-coordinate of $y_1 = \sin(x)$. Note, $y = \sin(x) - \cos(x)$ is a periodic function since it is the difference of two periodic functions.

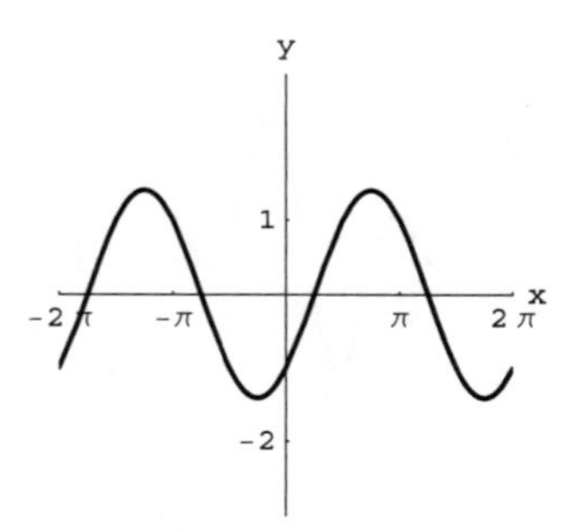

17. For each x-coordinate, the y-coordinate of $y = \sin(x) + \cos(2x)$ is obtained by adding the y-coordinates of $y_1 = \sin x$ and $y_2 = \cos(2x)$. Note, $y = \sin(x) + \cos(2x)$ is a periodic function since it is the sum of two periodic functions.

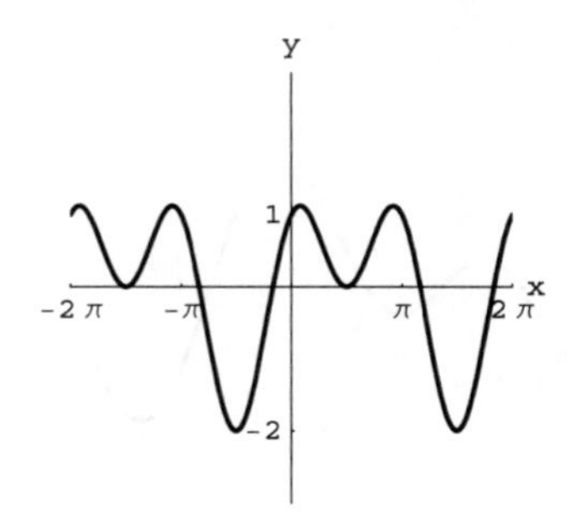

19. For each x-coordinate, the y-coordinate of $y = 2\sin(x) - \cos(2x)$ is obtained by subtracting the y-coordinate of $y_2 = \cos(2x)$ from the y-coordinate of $y_1 = 2\sin(x)$. Note, $y = 2\sin(x) - \cos(2x)$ is a periodic function since it is the difference of two periodic functions.

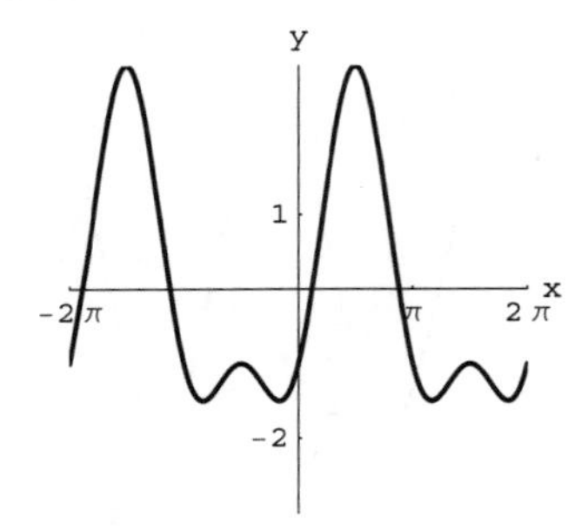

21. For each x-coordinate, the y-coordinate of $y = \sin(x) + \sin(2x)$ is obtained by adding the y-coordinates of $y_1 = \sin x$ and $y_2 = \sin(2x)$. Note, $y = \sin(x) + \sin(2x)$ is a periodic function since it is the sum of two periodic functions.

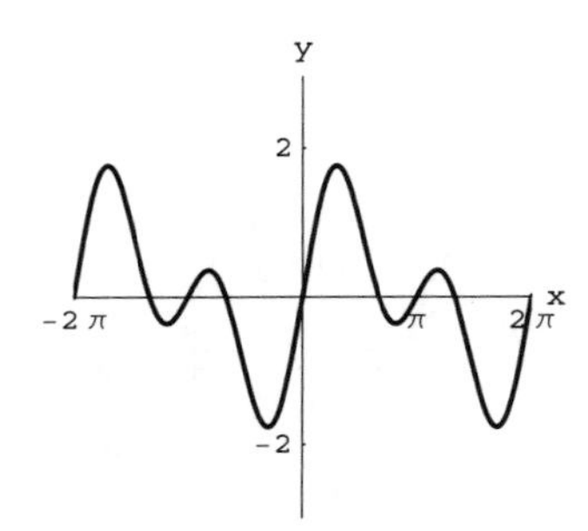

23. For each x-coordinate, the y-coordinate of $y = \sin(x) + \cos\left(\frac{x}{2}\right)$ is obtained by adding the y-coordinates of $y_1 = \sin x$ and $y_2 = \cos\left(\frac{x}{2}\right)$. Note, $y = \sin(x) + \cos\left(\frac{x}{2}\right)$ is a periodic function since it is the sum of periodic functions.

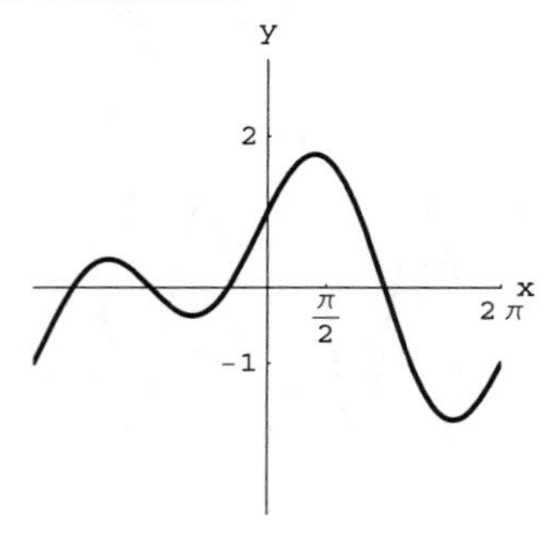

25. For each x-coordinate, the y-coordinate of $y = x + \cos(\pi x)$ is obtained by adding the y-coordinates of $y_1 = x$ and $y_2 = \cos(\pi x)$.

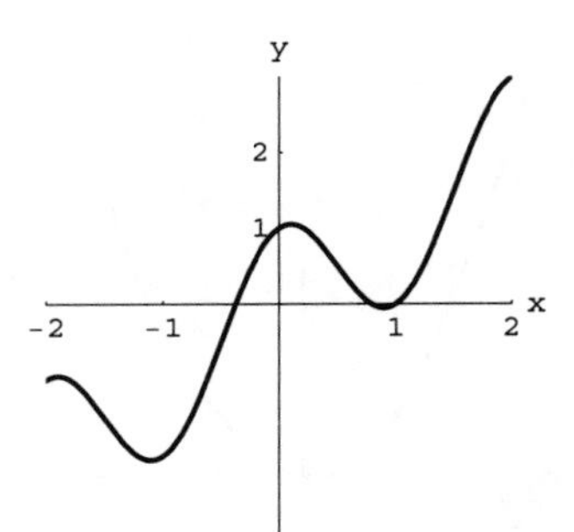

27. For each x-coordinate, the y-coordinate of $y = \frac{1}{x} + \cos(\pi x)$ is obtained by adding the y-coordinates of $y_1 = \frac{1}{x}$ and $y_2 = \cos(\pi x)$.

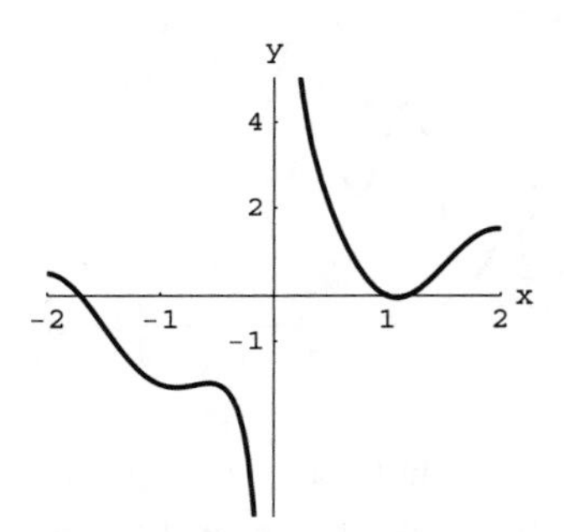

29. For each x-coordinate, the y-coordinate of $y = \sin(\pi x) - \cos(\pi x)$ is obtained by subtracting the y-coordinate of $y_2 = \cos(\pi x)$ from the y-coordinate of $y_1 = \sin(\pi x)$.

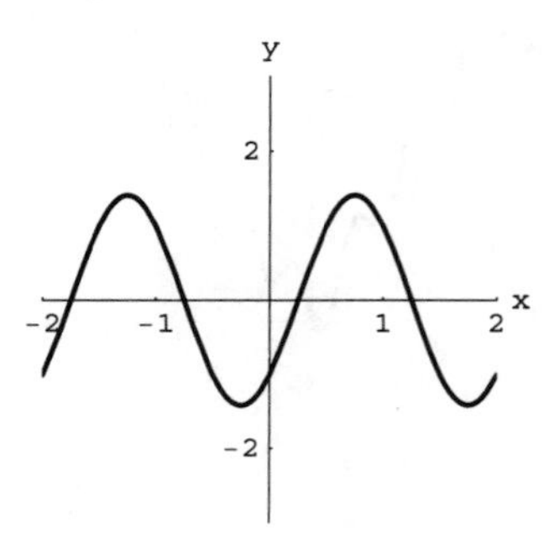

31. For each x-coordinate, the y-coordinate of $y = \sin(\pi x) + \sin(2\pi x)$ is obtained by adding the y-coordinates of $y_1 = \sin(\pi x)$ and $y_2 = \sin(2\pi x)$.

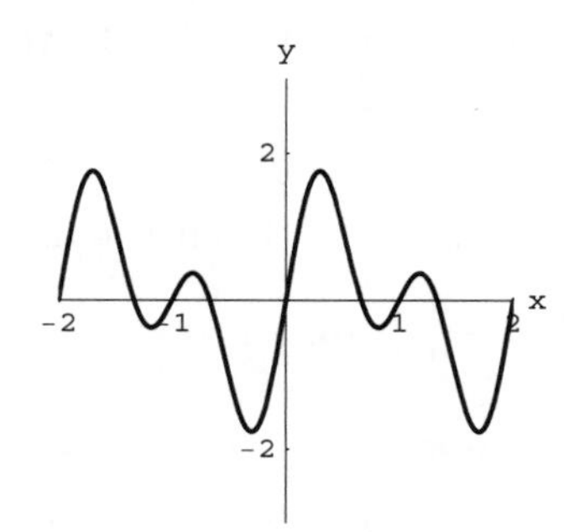

33 For each x-coordinate, the y-coordinate of $y = \cos\left(\frac{\pi}{2}x\right) + \sin(\pi x)$ is obtained by adding the y-coordinates of $y_1 = \cos\left(\frac{\pi}{2}x\right)$ and $y_2 = \sin(\pi x)$.

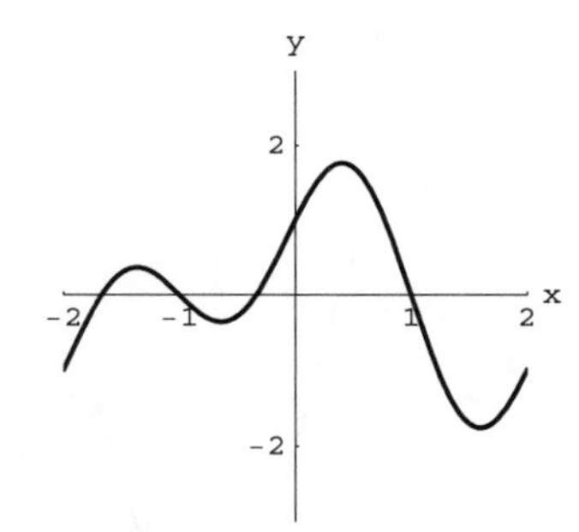

35 For each x-coordinate, the y-coordinate of $y = \sin(\pi x) + \sin\left(\frac{\pi}{2}x\right)$ is obtained by adding the y-coordinates of $y_1 = \sin(\pi x)$ and $y_2 = \sin\left(\frac{\pi}{2}x\right)$.

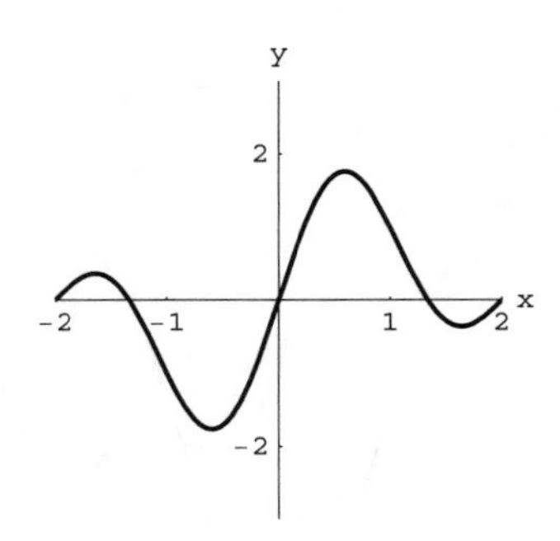

37. c, since the graph of of $y = x + \sin(6x)$ oscillates about the line $y = x$

39. a, since the graph of of $y = -x + \cos(0.5x)$ oscillates about the line $y = -x$

41. d, since the graph of of $y = \sin x + \cos(10x)$ is periodic and passes through $(0, 1)$

43. Since $x_\circ = -3$, $v_\circ = 4$, and $\omega = 1$, we obtain

$$\begin{aligned} x(t) &= \frac{v_\circ}{\omega}\sin(\omega t) + x_\circ \cos(\omega t) \\ &= 4\sin t - 3\cos t. \end{aligned}$$

After $t = 3$ sec, the location of the weight is

$$x(3) = 4\sin 3 - 3\cos 3 \approx 3.5 \text{ cm}.$$

The period and amplitude of

$$x(t) = 4\sin t - 3\cos t$$

are $2\pi \approx 6.3$ sec and $\sqrt{4^2 + 3^2} = 5$, respectively.

45. a) The graph of

$$P(x) = 1000(1.01)^x + 500\sin\left(\frac{\pi}{6}(x-4)\right) + 2000$$

for $1 \le x \le 60$ is given below

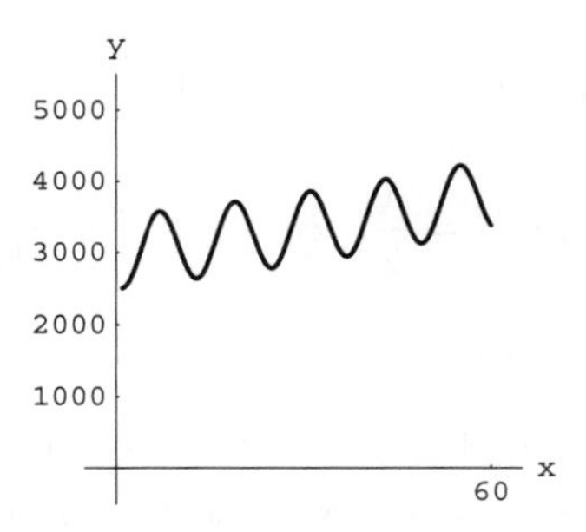

b) The graph for $1 \le x \le 600$ is given below

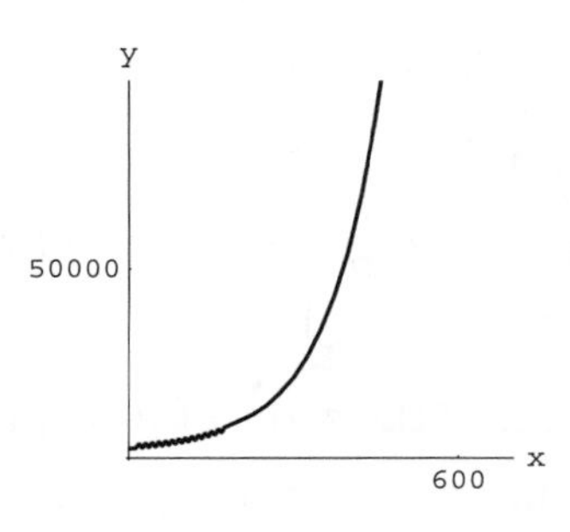

The graph looks like an exponential function.

47. By adding the ordinates of $y = x$ and $y = \sin(x)$, one can obtain the graph of $y = x + \sin(x)$ (which is given below).

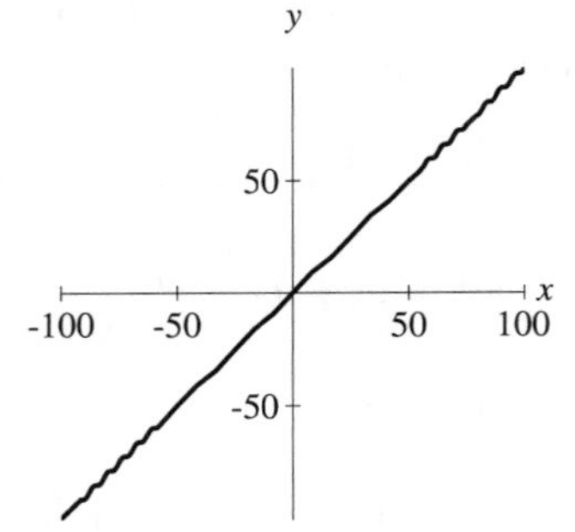

The graph above looks like the graph of $y = x$.

Review Exercises

1. 1

3. $\dfrac{\sin(\pi/4)}{\cos(\pi/4)} = \dfrac{\sqrt{2}/2}{\sqrt{2}/2} = 1$

5. $\dfrac{1}{\cos\pi} = \dfrac{1}{-1} = -1$

7. 0

9. Since $B = 1$, the period is $\dfrac{2\pi}{B} = \dfrac{2\pi}{1} = 2\pi$.

11. Since $B = 2$, the period is $\dfrac{\pi}{B} = \dfrac{\pi}{2}$.

13. Since $B = \pi$, the period is $\dfrac{2\pi}{B} = \dfrac{2\pi}{\pi} = 2$.

15. Since $B = \dfrac{1}{2}$, the period is $\dfrac{2\pi}{B} = \dfrac{2\pi}{1/2} = 4\pi$.

17. Domain $(-\infty, \infty)$, range $[-2, 2]$

19. To find the domain of $y = \tan(2x)$, solve $2x = \dfrac{\pi}{2} + k\pi$, and so the domain is $\left\{x \mid x \neq \dfrac{\pi}{4} + \dfrac{k\pi}{2}\right\}$. The range is $(-\infty, \infty)$.

21. Since the zeros of $y = \cos(x)$ are $x = \dfrac{\pi}{2} + k\pi$, the domain of $y = \sec(x) - 2$ is $\left\{x \mid x \neq \dfrac{\pi}{2} + k\pi\right\}$. The range is $(-\infty, -3] \cup [-1, \infty)$.

23. By using the zeros of $y = \sin(x)$ and by solving $\pi x = k\pi$, we get that the domain of $y = \cot(\pi x)$ is $\{x \mid x \neq k\}$. The range of $y = \cot(\pi x)$ is $(-\infty, \infty)$.

25. By solving $2x = \dfrac{\pi}{2} + k\pi$, we get that the asymptotes of $y = \tan(2x)$ are $x = \dfrac{\pi}{4} + \dfrac{k\pi}{2}$.

27. By solving $\pi x = k\pi$, we find that the asymptotes of $y = \cot(\pi x) + 1$ are $x = k$.

29. Solving $x - \dfrac{\pi}{2} = \dfrac{\pi}{2} + k\pi$, we get $x = \pi + k\pi$ or equivalently $x = k\pi$. Then the asymptotes of $y = \sec\left(x - \dfrac{\pi}{2}\right)$ are $x = k\pi$.

31. Solving $\pi x + \pi = k\pi$, we get $x = -1 + k$ or equivalently $x = k$. Then the asymptotes of $y = \csc(\pi x + \pi)$ are $x = k$.

33. Amplitude 3, since $B = 2$ the period is $\dfrac{2\pi}{B} = \dfrac{2\pi}{2}$ or equivalently π, phase shift is 0, and range is $[-3, 3]$. Five points are $(0, 0)$, $\left(\dfrac{\pi}{4}, 3\right)$, $\left(\dfrac{\pi}{2}, 0\right)$, $\left(\dfrac{3\pi}{4}, -3\right)$, and $(\pi, 0)$.

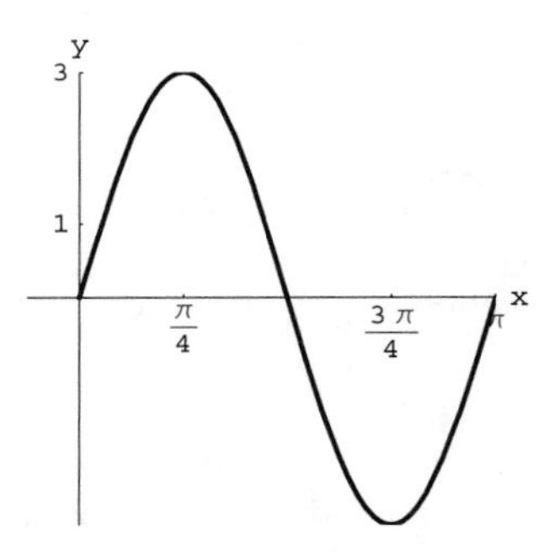

35. Amplitude 1, since $B = 1$ the period is $\dfrac{2\pi}{B} = \dfrac{2\pi}{1}$ or equivalently 2π, phase shift is $\dfrac{\pi}{6}$, and range is $[-1, 1]$. Five points are $\left(\dfrac{\pi}{6}, 1\right)$, $\left(\dfrac{2\pi}{3}, 0\right)$, $\left(\dfrac{7\pi}{6}, -1\right)$, $\left(\dfrac{5\pi}{3}, 0\right)$, and $\left(\dfrac{13\pi}{6}, 1\right)$.

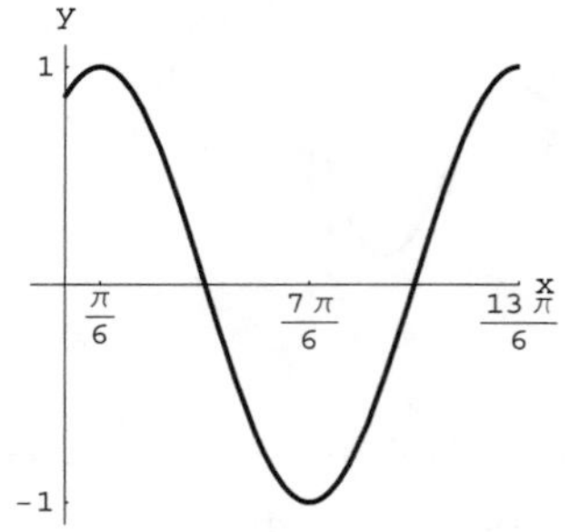

37. Amplitude 1, since $B = 1$ the period is $\frac{2\pi}{B} = \frac{2\pi}{1}$ or equivalently 2π, phase shift is $-\frac{\pi}{4}$, and range is $[-1+1, 1+1]$ or $[0, 2]$. Five points are $\left(-\frac{\pi}{4}, 2\right)$, $\left(\frac{\pi}{4}, 1\right)$, $\left(\frac{3\pi}{4}, 0\right)$, $\left(\frac{5\pi}{4}, 1\right)$, and $\left(\frac{7\pi}{4}, 2\right)$.

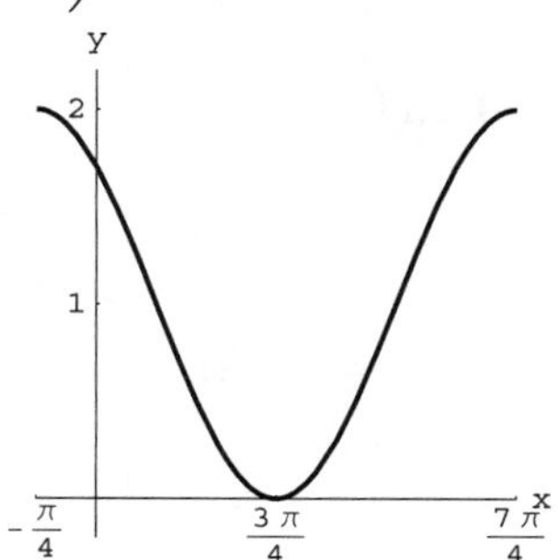

39. Amplitude 1, since $B = \frac{\pi}{2}$ the period is $\frac{2\pi}{B} = \frac{2\pi}{\pi/2}$ or equivalently 4, phase shift is 0, and range is $[-1, 1]$. Five points are $(0, 1)$, $(1, 0)$, $(2, -1)$, $(3, 0)$, and $(4, 1)$.

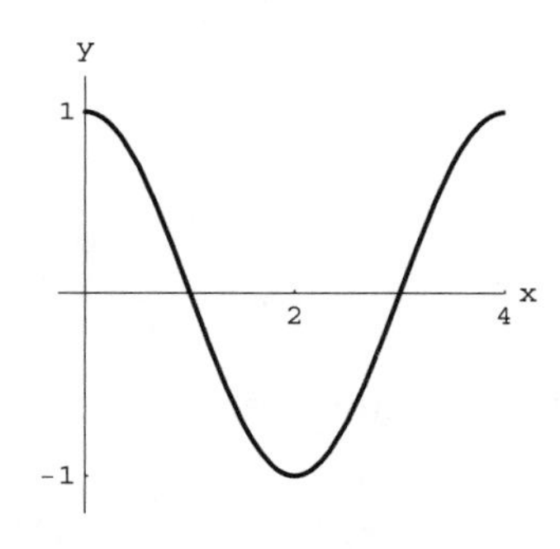

41. Note, $y = \sin\left(2\left(x + \frac{\pi}{2}\right)\right)$. Then amplitude is 1, since $B = 2$ the period is $\frac{2\pi}{B} = \frac{2\pi}{2}$ or equivalently π, phase shift is $-\frac{\pi}{2}$, and range is $[-1, 1]$. Five points are $\left(-\frac{\pi}{2}, 0\right)$, $\left(-\frac{\pi}{4}, 1\right)$, $(0, 0)$, $\left(\frac{\pi}{4}, -1\right)$, and $\left(\frac{\pi}{2}, 0\right)$.

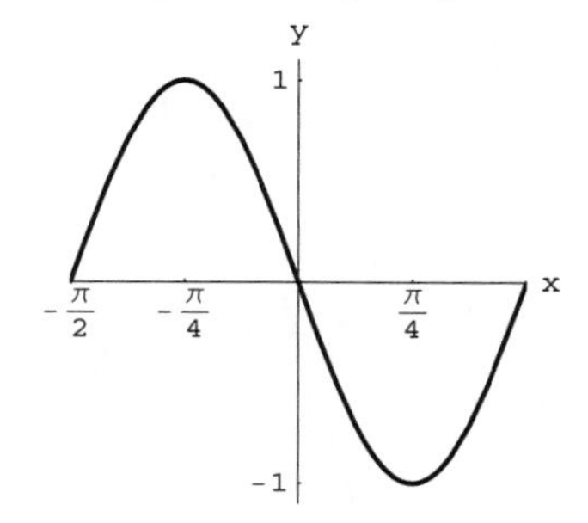

43. Amplitude $\frac{1}{2}$, since $B = 2$ the period is $\frac{2\pi}{B} = \frac{2\pi}{2}$ or equivalently π, phase shift is 0, and range is $\left[-\frac{1}{2}, \frac{1}{2}\right]$. Five points are $\left(0, -\frac{1}{2}\right)$, $\left(\frac{\pi}{4}, 0\right)$, $\left(\frac{\pi}{2}, \frac{1}{2}\right)$, $\left(\frac{3\pi}{4}, 0\right)$, and $\left(\pi, -\frac{1}{2}\right)$.

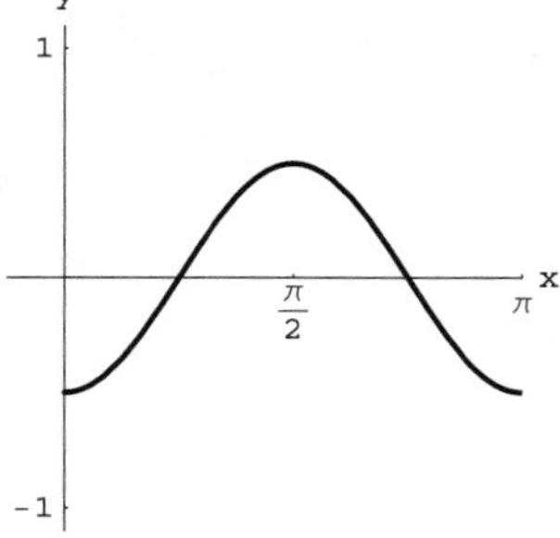

45. Note, $y = -2\sin\left(2\left(x + \frac{\pi}{6}\right)\right) + 1$.

Then amplitude is 2, since $B = 2$ the period is $\frac{2\pi}{B} = \frac{2\pi}{2}$ or π, phase shift is $-\frac{\pi}{6}$, and range is $[-2+1, 2+1]$ or $[-1, 3]$. Five points are $\left(-\frac{\pi}{6}, 1\right)$, $\left(\frac{\pi}{12}, -1\right)$, $\left(\frac{\pi}{3}, 1\right)$, $\left(\frac{7\pi}{12}, 3\right)$, and $\left(\frac{5\pi}{6}, 1\right)$.

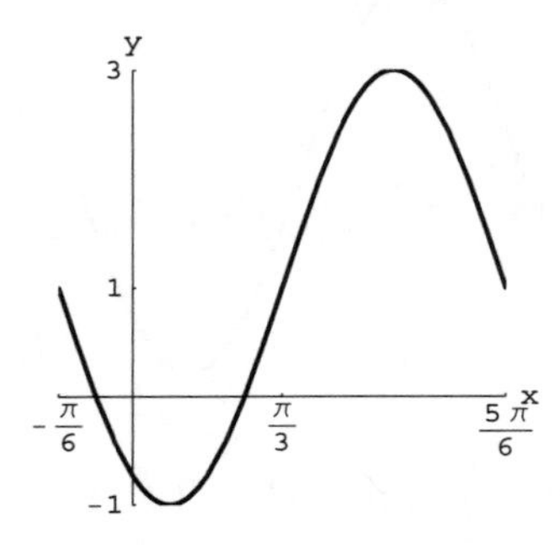

47. Note, $y = \frac{1}{3}\cos\left(\pi\left(x + 1\right)\right) + 1$.

Then amplitude is $\frac{1}{3}$, since $B = \pi$ the period is $\frac{2\pi}{B} = \frac{2\pi}{\pi}$ or 2, phase shift is -1, and range is $\left[-\frac{1}{3} + 1, \frac{1}{3} + 1\right]$ or $\left[\frac{2}{3}, \frac{4}{3}\right]$. Five points are $\left(-1, \frac{4}{3}\right)$, $\left(-\frac{1}{2}, 1\right)$, $\left(0, \frac{2}{3}\right)$, $\left(\frac{1}{2}, 1\right)$, and $\left(1, \frac{4}{3}\right)$.

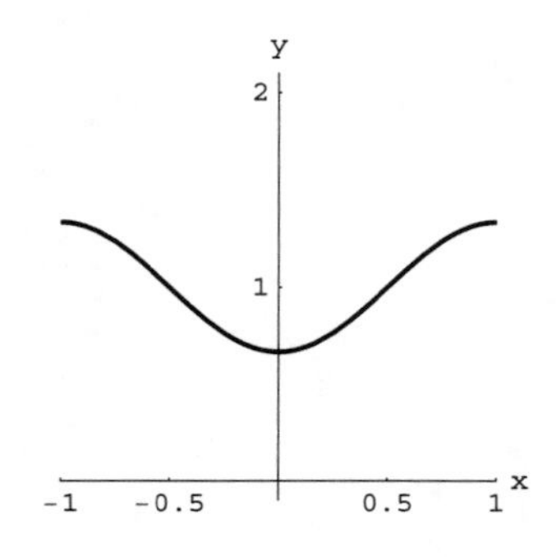

49. Note, $A = 10$, period is 8 and so $B = \frac{\pi}{4}$, and the phase shift can be $C = -2$.

Then $y = 10\sin\left(\frac{\pi}{4}\left(x + 2\right)\right)$.

51. Note, $A = 20$, period is 4 and so $B = \frac{\pi}{2}$, and the phase shift can be $C = 1$, and vertical shift is 10 units up or $D = 10$.

Then $y = 20\sin\left(\frac{\pi}{2}\left(x - 1\right)\right) + 10$.

53. Since $B = 3$, the period is $\frac{\pi}{B} = \frac{\pi}{3}$. To find the asymptotes, let $3x = \frac{\pi}{2} + k\pi$. Then the asymptotes are $x = \frac{\pi}{6} + \frac{k\pi}{3}$. The range is $(-\infty, \infty)$.

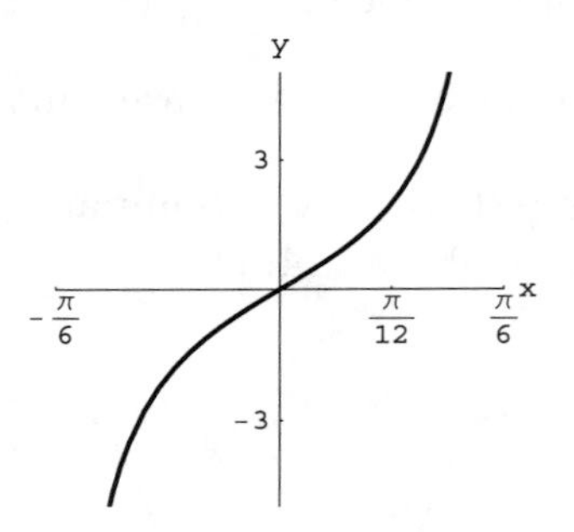

55. Since $B = 2$, the period is $\frac{\pi}{B} = \frac{\pi}{2}$.

To find the asymptotes, let $2x + \pi = \frac{\pi}{2} + k\pi$.

Then $x = -\frac{\pi}{4} + \frac{k\pi}{2}$ or equivalently we get $x = \frac{\pi}{4} + \frac{k\pi}{2}$, which are the asymptotes. The range is $(-\infty, \infty)$.

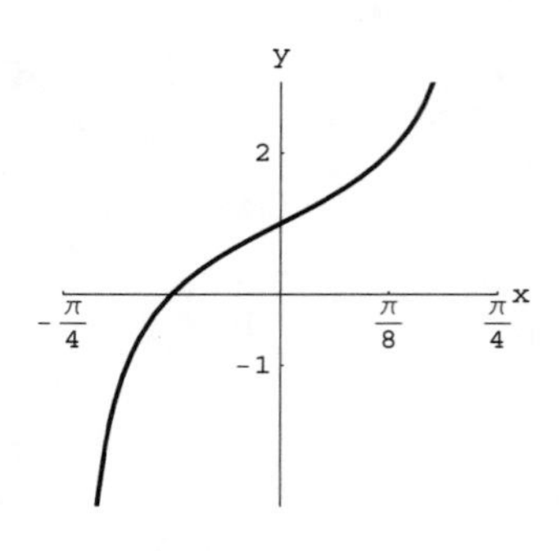

57. Since $B = \frac{1}{2}$, the period is $\frac{2\pi}{B} = \frac{2\pi}{1/2}$ or 4π.

To find the asymptotes, let $\frac{1}{2}x = \frac{\pi}{2} + k\pi$.

Then $x = \pi + 2k\pi$ which are the asymptotes.
The range is $(-\infty, -1] \cup [1, \infty)$.

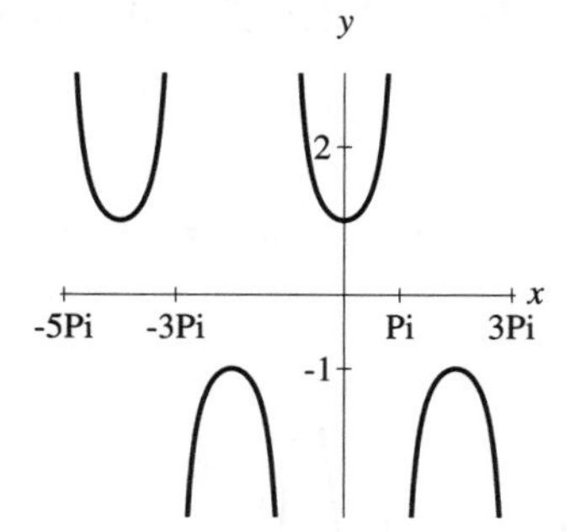

59. Since $B = 2$, the period is $\frac{\pi}{B} = \frac{\pi}{2}$.

To find the asymptotes, let $2x = k\pi$.

Then the asymptotes are $x = \frac{k\pi}{2}$.
The range is $(-\infty, \infty)$.

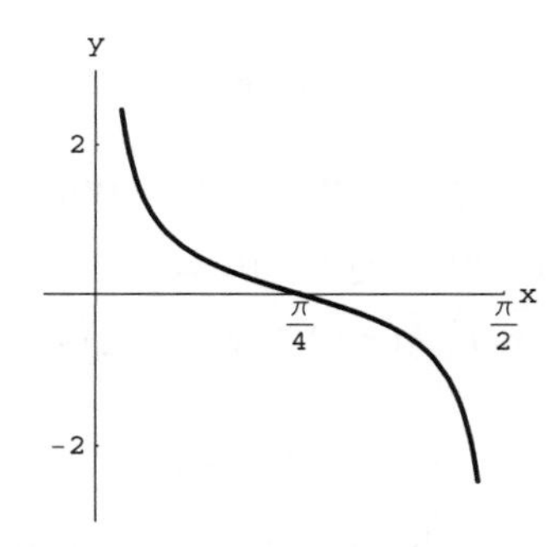

61. Since $B = 2$, the period is $\frac{\pi}{B} = \frac{\pi}{2}$. To find the asymptotes, let $2x + \frac{\pi}{3} = k\pi$. Then $x = -\frac{\pi}{6} + \frac{k\pi}{2}$ or equivalently $x = \frac{\pi}{3} + \frac{k\pi}{2}$, which are the asymptotes.
The range is $(-\infty, \infty)$.

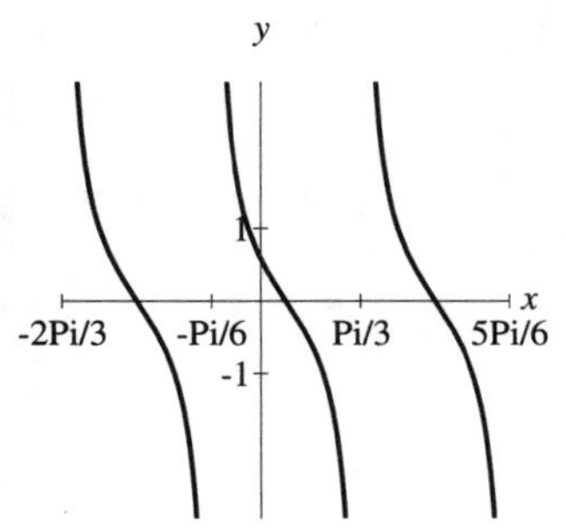

63. Since $B = 2$, the period is $\frac{2\pi}{B} = \frac{2\pi}{2}$ or π.
To obtain the asymptotes, let $2x + \pi = k\pi$ or equivalently $2x = k\pi$. Then the asymptotes are

$$x = \frac{k\pi}{2}.$$

The range is

$$(-\infty, -1/3] \cup [1/3, \infty).$$

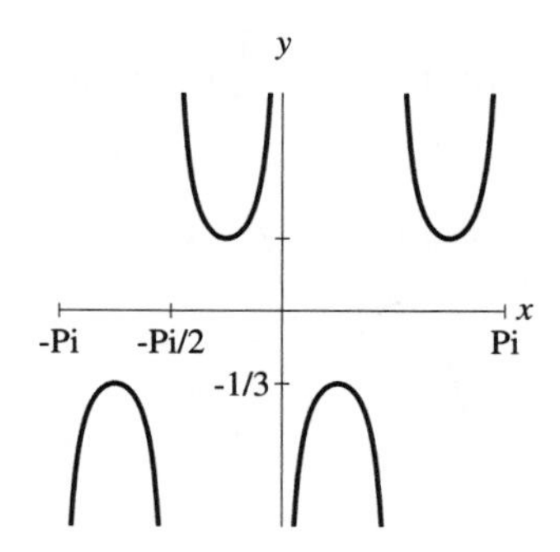

65. Since $B = 1$, the period is $\frac{2\pi}{B} = \frac{2\pi}{1}$ or 2π.

To find the asymptotes, let $x + \frac{\pi}{4} = \frac{\pi}{2} + k\pi$.

Then the asymptotes are

$$x = \frac{\pi}{4} + k\pi.$$

The range is $(-\infty, -2 - 1] \cup [2 - 1, \infty)$ or

$$(-\infty, -3] \cup [1, \infty).$$

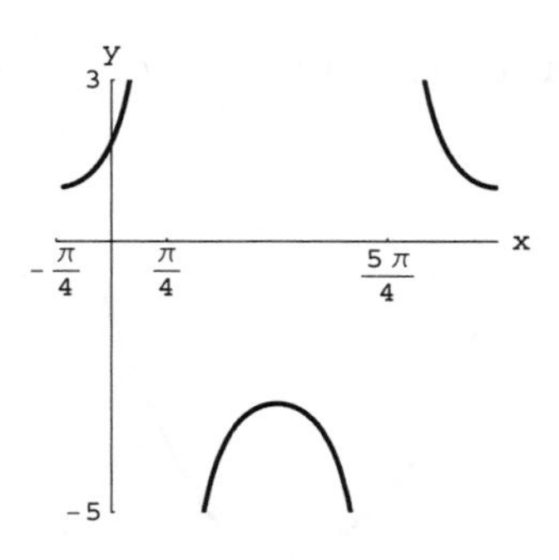

67. For each x-coordinate, the y-coordinate of

$$y = \frac{1}{2}x + \sin x$$

is obtained by adding the y-coordinates of

$y_1 = \frac{1}{2}x$ and $y_2 = \sin x$.

For instance, $\left(\pi/2, \frac{1}{2} \cdot \pi/2 + \sin(\pi/2)\right)$ or

$$(\pi/2, \pi/4 + 1)$$

is a point on the graph of $y = \frac{1}{2}x + \sin(x)$.

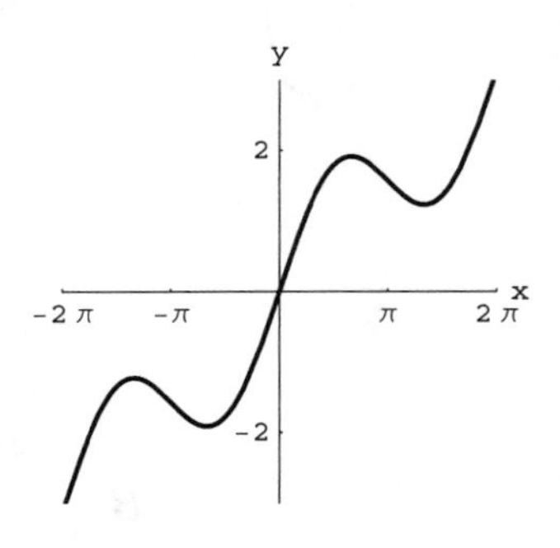

69. For each x-coordinate, the y-coordinate of

$$y = x^2 - \sin x$$

is obtained by subtracting the y-coordinate of $y_2 = \sin x$ from the y-coordinate of $y_1 = x^2$.

For instance, $(\pi, \pi^2 - \sin \pi)$ or

$$\left(\pi, \pi^2\right)$$

is a point on the graph of $y = x^2 - \sin x$.

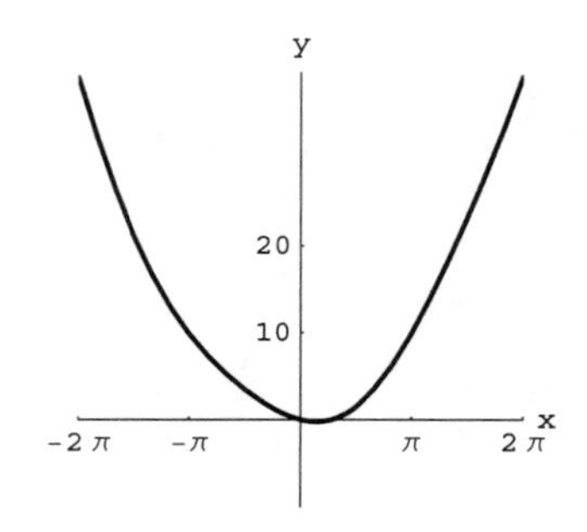

71. For each x-coordinate, the y-coordinate of

$$y = \sin x - \sin 2x$$

is obtained by subtracting the y-coordinate of $y_2 = \sin 2x$ from the y-coordinate of $y_1 = \sin x$.

For instance, $(\pi/2, \sin(\pi/2) - \sin(2 \cdot \pi/2))$ or

$$(\pi/2, 1)$$

is a point on the graph of $y = \sin x - \sin 2x$.

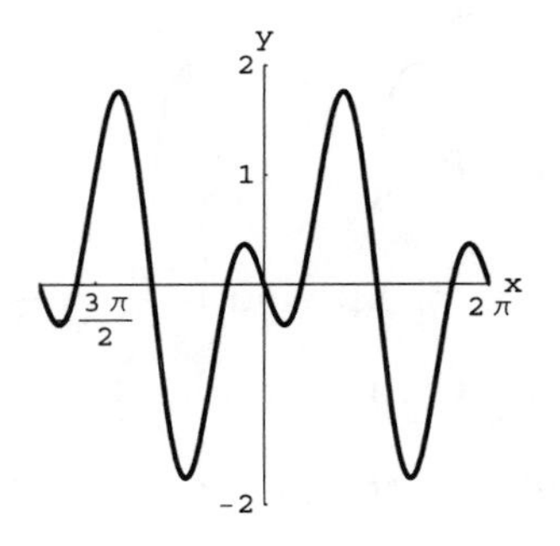

73. For each x-coordinate, the y-coordinate of

$$y = 3\sin x + \sin 2x$$

is obtained by adding the y-coordinates of

$y_1 = 3\sin x$ and $y_2 = \sin 2x$.

For instance, $(\pi/2, 3\sin(\pi/2) + \sin(2 \cdot \pi/2))$ or

$$(\pi/2, 3)$$

is a point on the graph of $y = 3\sin x + \sin 2x$.

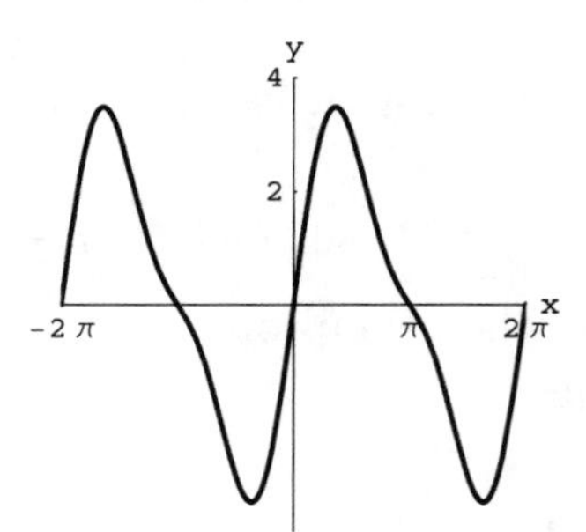

75. The period is $\dfrac{1}{92.3 \times 10^6} \approx 1.08 \times 10^{-8}$ sec

77. Since the period is 20 minutes, $\frac{2\pi}{b} = 20$ or $b = \frac{\pi}{10}$. Since the depth is between 12 ft and 16 ft, the vertical upward shift is 14 and $a = 2$. Since the depth is 16 ft at time $t = 0$, one can assume there is a left shift of 5 minutes. An equation is

$$y = 2\sin\left(\frac{\pi}{10}(x+5)\right) + 14$$

and its graph is given below.

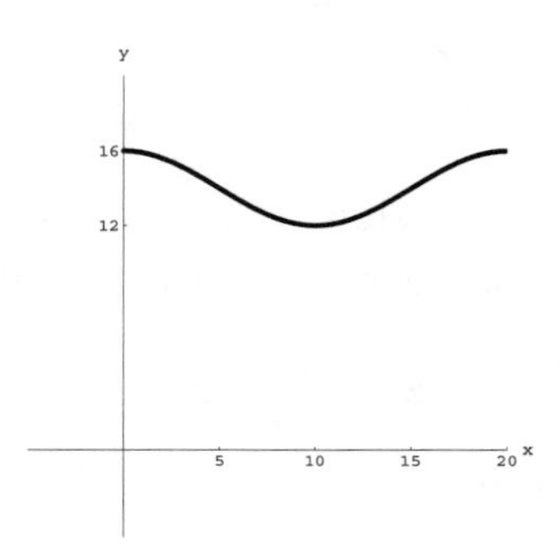

Chapter 2 Test

1. Period $2\pi/3$, range $[-1, 1]$, amplitude 1, some points are $(0, 0)$, $\left(\frac{\pi}{6}, 1\right)$, $\left(\frac{\pi}{3}, 0\right)$, $\left(\frac{\pi}{2}, -1\right)$, $\left(\frac{2\pi}{3}, 0\right)$

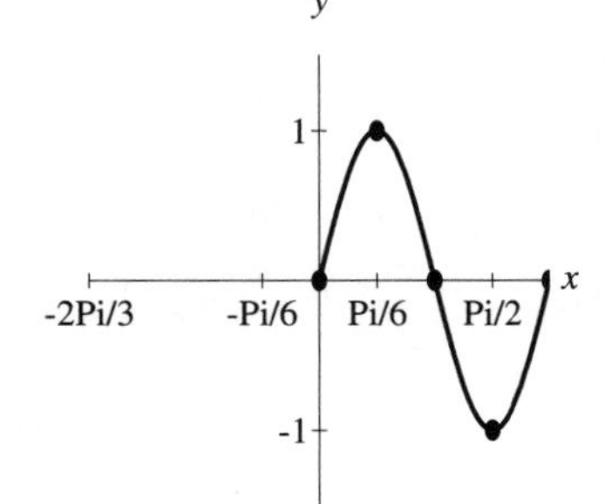

2. Period 2π, range $[-1, 1]$, amplitude 1, some points are $\left(-\frac{\pi}{4}, 1\right)$, $\left(\frac{\pi}{4}, 0\right)$, $\left(\frac{3\pi}{4}, -1\right)$, $\left(\frac{5\pi}{4}, 0\right)$, $\left(\frac{7\pi}{4}, 1\right)$

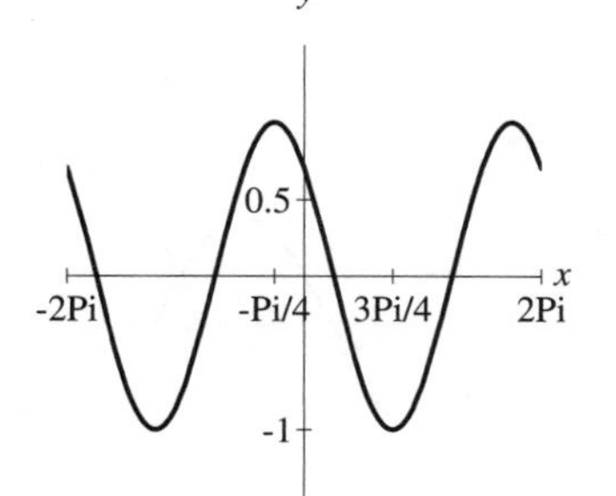

3. Since $B = \frac{\pi}{2}$, the period is 4. The range is $[-1, 1]$ and amplitude is 1. Some points are $(0, 0)$, $(1, 1)$, $(2, 0)$, $(3, -1)$, $(4, 0)$

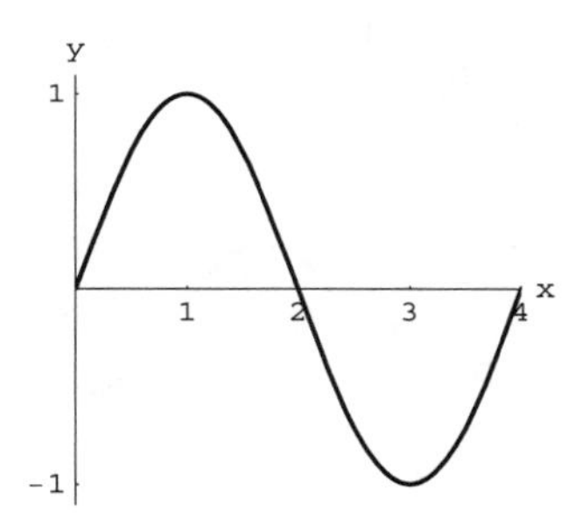

4. Period 2π, range $[-2, 2]$, amplitude 2, some points are $(\pi, -2)$, $\left(\frac{3\pi}{2}, 0\right)$, $(2\pi, 2)$, $\left(\frac{5\pi}{2}, 0\right)$, $(3\pi, -2)$

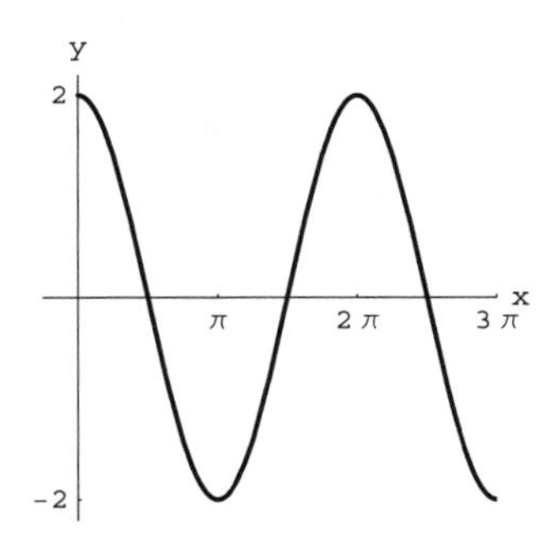

5. Period 2π, range $\left[-\frac{1}{2}, \frac{1}{2}\right]$, amplitude $\frac{1}{2}$, some points are $\left(\frac{\pi}{2}, 0\right)$, $\left(\pi, -\frac{1}{2}\right)$, $\left(\frac{3\pi}{2}, 0\right)$, $\left(2\pi, \frac{1}{2}\right)$, $\left(\frac{5\pi}{2}, 0\right)$

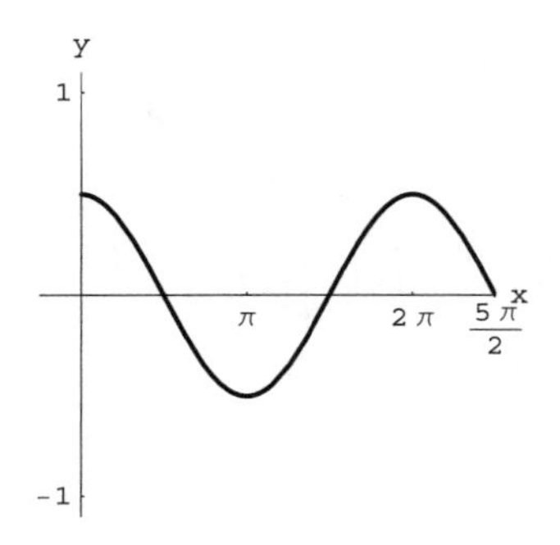

6. Period π since $B = 2$, range is $[-1+3, 1+3]$ or $[2, 4]$, amplitude is 1, some points are $\left(\frac{\pi}{6}, 2\right)$, $\left(\frac{5\pi}{12}, 3\right)$, $\left(\frac{2\pi}{3}, 4\right)$, $\left(\frac{11\pi}{12}, 3\right)$, $\left(\frac{7\pi}{6}, 2\right)$

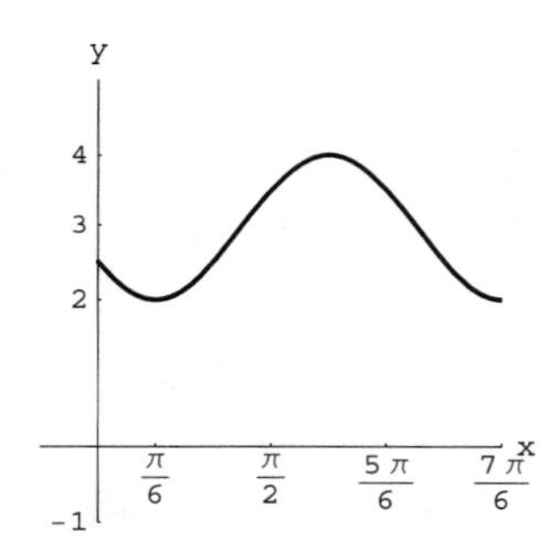

7. Note, $A = 4$, period is 12 and so $B = \frac{\pi}{6}$, phase shift is $C = 3$, and the vertical shift downward is 2 units. Then $y = 4\sin\left(\frac{\pi}{6}(x-3)\right) - 2$.

8. The period is $\frac{\pi}{3}$ since $B = 3$, by setting $3x = \frac{\pi}{2} + k\pi$ we get that the asymptotes are $x = \frac{\pi}{6} + \frac{k\pi}{3}$, and the range is $(-\infty, \infty)$.

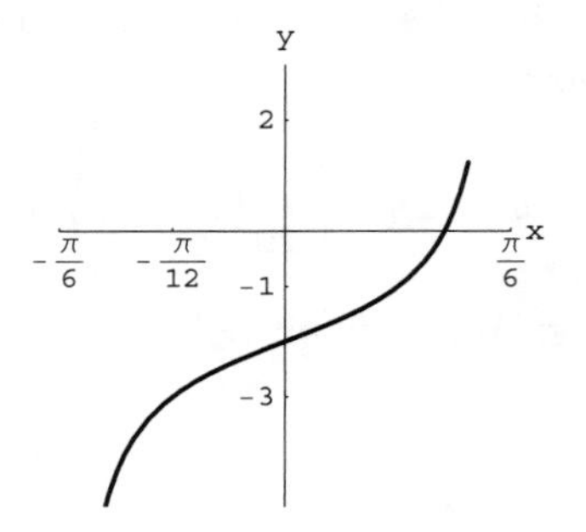

9. The period is π since $B = 1$, by setting $x + \frac{\pi}{2} = k\pi$ we get that the asymptotes are $x = -\frac{\pi}{2} + k\pi$, $x = -\frac{\pi}{2} + \pi + k\pi$, or $x = \frac{\pi}{2} + k\pi$, and the range is $(-\infty, \infty)$.

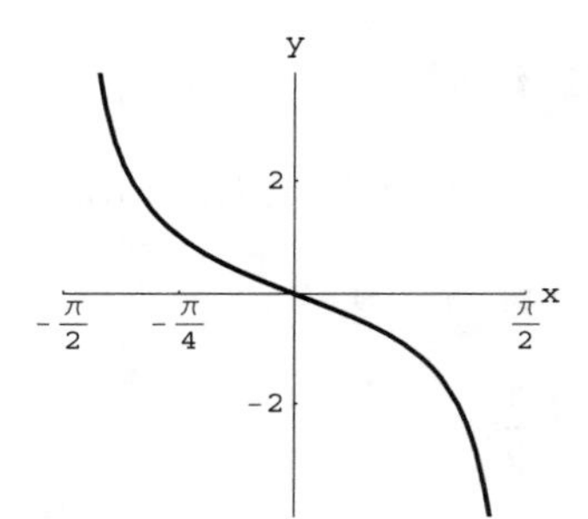

10. The period is $\frac{2\pi}{1}$ or 2π, by setting $x - \pi = \frac{\pi}{2} + k\pi$ we get $x = \frac{3\pi}{2} + k\pi$ and equivalently the asymptotes are $x = \frac{\pi}{2} + k\pi$, and the range is $(-\infty, -2] \cup [2, \infty)$.

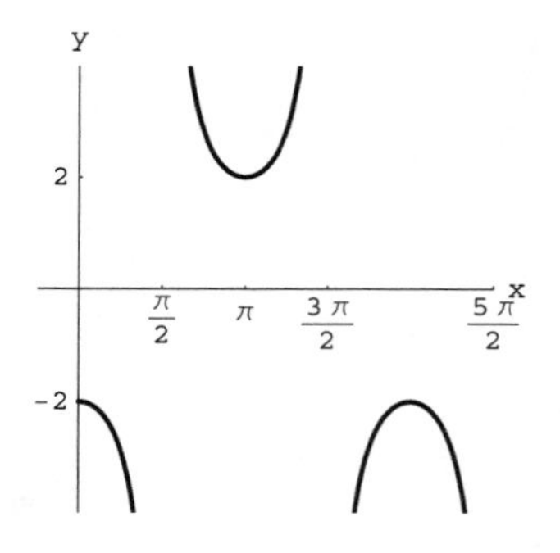

11. The period is $\frac{2\pi}{1}$ or 2π, by setting $x - \frac{\pi}{2} = k\pi$ we get that the asymptotes are

$$x = \frac{\pi}{2} + k\pi$$

and the range is $(-\infty, -1+1] \cup [1+1, \infty)$ or

$$(-\infty, 0] \cup [2, \infty).$$

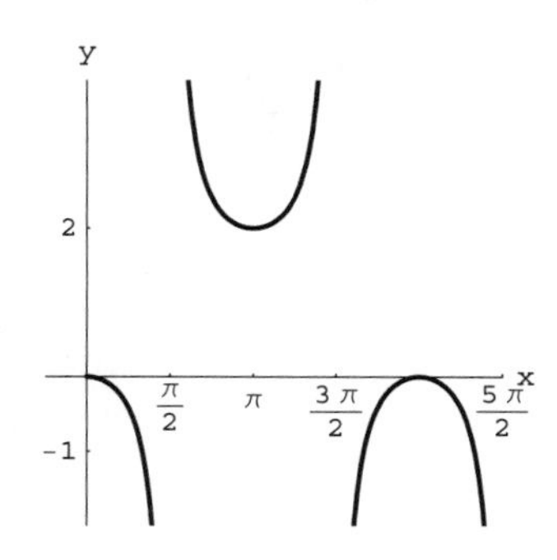

12. For each x-coordinate, the y-coordinate of

$$y = 3\sin x + \cos 2x$$

is obtained by adding the y-coordinates of $y_1 = 3\sin x$ and $y_2 = \cos 2x$.
For instance, the point

$$(\pi/2, 3\sin(\pi/2) + \cos(2 \cdot \pi/2)) = (\pi/2, 3 + (-1))$$

or equivalently

$$(\pi/2, 2)$$

is a point on the graph of $y = 3\sin x + \cos 2x$.

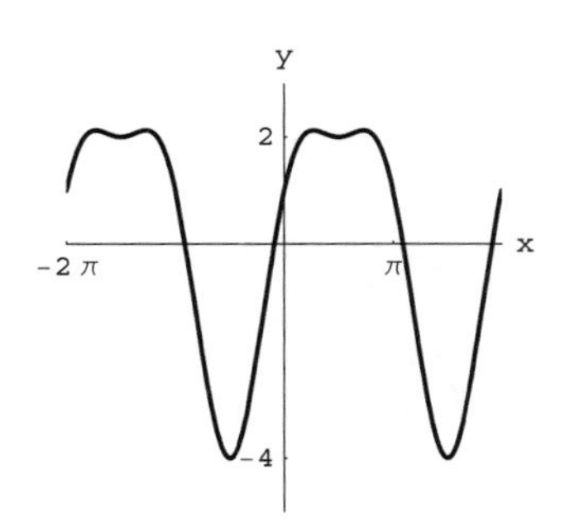

13. Since the pH oscillates between 7.2 and 7.8, there is a vertical upward shift of 7.5 and

$$a = 0.3.$$

Note, the period is 4 days. Thus, $\frac{2\pi}{b} = 4$ or

$$b = \frac{\pi}{2}.$$

Since the pH is 7.2 on day 13, the pH is 7.5 on day 14. We can assume a right shift of 14 days. Hence, an equation is

$$y = 0.3\sin\left(\frac{\pi}{2}(x - 14)\right) + 7.5.$$

A graph of one cycle is given below.

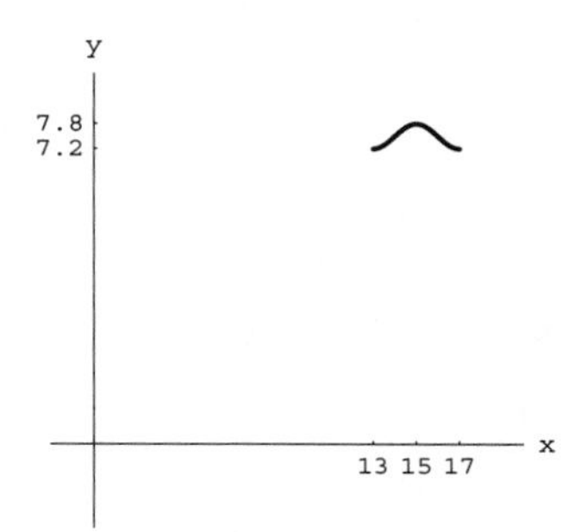

Tying It All Together Chapters P-2

1.

θ deg	0	30	45	60	90	120	135	150	180
θ rad	0	$\frac{\pi}{6}$	$\frac{\pi}{4}$	$\frac{\pi}{3}$	$\frac{\pi}{2}$	$\frac{2\pi}{3}$	$\frac{3\pi}{4}$	$\frac{5\pi}{6}$	π
$\sin\theta$	0	$\frac{1}{2}$	$\frac{\sqrt{2}}{2}$	$\frac{\sqrt{3}}{2}$	1	$\frac{\sqrt{3}}{2}$	$\frac{\sqrt{2}}{2}$	$\frac{1}{2}$	0
$\cos\theta$	1	$\frac{\sqrt{3}}{2}$	$\frac{\sqrt{2}}{2}$	$\frac{1}{2}$	0	$-\frac{1}{2}$	$-\frac{\sqrt{2}}{2}$	$-\frac{\sqrt{3}}{2}$	-1
$\tan\theta$	0	$\frac{\sqrt{3}}{3}$	1	$\sqrt{3}$	und	$-\sqrt{3}$	-1	$-\frac{\sqrt{3}}{3}$	0
$\csc\theta$	und	2	$\sqrt{2}$	$\frac{2\sqrt{3}}{3}$	1	$\frac{2\sqrt{3}}{3}$	$\sqrt{2}$	2	und
$\sec\theta$	1	$\frac{2\sqrt{3}}{3}$	$\sqrt{2}$	2	und	-2	$-\sqrt{2}$	$-\frac{2\sqrt{3}}{3}$	-1
$\cot\theta$	und	$\sqrt{3}$	1	$\frac{\sqrt{3}}{3}$	0	$-\frac{\sqrt{3}}{3}$	-1	$-\sqrt{3}$	und

2.

θ rad	π	$\frac{7\pi}{6}$	$\frac{5\pi}{4}$	$\frac{4\pi}{3}$	$\frac{3\pi}{2}$	$\frac{5\pi}{3}$	$\frac{7\pi}{4}$	$\frac{11\pi}{6}$	2π
θ deg	180	210	225	240	270	300	315	330	360
$\sin\theta$	0	$-\frac{1}{2}$	$-\frac{\sqrt{2}}{2}$	$-\frac{\sqrt{3}}{2}$	-1	$-\frac{\sqrt{3}}{2}$	$-\frac{\sqrt{2}}{2}$	$-\frac{1}{2}$	0
$\cos\theta$	-1	$-\frac{\sqrt{3}}{2}$	$-\frac{\sqrt{2}}{2}$	$-\frac{1}{2}$	0	$\frac{1}{2}$	$\frac{\sqrt{2}}{2}$	$\frac{\sqrt{3}}{2}$	1
$\tan\theta$	0	$\frac{\sqrt{3}}{3}$	1	$\sqrt{3}$	und	$-\sqrt{3}$	-1	$-\frac{\sqrt{3}}{3}$	0
$\csc\theta$	und	-2	$-\sqrt{2}$	$-\frac{2\sqrt{3}}{3}$	-1	$-\frac{2\sqrt{3}}{3}$	$-\sqrt{2}$	-2	und
$\sec\theta$	-1	$-\frac{2\sqrt{3}}{3}$	$-\sqrt{2}$	-2	und	2	$\sqrt{2}$	$\frac{2\sqrt{3}}{3}$	1
$\cot\theta$	und	$\sqrt{3}$	1	$\frac{\sqrt{3}}{3}$	0	$-\frac{\sqrt{3}}{3}$	-1	$-\sqrt{3}$	und

3. Domain $(-\infty, \infty)$, range $[-1, 1]$, and since $B = 2$ the period is $\dfrac{2\pi}{B} = \dfrac{2\pi}{2}$ or π.

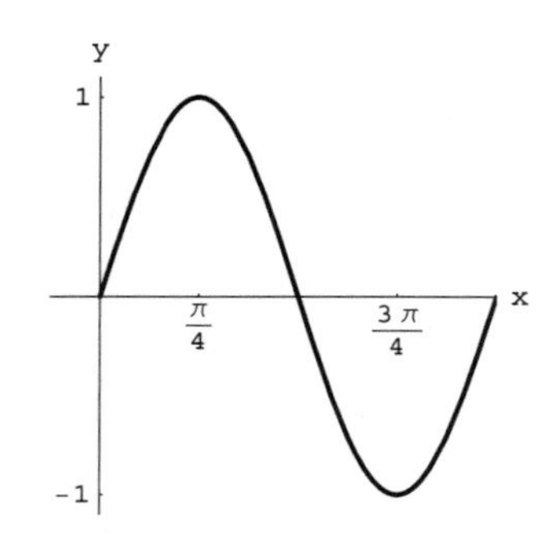

4. Domain $(-\infty, \infty)$, range $[-1, 1]$, and since $B = 2$ the period is $\dfrac{2\pi}{B} = \dfrac{2\pi}{2}$ or π.

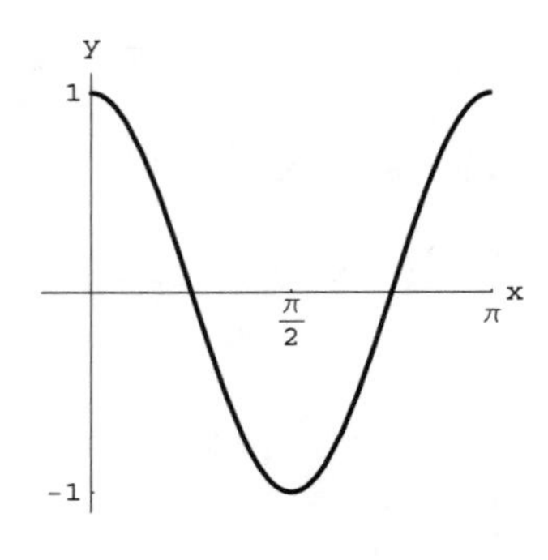

5. By setting $2x \neq \dfrac{\pi}{2} + k\pi$, the domain is

$$\left\{x \,:\, x \neq \frac{\pi}{4} + \frac{k\pi}{2}\right\},$$

range is $(-\infty, \infty)$, and the period is $\dfrac{\pi}{B} = \dfrac{\pi}{2}$.

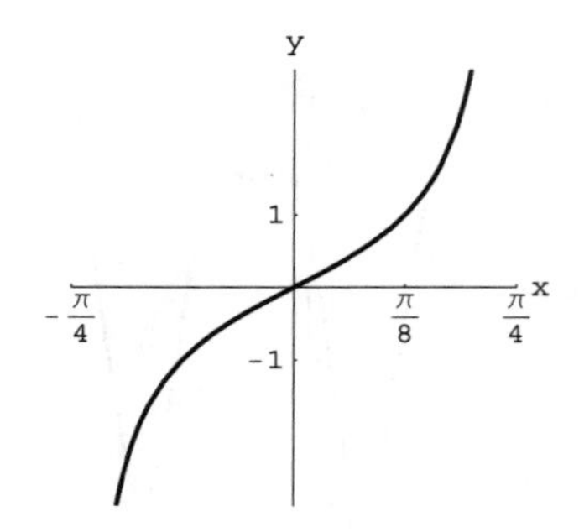

6. By setting $2x \neq \frac{\pi}{2} + k\pi$, we find that the domain is $\left\{x : x \neq \frac{\pi}{4} + \frac{k\pi}{2}\right\}$, the range is $(-\infty, -1] \cup [1, \infty)$, and the period is $\frac{2\pi}{B} = \frac{2\pi}{2}$ or π.

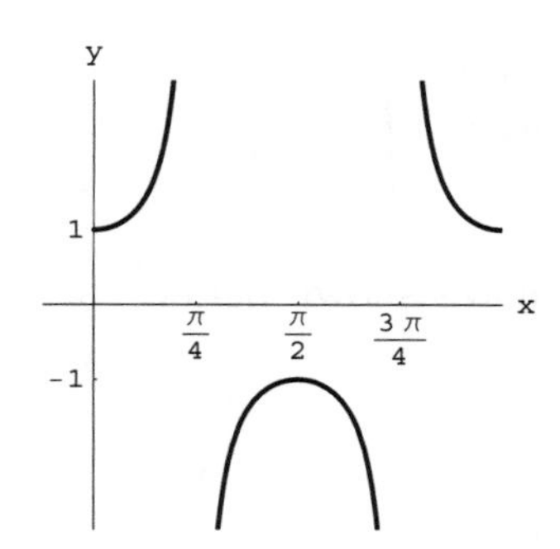

7. By setting $2x \neq k\pi$, we find that the domain is $\left\{x : x \neq \frac{k\pi}{2}\right\}$, the range is $(-\infty, -1] \cup [1, \infty)$, and the period is $\frac{2\pi}{B} = \frac{2\pi}{2}$ or π.

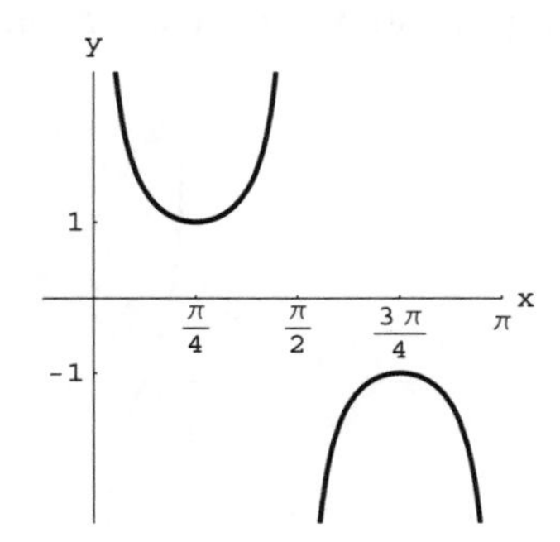

8. Domain $(-\infty, \infty)$, range $[-1, 1]$, and since $B = \pi$ the period is $\frac{2\pi}{B} = \frac{2\pi}{\pi}$ or 2.

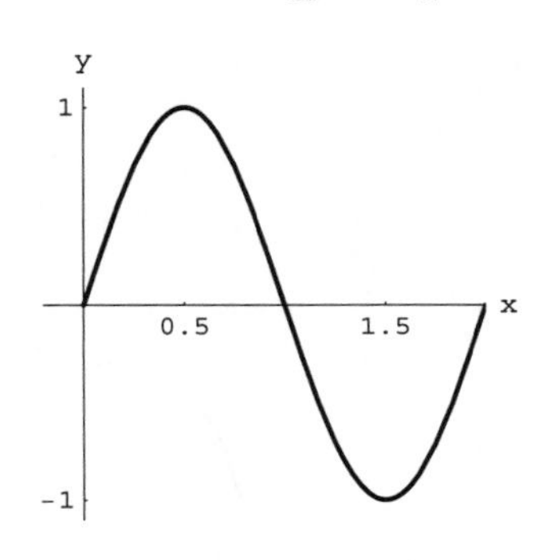

9. Domain $(-\infty, \infty)$, range $[-1, 1]$, and since $B = \pi$ the period is $\frac{2\pi}{B} = \frac{2\pi}{\pi}$ or 2.

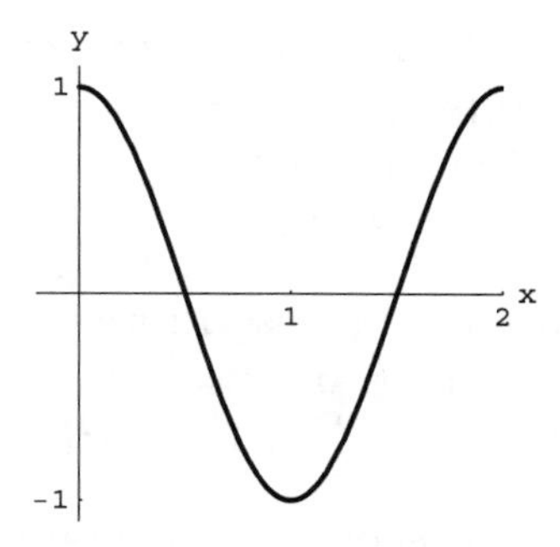

10. By setting $\pi x \neq \frac{\pi}{2} + k\pi$, the domain is $\left\{x : x \neq \frac{1}{2} + k\right\}$, range is $(-\infty, \infty)$, and the period is $\frac{\pi}{B} = \frac{\pi}{\pi}$ or 1.

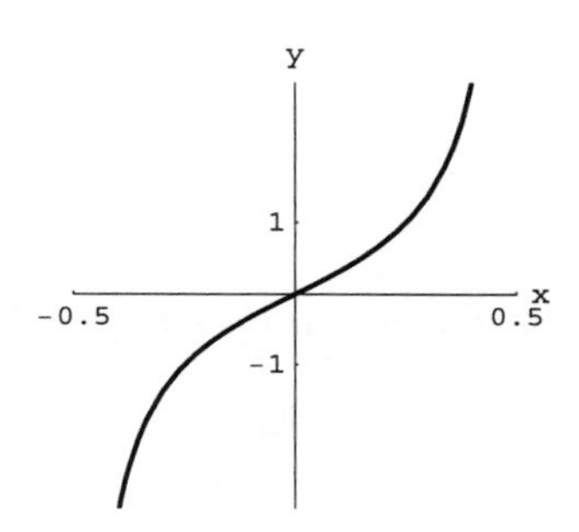

11. Odd, since $\sin(-x) = -\sin(x)$

12. Even, since $\cos(-x) = \cos(x)$

13. Odd, since $\tan(-x) = \frac{\sin(-x)}{\cos(-x)} = \frac{-\sin(x)}{\cos(x)} = -\tan(x)$, i.e., $\tan(-x) = -\tan(x)$

14. Odd, since $\cot(-x) = \frac{\cos(-x)}{\sin(-x)} = \frac{\cos(x)}{-\sin(x)} = -\cot(x)$, i.e., $\cot(-x) = -\cot(x)$

15. Even, since $\sec(-x) = \frac{1}{\cos(-x)} = \frac{1}{\cos(x)} = \sec(x)$, i.e., $\sec(-x) = \sec(x)$

16. Odd, since $\csc(-x) = \dfrac{1}{\sin(-x)} = \dfrac{1}{-\sin(x)} = -\csc(x)$, i.e., $\csc(-x) = -\csc(x)$

17. Even, since by Exercise 11 we get $\sin^2(-x) = (\sin(-x))^2 = (-\sin(x))^2 = (\sin(x))^2$, i.e., $\sin^2(-x) = \sin^2(x)$

18. Even, since by Exercise 12 we get $\cos^2(-x) = (\cos(-x))^2 = (\cos(x))^2$, i.e., $\cos^2(-x) = \cos^2(x)$

19. Increasing, as shown by the graph below.

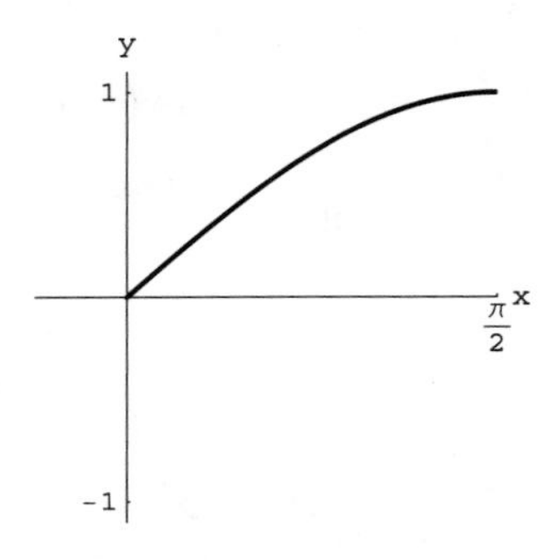

20. Decreasing, as shown by the graph below.

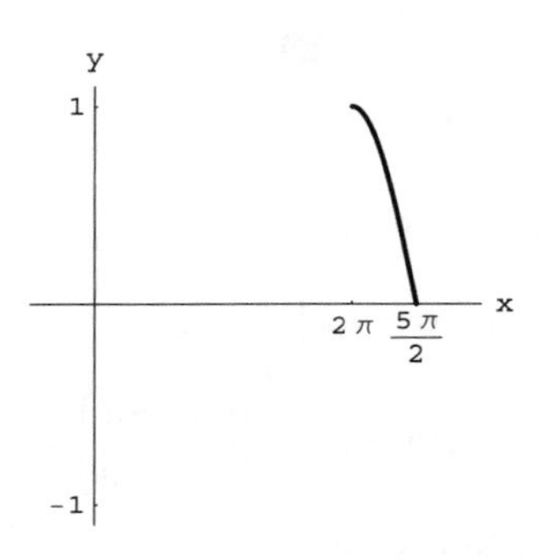

21. Increasing, as shown below.

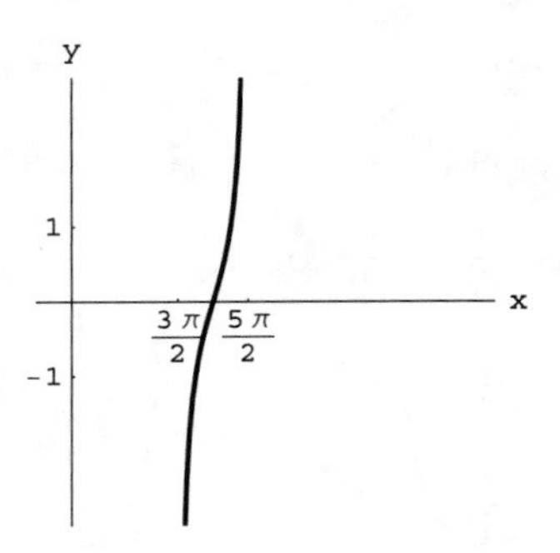

22. Decreasing, as shown below.

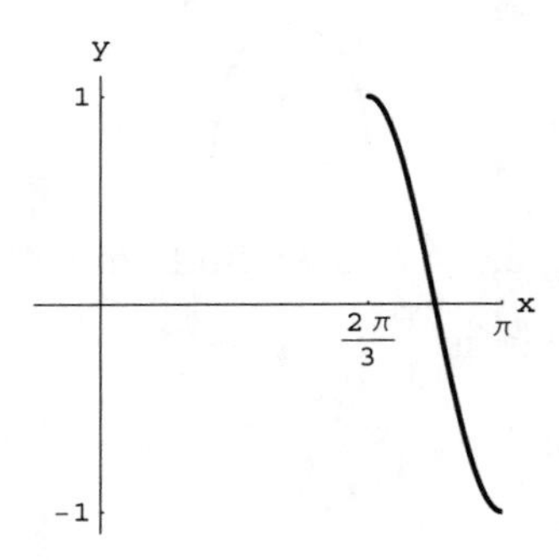

23. Decreasing, as shown by the graph below.

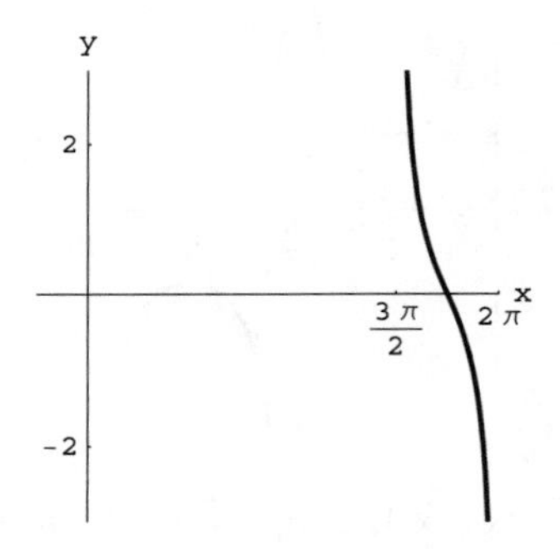

24. Increasing, as shown by the graph below.

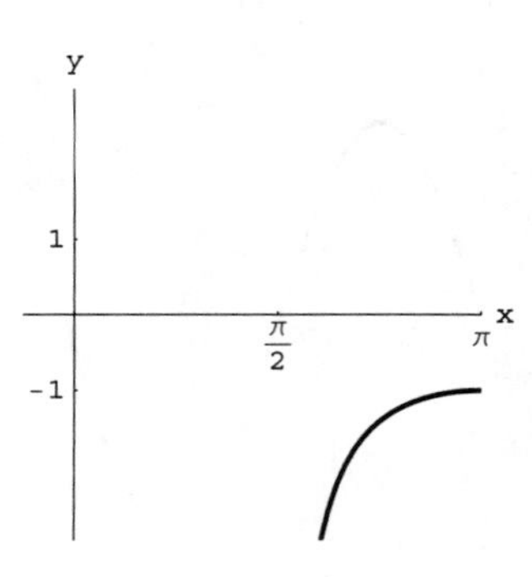

For Thought

1. False, for example $\sin 0 = 0$ but $\cos 0 = 1$.

2. True, since $\dfrac{\sin(x)}{\cos(x)} \cdot \dfrac{\cos(x)}{\sin(x)} = 1$.

3. True, since $\dfrac{2\sin(x)\cos(x)}{\sin(x)} = 2\cos(x)$.

4. False, since $(\sin x + \cos x)^2 =$ $1 + 2\sin(x)\cos(x) \neq 1 = \sin^2 x + \cos^2 x$.

5. False, since $\tan 1 = \sqrt{\sec^2(1) - 1}$.

6. False, since $\sin^2(-6) = (-\sin(6))^2 = \sin^2(6)$.

7. False, since $\cos(-9) = \cos(9)$ implies $\cos^3(-9) = \cos^3(9)$.

8. True, since $\sin(x)\csc(x) = 1$.

9. False, since $\sin^2(x) + \cos^2(x) = 1$.

10. True, since $(1 - \sin x)(1 + \sin x) = (1 - \sin^2 x)$.

3.1 Exercises

1. $\dfrac{1/\cos(x)}{\sin(x)/\cos(x)} = \dfrac{1}{\cos(x)} \cdot \dfrac{\cos(x)}{\sin(x)} =$ $\dfrac{1}{\sin(x)} = \csc(x)$

3.
$$\frac{\sin(x)}{1/\sin(x)} + \frac{\cos(x)}{1/\cos(x)} = \sin^2(x) + \cos^2(x) = 1$$

5. $1 - \sin^2\alpha = \cos^2\alpha$

7. $(\sin\beta + 1)(\sin\beta - 1) = \sin^2\beta - 1 = -\cos^2\beta$

9.
$$\frac{1 + \cos\alpha \cdot \dfrac{\sin\alpha}{\cos\alpha} \cdot \dfrac{1}{\sin\alpha}}{1/\sin\alpha} = \frac{1+1}{1/\sin\alpha} = 2\sin\alpha$$

11. Since $\cot^2 x = \csc^2 x - 1$, we get $\cot x = \pm\sqrt{\csc^2 x - 1}$.

13. $\sin(x) = \dfrac{1}{\csc(x)} = \dfrac{1}{\pm\sqrt{1 + \cot^2(x)}}$

15. Since $\cot^2(x) = \csc^2(x) - 1$, we obtain $\tan(x) = \dfrac{1}{\cot(x)} = \dfrac{1}{\pm\sqrt{\csc^2(x) - 1}}$

17. Since $\sec\alpha = \sqrt{1 + \left(\dfrac{1}{2}\right)^2} = \dfrac{\sqrt{5}}{2}$, we get $\cos\alpha = \dfrac{2}{\sqrt{5}} = \dfrac{2\sqrt{5}}{5}$, $\sin\alpha = \sqrt{1 - \left(\dfrac{2}{\sqrt{5}}\right)^2} =$ $\dfrac{1}{\sqrt{5}}$ or $\sin\alpha = \dfrac{\sqrt{5}}{5}$, $\csc\alpha = \sqrt{5}$, and $\cot\alpha = 2$.

19. Since $\sin\alpha = -\sqrt{1 - \left(-\dfrac{\sqrt{3}}{5}\right)^2} =$ $-\sqrt{1 - \dfrac{3}{25}} = -\dfrac{\sqrt{22}}{5}$, we find

$\csc\alpha = -\dfrac{5}{\sqrt{22}}$ or $\csc\alpha = -\dfrac{5\sqrt{22}}{22}$,

$\sec\alpha = -\dfrac{5}{\sqrt{3}} = -\dfrac{5\sqrt{3}}{3}$,

$\tan\alpha = \dfrac{-\sqrt{22}/5}{-\sqrt{3}/5} = \dfrac{\sqrt{22}}{\sqrt{3}} = \dfrac{\sqrt{66}}{3}$, and

$\cot\alpha = \dfrac{\sqrt{3}}{\sqrt{22}} = \dfrac{\sqrt{66}}{22}$.

21. Since α is in Quadrant IV, we get

$\csc\alpha = -\sqrt{1 + \left(-\dfrac{1}{3}\right)^2} = -\dfrac{\sqrt{10}}{3}$,

$\sin\alpha = -\dfrac{3}{\sqrt{10}} = -\dfrac{3\sqrt{10}}{10}$,

$\cos\alpha = \sqrt{1 - \left(-\dfrac{3}{\sqrt{10}}\right)^2} =$

$\sqrt{1 - \dfrac{9}{10}} = \dfrac{1}{\sqrt{10}} = \dfrac{\sqrt{10}}{10}$,

$\sec\alpha = \sqrt{10}$, and $\tan\alpha = -3$.

23. $(-\sin x) \cdot (-\cot x) = \sin(x) \cdot \dfrac{\cos x}{\sin x} = \cos(x)$

25. $\sin(y) + (-\sin(y)) = 0$

27. $\dfrac{\sin(x)}{\cos(x)} + \dfrac{-\sin(x)}{\cos(x)} = 0$

29. $(1 + \sin\alpha)(1 - \sin\alpha) = 1 - \sin^2\alpha = \cos^2\alpha$

31. $(-\sin\beta)(\cos\beta)(1/\sin\beta) = -\cos\beta$

33. Odd, since $\sin(-y) = -\sin(y)$ for any y, even if $y = 2x$.

35. Neither, since $f(-\pi/6) \neq f(\pi/6)$ and $f(-\pi/6) \neq -f(\pi/6)$.

37. Even, since $\sec^2(-t) - 1 = \sec^2(t) - 1$.

39 Even, $f(-\alpha) = 1+\sec(-\alpha) = 1+\sec(\alpha) = f(\alpha)$

41. Even, $f(-x) = \dfrac{\sin(-x)}{-x} = \dfrac{-\sin(x)}{-x} = f(x)$

43. Odd, $f(-x) = -x + \sin(-x) = -x - \sin(x) = -f(x)$

45. If $\gamma = \pi/3$, then $(\sin(\pi/3) + \cos(\pi/3))^2 = \left(\dfrac{\sqrt{3}}{2} + \dfrac{1}{2}\right)^2 = \dfrac{(\sqrt{3}+1)^2}{4} = \dfrac{4+2\sqrt{3}}{4} \neq 1$ while $\sin^2(\pi/3) + \cos^2(\pi/3) = 1$. Thus, it is not an identity.

47. If $\beta = \pi/6$, then

$$(1 + \sin(\pi/6))^2 = \left(1 + \frac{1}{2}\right)^2 = \left(\frac{3}{2}\right)^2 = \frac{9}{4}$$

and

$$1 + \sin^2(\pi/6) = 1 + \left(\frac{1}{2}\right)^2 = \frac{5}{4}.$$

Thus, it is not an identity.

49. If $\alpha = 7\pi/6$, then $\sin(7\pi/6) = -\dfrac{1}{2}$ while $\sqrt{1 - \cos^2\left(\dfrac{7\pi}{6}\right)}$ is a positive number. Thus, it is not an identity.

51. If $y = \pi/6$, then $\sin(\pi/6) = \dfrac{1}{2}$ while $\sin(-\pi/6) = -\dfrac{1}{2}$. Thus, it is not an identity.

53. If $y = \pi/6$, then $\cos^2(\pi/6) - \sin^2(\pi/6) = \left(\dfrac{\sqrt{3}}{2}\right)^2 - \left(\dfrac{1}{2}\right)^2 = \dfrac{3}{4} - \dfrac{1}{4} = \dfrac{1}{2}$ and

$$\sin(2 \cdot \pi/6) = \sin(\pi/3) = \frac{\sqrt{3}}{2}.$$

Thus, it is not an identity.

55. $1 - \dfrac{1}{\cos^2(x)} = 1 - \sec^2(x) = -\tan^2(x)$

57. $\dfrac{-(\tan^2 t + 1)}{\sec^2 t} = \dfrac{-\sec^2 t}{\sec^2 t} = -1$

59. $\dfrac{(1 - \cos^2 w) - \cos^2 w}{1 - 2\cos^2 w} = \dfrac{1 - 2\cos^2 w}{1 - 2\cos^2 w} = 1$

61. $\dfrac{\tan x(\tan^2 x - \sec^2 x)}{-\cot x} = \dfrac{\tan x(-1)}{-\cot x} = \tan^2 x$

63.
$$\frac{1}{\sin^3 x} - \frac{\cos^2(x)/\sin^2(x)}{\sin x} = \frac{1}{\sin^3 x} - \frac{\cos^2(x)}{\sin^3 x} = \frac{1 - \cos^2 x}{\sin^3 x} = \frac{\sin^2 x}{\sin^3 x} = \frac{1}{\sin x} = \csc x$$

65. $(\sin^2 x - \cos^2 x)(\sin^2 x + \cos^2 x) = (\sin^2 x - \cos^2 x)(1) = \sin^2 x - \cos^2 x$

For Thought

1. True, $\dfrac{\sin x}{1/\sin x} = \sin x \cdot \dfrac{\sin x}{1} = \sin^2 x$.

2. False, if $x = \pi/3$ then $\dfrac{\cot(\pi/3)}{\tan(\pi/3)} = \dfrac{\sqrt{3}/3}{\sqrt{3}} = \dfrac{1}{3}$ and $\tan^2(\pi/3) = (\sqrt{3})^2 = 3$.

3. True, $\dfrac{1/\cos x}{1/\sin x} = \dfrac{1}{\cos x} \cdot \dfrac{\sin x}{1} = \dfrac{\sin x}{\cos x} = \tan x$.

4. True, $\sin x \cdot \dfrac{1}{\cos x} = \dfrac{\sin x}{\cos x} = \tan x$.

5. True, $\dfrac{\cos x}{\cos x} + \dfrac{\sin x}{\cos x} = 1 + \tan x$.

6. False, if $x = \pi/4$ then

$$\sec(\pi/4) + \frac{\sin(\pi/4)}{\cos(\pi/4)} = \sqrt{2} + 1$$

and

$$\frac{1 + \sin(\pi/4)\cos(\pi/4)}{\cos(\pi/4)} = \frac{1 + (\sqrt{2}/2)(\sqrt{2}/2)}{\sqrt{2}/2} = \frac{1 + (1/2)}{\sqrt{2}/2} = \frac{3}{2} \cdot \frac{2}{\sqrt{2}} = \frac{3}{\sqrt{2}}.$$

7. True, $\dfrac{1 + \sin x}{1 - \sin^2 x} = \dfrac{1 + \sin x}{(1 - \sin x)(1 + \sin x)} = \dfrac{1}{1 - \sin x}$.

8. True, since $\tan x \cdot \cot x = \tan x \cdot \dfrac{1}{\tan x} = 1$.

9. False, if $x = \pi/3$ then $(1 - \cos(\pi/3))^2 = (1 - 1/2)^2 = (1/2)^2 = 1/4$ and $\sin^2(\pi/3) = (\sqrt{3}/2)^2 = 3/4$.

10. False, if $x = \pi/6$ then $(1 - \csc(\pi/6))(1 + \csc(\pi/6)) = (1-2)(1+2) = -3$ and $\cot^2(\pi/6) = (\sqrt{3})^2 = 3$.

3.2 Exercises

1. D, since $\cos x \tan x = \cos x \cdot \dfrac{\sin x}{\cos x} = \sin x$.

3. A, since $\csc^2 x - \cot^2 x = 1$.

5. B, for $1 - \sec^2 x = -\tan^2 x$.

7. H, since $\dfrac{\csc x}{\csc x} - \dfrac{\sin x}{\csc x} = 1 - \sin^2 x = \cos^2 x$.

9. G, for $\csc^2 x = 1 + \cot^2 x$.

11. $\sin^2(\alpha) - 1 = -\cos^2 \alpha$

13. $2\cos^2 \beta - \cos\beta - 1$

15. $\csc^2 x + 2\csc x \sin x + \sin^2 x = \csc^2 x + 2 + \sin^2 x$

17. $4\sin^2\theta - 1$

19. $9\sin^2\theta + 12\sin\theta + 4$

21. $4\sin^4 y - 4\sin^2 y \csc^2 y + \csc^4 y = 4\sin^4 y - 4 + \csc^4 y$

23. $(2\sin\gamma + 1)(\sin\gamma - 3)$

25. $(\tan\alpha - 4)(\tan\alpha - 2)$

27. $(2\sec\beta + 1)^2$

29. $(\tan\alpha - \sec\beta)(\tan\alpha + \sec\beta)$

31. $\cos\beta(\sin^2\beta + \sin\beta - 2) = \cos\beta(\sin\beta + 2)(\sin\beta - 1)$

33. $(2\sec^2 x - 1)^2$

35. $\cos\alpha(\sin\alpha + 1) + (\sin\alpha + 1) = (\cos\alpha + 1)(\sin\alpha + 1)$

37. Rewrite the left side of the equation.

$$\begin{aligned}\sin(x)\cot(x) &= \\ \sin(x) \cdot \frac{\cos(x)}{\sin(x)} &= \\ \cos x\end{aligned}$$

39. Rewrite the left side of the equation.

$$\begin{aligned}1 - \sec(x)\cos^3(x) &= \\ 1 - \frac{1}{\cos(x)} \cdot \cos^3(x) &= \\ 1 - \cos^2(x) &= \\ \sin^2(x)\end{aligned}$$

41.

$$\begin{aligned}1 + \sec^2(x)\sin^2(x) &= \\ 1 + \frac{1}{\cos^2(x)}\sin^2(x) &= \\ 1 + \frac{\sin^2(x)}{\cos^2(x)} &= \\ 1 + \tan^2(x) &= \\ \sec^2(x)\end{aligned}$$

43.

$$\begin{aligned}\frac{\sin^3(x) + \sin(x)\cos^2(x)}{\cos(x)} &= \\ \frac{\sin(x)[\sin^2(x) + \cos^2(x)]}{\cos(x)} &= \\ \frac{\sin(x)[1]}{\cos(x)} &= \\ \tan(x)\end{aligned}$$

45.

$$\begin{aligned}\frac{\sin(x)}{\csc(x)} + \frac{\cos(x)}{\sec(x)} &= \\ \frac{\sin(x)}{1/\sin(x)} + \frac{\cos(x)}{1/\cos(x)} &= \\ \sin^2(x) + \cos^2(x) &= \\ 1\end{aligned}$$

47. Rewrite the left side of the equation.

$$\begin{aligned}\tan(x)\cos(x) + \csc(x)\sin^2(x) &= \\ \sin x + \sin x &= \\ 2\sin x\end{aligned}$$

49.

$$\begin{aligned}(1+\sin\alpha)^2+\cos^2\alpha &= \\ 1+2\sin\alpha+\sin^2\alpha+\cos^2\alpha &= \\ 2+2\sin\alpha\end{aligned}$$

51.

$$\begin{aligned}2-\csc(\beta)\sin(\beta) &= \\ 2-1 &= \\ 1 &= \\ \sin^2(\beta)+\cos^2(\beta)\end{aligned}$$

53.

$$\begin{aligned}\tan x+\cot x &= \\ \frac{\sin x}{\cos x}+\frac{\cos x}{\sin x} &= \\ \frac{\sin^2 x+\cos^2 x}{\sin(x)\cos(x)} &= \\ \frac{1}{\sin(x)\cos(x)} &= \\ \sec(x)\csc(x)\end{aligned}$$

55.

$$\begin{aligned}\frac{\sec(x)}{\tan(x)}-\frac{\tan(x)}{\sec(x)} &= \\ \frac{\sec^2(x)-\tan^2(x)}{\tan(x)\sec(x)} &= \\ \frac{1}{\tan(x)\sec(x)} &= \\ \cot(x)\cos(x)\end{aligned}$$

57. Rewrite the right side of the equation.

$$\begin{aligned}&= \frac{\csc x}{\csc x-\sin x} \\ &= \frac{\csc x}{\csc x-\sin x}\cdot\frac{\sin x}{\sin x} \\ &= \frac{1}{1-\sin^2 x} \\ &= \frac{1}{\cos^2 x} \\ &\sec^2 x\end{aligned}$$

59. Rewrite the right side of the equation.

$$\begin{aligned}&= \frac{\cos(-x)-\csc(-x)}{\cos x} \\ &= \frac{\cos x+\csc x}{\cos x} \\ &= \frac{\cos x}{\cos x}+\frac{\csc x}{\cos x} \\ &1+\csc x\sec x\end{aligned}$$

61. Rewrite the left side of the equation.

$$\begin{aligned}\frac{1}{\csc\theta-\cot\theta} &= \\ \frac{1}{\csc\theta-\cot\theta}\cdot\frac{\sin\theta}{\sin\theta} &= \\ \frac{\sin\theta}{1-\cos\theta} &= \\ \frac{\sin\theta}{1-\cos\theta}\cdot\frac{1+\cos\theta}{1+\cos\theta} &= \\ \frac{\sin\theta(1+\cos\theta)}{1-\cos^2\theta} &= \\ \frac{\sin\theta(1+\cos\theta)}{\sin^2\theta} &= \\ \frac{1+\cos\theta}{\sin\theta}\end{aligned}$$

63. Rewrite the right side of the equation.

$$\begin{aligned}&= \frac{1+\sin(y)}{1-\sin(y)} \\ &= \frac{1+\sin(y)}{1-\sin(y)}\cdot\frac{\csc(y)}{\csc(y)} \\ &= \frac{\csc(y)+1}{\csc(y)-1}\end{aligned}$$

65. Rewrite the left side of the equation.

$$\begin{aligned}\frac{\cot x+\tan x}{\csc x} &= \\ \frac{\frac{\cos x}{\sin x}+\frac{\sin x}{\cos x}}{\frac{1}{\sin x}} &= \\ \frac{\frac{\cos^2 x+\sin^2 x}{\sin x\cos x}}{\frac{1}{\sin x}} &=\end{aligned}$$

$$\frac{\frac{1}{\sin x \cos x}}{\frac{1}{\sin x}} =$$

$$\frac{1}{\cos x}$$

67. Rewrite the left side of the equation.

$$\frac{1-\sin(-x))^2}{1-\sin(-x)} =$$

$$\frac{1-(-\sin x)^2}{1+\sin x} =$$

$$\frac{1-\sin^2 x}{1+\sin x} =$$

$$\frac{(1-\sin x)(1+\sin x)}{1+\sin x} =$$

$$1-\sin(x)$$

69. Rewrite the left side of the equation.

$$\frac{1-\cot^2 w+\cos^2 w \cot^2 w}{\csc^2 w} =$$

$$\frac{1-\cot^2 w\,(1-\cos^2 w)}{\csc^2 w} =$$

$$\frac{1-\cot^2 w \sin^2 w}{\csc^2 w} =$$

$$\frac{1-\frac{\cos^2 w}{\sin^2 w}\sin^2 w}{\csc^2 w} =$$

$$\frac{1-\cos^2 w}{\csc^2 w} =$$

$$\frac{\sin^2 w}{\csc^2 w} =$$

$$\sin^4 w$$

71.

$$\ln(\sec\theta) =$$

$$\ln((\cos\theta)^{-1}) =$$

$$-\ln(\cos\theta)$$

73. Rewrite the left side of the equation.

$$\ln|\sec\alpha+\tan\alpha| =$$

$$\ln\left|(\sec\alpha+\tan\alpha)\cdot\frac{\sec\alpha-\tan\alpha}{\sec\alpha-\tan\alpha}\right| =$$

$$\ln\left|\frac{\sec^2\alpha-\tan^2\alpha}{\sec\alpha-\tan\alpha}\right| =$$

$$\ln\left|\frac{1}{\sec\alpha-\tan\alpha}\right| =$$

$$\ln\left|(\sec\alpha-\tan\alpha)^{-1}\right| =$$

$$-\ln|\sec\alpha-\tan\alpha|$$

75. It is an identity since

$$\frac{\sin\theta}{\sin\theta}+\frac{\cos\theta}{\sin\theta} =$$

$$1+\cot\theta.$$

The graphs of $y=\frac{\sin\theta+\cos\theta}{\sin\theta}$ and $y=1+\cot\theta$ are shown to be identical.

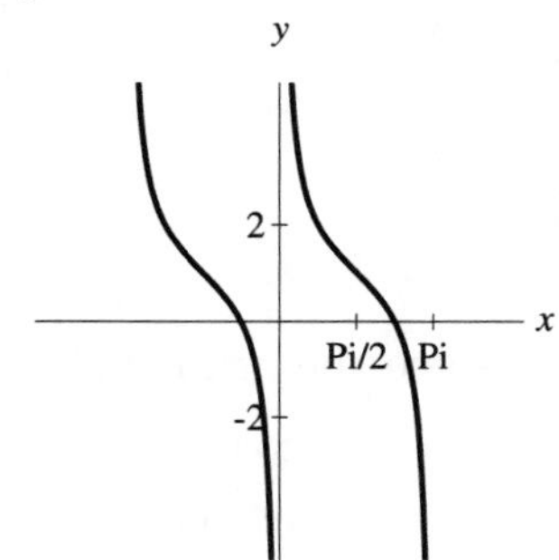

77. It is not an identity since the graphs of $y=(\sin x+\csc x)^2$ and $y=\sin^2 x+\csc^2 x$ do not coincide as shown.

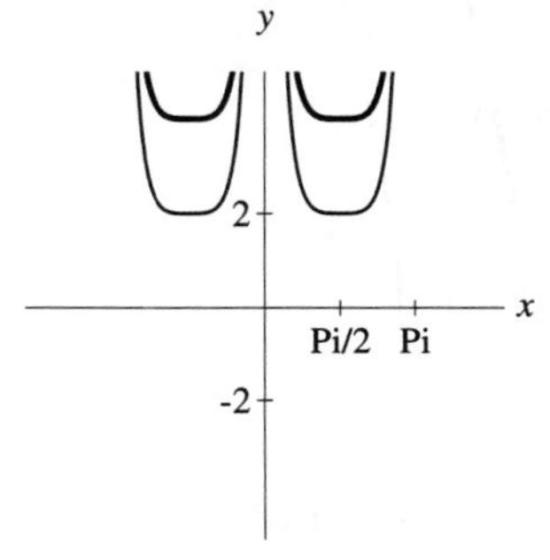

79. It is an identity. Re-arranging the numerator in the right-hand side, we find

$$= \frac{1+\cos x - \cos^2 x}{\sin x}$$
$$= \frac{1-\cos^2 x + \cos x}{\sin x}$$
$$= \frac{\sin^2 x + \cos x}{\sin x}$$
$$= \frac{\sin^2 x}{\sin x} + \frac{\cos x}{\sin x}$$
$$\sin x + \cot x.$$

The graphs of $y = \cot x + \sin x$ and $y = \dfrac{1+\cos x - \cos^2 x}{\sin x}$ are shown to be identical.

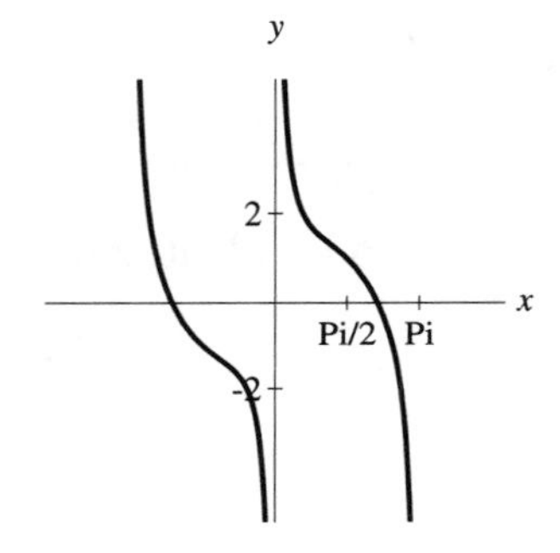

81. It is not an identity since the graphs of $y = \dfrac{\sin x}{\cos x} - \dfrac{\cos x}{\sin x}$ and $y = \dfrac{2\cos^2 x - 1}{\sin x \cos x}$ are not the same as shown.

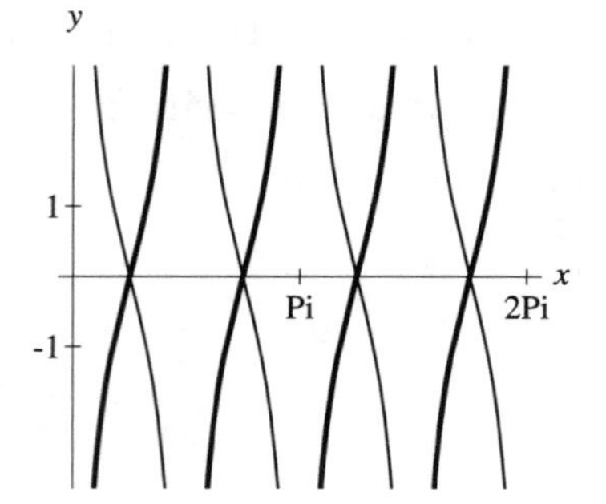

83. It is an identity.

$$\frac{\cos(-x)}{1-\sin(x)} =$$
$$\frac{\cos(x)}{1-\sin(x)} =$$
$$\frac{\cos x}{1-\sin(x)} \cdot \frac{1+\sin x}{1+\sin x} =$$
$$\frac{\cos x(1+\sin x)}{1-\sin^2 x} =$$
$$\frac{\cos x(1+\sin x)}{\cos^2 x} =$$
$$\frac{1+\sin x}{\cos x} =$$
$$\frac{1-\sin(-x)}{\cos x} =$$

The graphs of $y = \dfrac{\cos(-x)}{1-\sin x}$ and $y = \dfrac{1-\sin(-x)}{\cos x}$ are shown to be identical.

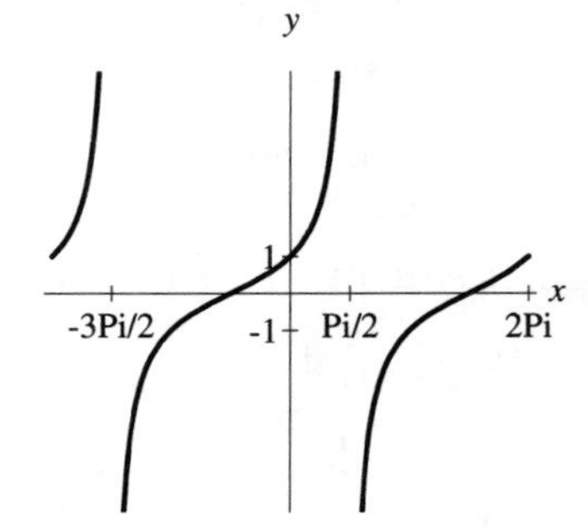

For Thought

1. False, since $\frac{\pi}{4} + \frac{\pi}{2} = \frac{3\pi}{4}$.

2. False, since $\frac{\pi}{4} - \frac{\pi}{3} = -\frac{\pi}{12}$.

3. False, the right-hand side should be $\cos(5°)$.

4. True, by the sum identity for cosine.

5. True, since $\cos\left(\frac{\pi}{2} - \alpha\right) = \sin \alpha$.

6. True, since $\cos(-x) = \cos(x)$.

7. False, rather $\sin(\pi/7 - 3) = -\sin(3 - \pi/7)$.

8. False, since $\sin(x - \pi/2) \neq \cos(x)$ if $x = 0$.

9. True, for $\sec(\pi/3) = \sec(\pi/2 - \pi/6) = \csc(\pi/6)$.

10. True, since the cofunction identity for tangent is applied to $90° - 68°29'55" = 21°30'5"$.

3.3 Exercises

1. $\dfrac{7\pi}{12}$

3. $\dfrac{13\pi}{12}$

5. $\dfrac{3\pi}{12} - \dfrac{4\pi}{12} = -\dfrac{\pi}{12}$

7. $\dfrac{3\pi}{12} - \dfrac{2\pi}{12} = \dfrac{\pi}{12}$

9. $30° + 45°$

11. $120° + 45°$

13. $\dfrac{\pi}{3} - \dfrac{\pi}{4}$

15. $\dfrac{\pi}{3} + \dfrac{\pi}{4}$

17. $\cos(60° - 45°) =$
$\cos(60°)\cos(45°) + \sin(60°)\sin(45°) =$
$\dfrac{1}{2} \cdot \dfrac{\sqrt{2}}{2} + \dfrac{\sqrt{3}}{2} \cdot \dfrac{\sqrt{2}}{2} = \dfrac{\sqrt{2}+\sqrt{6}}{4}$

19. $\cos(60° + 45°) =$
$\cos(60°)\cos(45°) - \sin(60°)\sin(45°) =$
$\dfrac{1}{2} \cdot \dfrac{\sqrt{2}}{2} - \dfrac{\sqrt{3}}{2} \cdot \dfrac{\sqrt{2}}{2} = \dfrac{\sqrt{2}-\sqrt{6}}{4}$

21. $\cos\left(\dfrac{5\pi}{12}\right) = \cos\left(\dfrac{\pi}{6} + \dfrac{\pi}{4}\right) =$
$\cos\left(\dfrac{\pi}{6}\right)\cos\left(\dfrac{\pi}{4}\right) - \sin\left(\dfrac{\pi}{6}\right)\sin\left(\dfrac{\pi}{4}\right) =$
$\dfrac{\sqrt{3}}{2} \cdot \dfrac{\sqrt{2}}{2} - \dfrac{1}{2} \cdot \dfrac{\sqrt{2}}{2} = \dfrac{\sqrt{6}-\sqrt{2}}{4}$

23. $\cos\left(\dfrac{13\pi}{12}\right) = \cos\left(\dfrac{3\pi}{4} + \dfrac{\pi}{3}\right) =$
$\cos\left(\dfrac{3\pi}{4}\right)\cos\left(\dfrac{\pi}{3}\right) - \sin\left(\dfrac{3\pi}{4}\right)\sin\left(\dfrac{\pi}{3}\right) =$
$-\dfrac{\sqrt{2}}{2} \cdot \dfrac{1}{2} - \dfrac{\sqrt{2}}{2} \cdot \dfrac{\sqrt{3}}{2} = -\dfrac{\sqrt{2}+\sqrt{6}}{4}$

25. $\cos\left(-\dfrac{\pi}{12}\right) = \cos\left(\dfrac{\pi}{4} - \dfrac{\pi}{3}\right) =$
$\cos\left(\dfrac{\pi}{4}\right)\cos\left(\dfrac{\pi}{3}\right) + \sin\left(\dfrac{\pi}{4}\right)\sin\left(\dfrac{\pi}{3}\right) =$
$\dfrac{\sqrt{2}}{2} \cdot \dfrac{1}{2} + \dfrac{\sqrt{2}}{2} \cdot \dfrac{\sqrt{3}}{2} = \dfrac{\sqrt{2}+\sqrt{6}}{4}$

27. $\cos\left(-\dfrac{13\pi}{12}\right) = \cos\left(\dfrac{13\pi}{12}\right) =$
$\cos\left(\dfrac{3\pi}{4} + \dfrac{\pi}{3}\right) =$
$\cos\left(\dfrac{3\pi}{4}\right)\cos\left(\dfrac{\pi}{3}\right) - \sin\left(\dfrac{3\pi}{4}\right)\sin\left(\dfrac{\pi}{3}\right) =$
$-\dfrac{\sqrt{2}}{2} \cdot \dfrac{1}{2} - \dfrac{\sqrt{2}}{2} \cdot \dfrac{\sqrt{3}}{2} = -\dfrac{\sqrt{2}+\sqrt{6}}{4}$

29. $\cos(23° - 67°) = \cos(-44°) = \cos(44°)$

31. $\cos(5 + 6) = \cos(11)$

33. $\cos(2k - k) = \cos(k)$

35. $\cos\left(\dfrac{\pi}{2} - \dfrac{\pi}{5}\right) = \cos\left(\dfrac{3\pi}{10}\right)$

37. The problem can be expressed as
$\cos\left(-\dfrac{\pi}{5}\right)\cos\left(-\dfrac{\pi}{3}\right) + \sin\left(-\dfrac{\pi}{5}\right)\sin\left(-\dfrac{\pi}{3}\right) =$
$\cos\left(-\dfrac{\pi}{5} - \left(-\dfrac{\pi}{3}\right)\right) = \cos\left(\dfrac{2\pi}{15}\right)$

39. e, $\sin(20°) = \cos(90° - 20°) = \cos(70°)$

41. c, $\cos(90°) = 0 = \sin(0°)$

43. a, $\sec(\pi/6) = \csc(\pi/2 - \pi/6) = \csc(\pi/3)$

45. g, $\sin(5\pi/12) = \cos(\pi/2 - 5\pi/12) = \cos(\pi/12)$

47. g, $\cos(44°) = \sin(90° - 44°) = \sin(46°)$

49. c, $\cot(134°) = \tan(90° - 134°) = -\tan(44°)$

51. f, $\sec(1) = \csc\left(\dfrac{\pi}{2} - 1\right) = \csc\left(\dfrac{\pi - 2}{2}\right)$

53. a, $\csc(\pi/2) = 1 = \cos(0)$

55. Note, $\cos(61°) = \sin(29°)$. Rewriting we get
$\cos(14°)\cos(29°) + \sin(14°)\sin(29°) =$
$\cos(14° - 29°) = \cos(-15°) = \cos(15°)$.

57. Note, $\cos(-10°) = \cos(10°)$ and
$\cos(70°) = \sin(20°)$. Rewriting we obtain
$\cos(-10°)\cos(20°) + \sin(-10°)\sin(20°) =$
$\cos(-10° - 20°) = \cos(-30°) =$
$\cos(30°) = \dfrac{\sqrt{3}}{2}$.

59. Rewriting we get
$\cos(\pi/2 - \alpha)\cos(-\alpha) - \sin(\alpha - \pi/2)\sin(-\alpha) =$
$\cos(\pi/2 - \alpha)\cos(-\alpha) + \sin(\pi/2 - \alpha)\sin(-\alpha) =$
$\cos((\pi/2 - \alpha) - (-\alpha)) = \cos(\pi/2) = 0$.

61. Rewriting we find
$\cos(-3k)\cos(-k) - \cos(\pi/2 - 3k)\sin(-k) =$
$\cos(3k)\cos(-k) - \sin(3k)\sin(-k) =$
$\cos(3k + (-k)) = \cos(2k)$.

63. Since α is in quadrant II and β is in quadrant I, we find

$$\cos\alpha = -\sqrt{1 - \left(\frac{3}{5}\right)^2} = -\sqrt{1 - \frac{9}{25}} =$$

$$-\sqrt{\frac{16}{25}} = -\frac{4}{5} \text{ and } \cos\beta = \sqrt{1 - \left(\frac{5}{13}\right)^2} =$$

$$\sqrt{1 - \frac{25}{169}} = \sqrt{\frac{144}{169}} = \frac{12}{13}.$$ Then

$\cos(\alpha + \beta) = \cos\alpha\cos\beta - \sin\alpha\sin\beta =$

$$-\frac{4}{5} \cdot \frac{12}{13} - \frac{3}{5} \cdot \frac{5}{13} = -\frac{63}{65}.$$

65. Since α is in quadrant IV and β is in quadrant II, we obtain

$$\cos\alpha = \sqrt{1 - \left(\frac{-7}{25}\right)^2} = \frac{24}{25}$$

and

$$\cos\beta = -\sqrt{1 - \left(\frac{8}{17}\right)^2} = -\frac{15}{17}.$$

Then

$\cos(\alpha + \beta) = \cos\alpha\cos\beta - \sin\alpha\sin\beta =$

$$\frac{24}{25} \cdot \left(\frac{-15}{17}\right) - \left(\frac{-7}{25}\right) \cdot \frac{8}{17} = -\frac{304}{425}.$$

67. Since α is in quadrant I and β is in quadrant II, we find

$$\cos\alpha = \sqrt{1 - \left(\frac{\sqrt{3}}{2}\right)^2} = \frac{1}{2}$$

and

$$\sin\beta = \sqrt{1 - \left(\frac{-\sqrt{2}}{2}\right)^2} = \frac{\sqrt{2}}{2}.$$

Then

$\cos(\alpha - \beta) = \cos\alpha\cos\beta + \sin\alpha\sin\beta =$

$$\frac{1}{2} \cdot \left(\frac{-\sqrt{2}}{2}\right) + \frac{\sqrt{3}}{2} \cdot \frac{\sqrt{2}}{2} = \frac{\sqrt{6} - \sqrt{2}}{4}.$$

69. Since α is in quadrant I and β is in quadrant III, we obtain

$$\cos\alpha = \sqrt{1 - \left(\frac{2}{3}\right)^2} = \sqrt{1 - \frac{4}{9}} =$$

$$\sqrt{\frac{5}{9}} = \frac{\sqrt{5}}{3} \text{ and } \cos\beta = -\sqrt{1 - \left(\frac{-1}{2}\right)^2} =$$

$$-\sqrt{1 - \frac{1}{4}} = -\sqrt{\frac{3}{4}} = -\frac{\sqrt{3}}{2}.$$ Thus,

$\cos(\alpha + \beta) = \cos\alpha\cos\beta - \sin\alpha\sin\beta =$

$$\frac{\sqrt{5}}{3} \cdot \frac{-\sqrt{3}}{2} - \frac{2}{3} \cdot \frac{-1}{2} = \frac{2 - \sqrt{15}}{6}.$$

71. $\cos(\pi/2 - (-\alpha)) = \sin(-\alpha) = -\sin\alpha$

73. $\cos 180^\circ \cos\alpha + \sin 180^\circ \sin\alpha =$
$(-1) \cdot \cos\alpha + 0 \cdot \sin\alpha = -\cos\alpha$

75. $\cos(3\pi/2)\cos\alpha + \sin(3\pi/2)\sin\alpha =$
$0 \cdot \cos\alpha + (-1) \cdot \sin\alpha = -\sin\alpha$

77. $\cos(90^\circ - (-\alpha)) = \sin(-\alpha) = -\sin(\alpha)$

79. Rewrite left side.

$$\begin{aligned} \cos(x - \pi/2) &= \\ \cos x\cos(\pi/2) + \sin x\sin(\pi/2) &= \\ \cos x \cdot 0 + \sin x \cdot 1 &= \\ \sin x & \end{aligned}$$

81. Substitute the sum and difference cosine identities into the left-hand side to get a difference of two squares.

$$\begin{aligned} \cos(\alpha + \beta)\cos(\alpha - \beta) &= \\ (\cos\alpha\cos\beta)^2 - (\sin\alpha\sin\beta)^2 &= \\ (1 - \sin^2\alpha)\cos^2\beta - \sin^2\alpha(1 - \cos^2\beta) &= \\ \cos^2\beta - \sin^2\alpha\cos^2\beta - \sin^2\alpha + \sin^2\alpha\cos^2\beta &= \\ \cos^2\beta - \sin^2\alpha & \end{aligned}$$

83.

$$\begin{aligned} \cos(x - y) + \cos(y - x) &= \\ \cos(x - y) + \cos(x - y) &= \\ 2\cos(x - y) &= \\ 2(\cos x\cos y + \sin x\sin y) &= \\ 2\cos x\cos y - 2\sin x\sin y & \end{aligned}$$

85. In the proof, multiply each te rm by $\cos(\alpha-\beta)$. Also, the sum and difference identities for cosine expresses $\cos(\alpha+\beta)\cos(\alpha-\beta)$ as a difference of two squares.

$$\frac{\cos(\alpha+\beta)}{\cos\alpha+\sin\beta} =$$

$$\frac{\cos(\alpha+\beta)}{\cos\alpha+\sin\beta}\cdot\frac{\cos(\alpha-\beta)}{\cos(\alpha-\beta)} =$$

$$\frac{\cos^2\alpha\cos^2\beta-\sin^2\alpha\sin^2\beta}{(\cos\alpha+\sin\beta)\cos(\alpha-\beta)} =$$

$$\frac{\cos^2\alpha(1-\sin^2\beta)-(1-\cos^2\alpha)\sin^2\beta}{(\cos\alpha+\sin\beta)\cos(\alpha-\beta)} =$$

$$\frac{\cos^2\alpha-\cos^2\alpha\sin^2\beta-\sin^2\beta+\cos^2\alpha\sin^2\beta}{(\cos\alpha+\sin\beta)\cos(\alpha-\beta)} =$$

$$\frac{\cos^2\alpha-\sin^2\beta}{(\cos\alpha+\sin\beta)\cos(\alpha-\beta)} =$$

$$\frac{(\cos\alpha-\sin\beta)(\cos\alpha+\sin\beta)}{(\cos\alpha+\sin\beta)\cos(\alpha-\beta)} =$$

$$\frac{\cos\alpha-\sin\beta}{\cos(\alpha-\beta)} =$$

$$\frac{\cos\alpha-\sin\beta}{\cos(\beta-\alpha)}$$

For Thought

1. True, by the difference identity for sine.

2. False, since the left side of the equation is $\sin 9$.

3. True, by the sum identity for sine.

4. False, since the equation is false if $\alpha=\beta=\frac{\pi}{4}$.

5. True, since $\sin 2\pi = 0 = \sin\pi$.

6. False, since $\sin(\pi/2)=1$ and $\sin(\pi/4)+\sin(\pi/4)=\sqrt{2}$.

7. False, since $\tan 10° \approx 0.176$ and $\tan(2°)+\tan(8°)\approx 0.175$.

8. True, by the sum identity for tangent.

9. True, by the difference identity for tangent.

10. True, since both sides of the equation (by the sum identity for tangent) are equal to $\tan(-7°)$.

3.4 Exercises

1. $\frac{3\pi}{12}+\frac{\pi}{12}=\frac{4\pi}{12}=\frac{\pi}{3}$

3. $\frac{9\pi}{12}-\frac{4\pi}{12}=\frac{5\pi}{12}$

5. $60°+45°$

7. $30°-45°$

9. $\sin(\pi/3+\pi/4)=$
$\sin(\pi/3)\cos(\pi/4)+\cos(\pi/3)\sin(\pi/4)=$
$\frac{\sqrt{3}}{2}\cdot\frac{\sqrt{2}}{2}+\frac{1}{2}\cdot\frac{\sqrt{2}}{2}=\frac{\sqrt{6}+\sqrt{2}}{4}$

11. $\tan(45°+30°)=\frac{\tan(45°)+\tan(30°)}{1-\tan(45°)\tan(30°)}=$
$\frac{1+\sqrt{3}/3}{1-1\cdot\sqrt{3}/3}\cdot\frac{3}{3}=\frac{3+\sqrt{3}}{3-\sqrt{3}}\cdot\frac{3+\sqrt{3}}{3+\sqrt{3}}=$
$\frac{12+6\sqrt{3}}{9-3}=2+\sqrt{3}$

13. $\sin(30°-45°)=$
$\sin(30°)\cos(45°)-\cos(30°)\sin(45°)=$
$\frac{1}{2}\cdot\frac{\sqrt{2}}{2}-\frac{\sqrt{3}}{2}\cdot\frac{\sqrt{2}}{2}=\frac{\sqrt{2}-\sqrt{6}}{4}$

15. $\tan(-13\pi/12)=-\tan(13\pi/12)=$
$-\tan\left(\frac{3\pi}{4}+\frac{\pi}{3}\right)=$
$-\frac{\tan(3\pi/4)+\tan(\pi/3)}{1-\tan(3\pi/4)\tan(\pi/3)}=$
$-\frac{-1+\sqrt{3}}{1-(-1)\sqrt{3}}=\frac{1-\sqrt{3}}{1+\sqrt{3}}\cdot\frac{1-\sqrt{3}}{1-\sqrt{3}}=$
$\frac{4-2\sqrt{3}}{-2}=-2+\sqrt{3}$

17. $\sin(23°+67°)=\sin(90°)=1$

19. $\sin(34°)\cos(13°)-\cos(34°)\sin(13°)=$
$\sin(34°-13°)=\sin(21°)$

21. $\sin\left(-\frac{\pi}{2}\right)\cos\left(-\frac{\pi}{5}\right)+\cos\left(-\frac{\pi}{2}\right)\sin\left(-\frac{\pi}{5}\right)=$
$\sin\left(-\frac{\pi}{2}+\left(-\frac{\pi}{5}\right)\right)=\sin\left(-\frac{7\pi}{10}\right)=$
$-\sin\left(\frac{7\pi}{10}\right)$

23. $\sin(14°)\cos(35°) + \cos(14°)\sin(35°) =$
$\sin(14° + 35°) = \sin(49°)$

25. $\tan\left(\frac{\pi}{9} + \frac{\pi}{6}\right) = \tan\left(\frac{5\pi}{18}\right)$

27. $\frac{\tan(\pi/7) + \tan(\pi/6)}{1 - \tan(\pi/7)\tan(\pi/6)} = \tan\left(\frac{\pi}{7} + \frac{\pi}{6}\right) =$
$\tan(13\pi/42)$

29. Since α is in quadrant II and β is in quadrant I, we obtain

$\cos\alpha = -\sqrt{1 - \left(\frac{3}{5}\right)^2} = -\sqrt{1 - \frac{9}{25}} =$

$-\sqrt{\frac{16}{25}} = -\frac{4}{5}$ and $\cos\beta = \sqrt{1 - \left(\frac{5}{13}\right)^2} =$

$\sqrt{1 - \frac{25}{169}} = \sqrt{\frac{144}{169}} = \frac{12}{13}$. Then
$\sin(\alpha + \beta) = \sin\alpha\cos\beta + \cos\alpha\sin\beta =$
$\frac{3}{5} \cdot \frac{12}{13} + \frac{-4}{5} \cdot \frac{5}{13} = \frac{16}{65}$.

31. Since α is in quadrant II and β is in quadrant III, we find

$$\cos\alpha = -\sqrt{1 - \left(\frac{7}{25}\right)^2} = -\frac{24}{25}$$

and

$$\cos\beta = -\sqrt{1 - \left(\frac{-8}{17}\right)^2} = -\frac{15}{17}.$$

Then
$\sin(\alpha + \beta) = \sin\alpha\cos\beta + \cos\alpha\sin\beta =$
$\frac{7}{25} \cdot \left(\frac{-15}{17}\right) + \frac{24}{25} \cdot \left(\frac{-8}{17}\right) = -\frac{297}{425}$.

33. Since α is in quadrant II and β is in quadrant I, we find

$$\cos\alpha = -\sqrt{1 - \left(\frac{\sqrt{3}}{2}\right)^2} = -\frac{1}{2}$$

and

$$\sin\beta = \sqrt{1 - \left(\frac{\sqrt{2}}{2}\right)^2} = \frac{\sqrt{2}}{2}.$$

Then

$\sin(\alpha - \beta) = \sin\alpha\cos\beta - \cos\alpha\sin\beta =$
$\frac{\sqrt{3}}{2} \cdot \frac{\sqrt{2}}{2} - \left(\frac{-1}{2}\right) \cdot \frac{\sqrt{2}}{2} = \frac{\sqrt{6} + \sqrt{2}}{4}$.

35. Since α is in quadrant I and β is in quadrant III, we find

$\cos\alpha = \sqrt{1 - \left(\frac{2}{3}\right)^2} = \sqrt{1 - \frac{4}{9}} =$

$\sqrt{\frac{5}{9}} = \frac{\sqrt{5}}{3}$ and $\cos\beta = -\sqrt{1 - \left(\frac{-1}{2}\right)^2} =$

$-\sqrt{1 - \frac{1}{4}} = -\sqrt{\frac{3}{4}} = -\frac{\sqrt{3}}{2}$. Thus,
$\sin(\alpha + \beta) = \sin\alpha\cos\beta + \cos\alpha\sin\beta =$
$\frac{2}{3} \cdot \frac{-\sqrt{3}}{2} + \frac{\sqrt{5}}{3} \cdot \frac{-1}{2} = \frac{-2\sqrt{3} - \sqrt{5}}{6}$.

37. $\sin\alpha\cos\pi - \cos\alpha\sin\pi =$
$\sin\alpha \cdot (-1) - \cos\alpha \cdot 0 = -\sin\alpha$

39. Since the period of $\sin x$ is 360°, we get

$$\sin(360° - \alpha) = \sin(-\alpha) = -\sin\alpha.$$

41. $\frac{\tan(\pi/4) + \tan\alpha}{1 - \tan(\pi/4)\tan\alpha} = \frac{1 + \tan\alpha}{1 - 1 \cdot \tan\alpha} =$
$\frac{1 + \tan\alpha}{1 - \tan\alpha}$

43. Since the period of $y = \tan x$ is π or 180°, we find

$$\tan(180° + \alpha) = \tan\alpha.$$

45. We rewrite the left side.

$$\begin{aligned} \sin(180° - \alpha) &= \\ \sin(180°)\cos\alpha - \cos(180°)\sin\alpha &= \\ 0 \cdot \cos\alpha - (-1)\sin\alpha &= \\ \sin\alpha &= \\ \frac{\sin^2\alpha}{\sin\alpha} &= \\ \frac{1 - \cos^2\alpha}{\sin\alpha} & \end{aligned}$$

47. Substitute the sum and difference identities for sine. Then you obtain a difference of two squares as shown:

$$\begin{aligned} \sin(\alpha+\beta)\sin(\alpha-\beta) &= \\ (\sin\alpha\cos\beta)^2-(\cos\alpha\sin\beta)^2 &= \\ \sin^2\alpha(1-\sin^2\beta)-(1-\sin^2\alpha)\sin^2\beta &= \\ \sin^2\alpha-\sin^2\alpha\sin^2\beta-\sin^2\beta+\sin^2\alpha\sin^2\beta &= \\ \sin^2\alpha-\sin^2\beta & \end{aligned}$$

49. Rewrite the left side of the identity.

$$\begin{aligned} \sin(x-y)-\sin(y-x) &= \\ \sin(x-y)+\sin(x-y) &= \\ 2\sin(x-y) &= \\ 2(\sin x\cos y-\cos x\sin y) &= \\ 2\sin x\cos y-2\cos x\sin y & \end{aligned}$$

51. Apply the cofunction identity for cotangent.

$$\begin{aligned} \tan(\pi/4+x) &= \\ \cot(\pi/2-(\pi/4+x)) &= \\ \cot(\pi/2-\pi/4-x) &= \\ \cot(\pi/4-x) & \end{aligned}$$

53. Use sum and difference identities, then divide each term by $\cos\alpha\cos\beta$.

$$\begin{aligned} \frac{\cos(\alpha-\beta)}{\sin(\alpha+\beta)} &= \\ \frac{\dfrac{\cos\alpha\cos\beta}{\cos\alpha\cos\beta}+\dfrac{\sin\alpha\sin\beta}{\cos\alpha\cos\beta}}{\dfrac{\sin\alpha\cos\beta}{\cos\alpha\cos\beta}+\dfrac{\cos\alpha\sin\beta}{\cos\alpha\cos\beta}} &= \\ \frac{1+\tan(\alpha)\tan(\beta)}{\tan(\alpha)+\tan(\beta)} & \end{aligned}$$

55. In the proof, divide the numerator and denominator by $\sin x\sin y$.

$$\begin{aligned} \frac{\sin(x+y)}{\sin(x-y)} &= \\ \frac{\sin(x)\cos(y)+\cos(x)\sin(y)}{\sin(x)\cos(y)-\cos(x)\sin(y)} &= \\ \frac{\dfrac{\sin(x)\cos(y)}{\sin(x)\sin(y)}+\dfrac{\cos(x)\sin(y)}{\sin(x)\sin(y)}}{\dfrac{\sin(x)\cos(y)}{\sin(x)\sin(y)}-\dfrac{\cos(x)\sin(y)}{\sin(x)\sin(y)}} &= \\ \frac{\cot(y)+\cot(x)}{\cot(y)-\cot(x)} & \end{aligned}$$

For Thought

1. True, $\dfrac{\sin(2\cdot 21^\circ)}{2}=\dfrac{2\sin(21^\circ)\cos(21^\circ)}{2}$ $=\sin(21^\circ)\cos(21^\circ)$.

2. True, by a cosine double angle identity

$$\cos(2\sqrt{2})=2\cos^2(\sqrt{2})-1.$$

3. False, $\sin\left(\dfrac{300^\circ}{2}\right)=\sqrt{\dfrac{1-\cos(300^\circ)}{2}}$.

4. True, $\sin\left(\dfrac{400^\circ}{2}\right)=-\sqrt{\dfrac{1-\cos(400^\circ)}{2}}$ $=-\sqrt{\dfrac{1-\cos(40^\circ)}{2}}$.

5. False, $\tan\left(\dfrac{7\pi/4}{2}\right)=-\sqrt{\dfrac{1-\cos(7\pi/4)}{1+\cos(7\pi/4)}}$.

6. True, $\tan\left(\dfrac{-\pi/4}{2}\right)=\dfrac{1-\cos(-\pi/4)}{\sin(-\pi/4)}=$ $\dfrac{1-\cos(\pi/4)}{\sin(-\pi/4)}$

7. False, if $x=\pi/4$ then $\dfrac{\sin(2\cdot\pi/4)}{2}=$ $\dfrac{\sin(\pi/2)}{2}=\dfrac{1}{2}$ and $\sin(\pi/4)=\sqrt{2}/2$.

8. False, since $\cos(2\pi/3)=-1/2$ while $\sqrt{\dfrac{1+\cos(2x)}{2}}$ is a non-negative number.

9. True, since $1-\cos x\geq 0$ we find

$$\sqrt{(1-\cos x)^2}=|1-\cos x|=1-\cos x.$$

10. True, α is in quadrant III or IV while, $\alpha/2$ is in quadrant II.

3.5 Exercises

1. $\sin(2\cdot 45°) = 2\sin(45°)\cos(45°) =$

$2\cdot\frac{\sqrt{2}}{2}\cdot\frac{\sqrt{2}}{2} = 2\cdot\frac{2}{4} = 1.$

3. $\tan(2\cdot 30°) = \frac{2\tan(30°)}{1-\tan^2(30°)} = \frac{2(\sqrt{3}/3)}{1-(\sqrt{3}/3)^2} =$

$\frac{2\sqrt{3}/3}{1-1/3} = \frac{2\sqrt{3}/3}{2/3} = \sqrt{3}$

5. $\sin\left(2\cdot\frac{3\pi}{4}\right) = 2\sin(3\pi/4)\cos(3\pi/4) =$

$2\cdot\frac{\sqrt{2}}{2}\cdot\frac{-\sqrt{2}}{2} = 2\cdot\frac{-2}{4} = -1$

7. $\tan\left(2\cdot\frac{2\pi}{3}\right) = \frac{2\tan(2\pi/3)}{1-\tan^2(2\pi/3)} = \frac{2(-\sqrt{3})}{1-(-\sqrt{3})^2}$

$= \frac{-2\sqrt{3}}{1-3} = \frac{-2\sqrt{3}}{-2} = \sqrt{3}$

9. $\sin\left(\frac{30°}{2}\right) = \sqrt{\frac{1-\cos(30°)}{2}} =$

$\sqrt{\frac{1-\sqrt{3}/2}{2}\cdot\frac{2}{2}} = \sqrt{\frac{2-\sqrt{3}}{4}} = \frac{\sqrt{2-\sqrt{3}}}{2}$

11. $\tan\left(\frac{30°}{2}\right) = \frac{1-\cos(30°)}{\sin(30°)} =$

$\frac{1-\sqrt{3}/2}{1/2}\cdot\frac{2}{2} = 2-\sqrt{3}$

13. $\cos\left(\frac{\pi/4}{2}\right) = \sqrt{\frac{1+\cos(\pi/4)}{2}} =$

$\sqrt{\frac{1+\sqrt{2}/2}{2}\cdot\frac{2}{2}} = \sqrt{\frac{2+\sqrt{2}}{4}} = \frac{\sqrt{2+\sqrt{2}}}{2}$

15. $\sin\left(\frac{45°}{2}\right) = \sqrt{\frac{1-\cos(45°)}{2}} =$

$\sqrt{\frac{1-\sqrt{2}/2}{2}\cdot\frac{2}{2}} = \sqrt{\frac{2-\sqrt{2}}{4}} = \frac{\sqrt{2-\sqrt{2}}}{2}$

17. $\cos(7\pi/8) = -\sqrt{\frac{1+\cos(7\pi/4)}{2}} =$

$-\sqrt{\frac{1+\frac{\sqrt{2}}{2}}{2}} = -\sqrt{\frac{2+\sqrt{2}}{4}} = -\frac{\sqrt{2+\sqrt{2}}}{2}$

19. Positive, 118.5° is in quadrant II

21. Negative, 100° is in quadrant II

23. Negative, $-5\pi/12$ is in quadrant IV

25. $\sin(2\cdot 13°) = \sin 26°$

27. $\cos(2\cdot 22.5°) = \cos 45° = \frac{\sqrt{2}}{2}$

29. $\frac{1}{2}\cdot\frac{2\tan 15°}{1-\tan^2 15°} = \frac{1}{2}\cdot\tan(2\cdot 15°) =$

$\frac{1}{2}\cdot\tan 30° = \frac{1}{2}\cdot\frac{\sqrt{3}}{3} = \frac{\sqrt{3}}{6}$

31. $\frac{1}{2}\cdot\frac{2\tan 30°}{1-\tan^2 30°} = \frac{1}{2}\cdot\tan(2\cdot 30°) =$

$\frac{1}{2}\cdot\tan 60° = \frac{1}{2}\cdot\sqrt{3} = \frac{\sqrt{3}}{2}$

33. $2\sin\left(\frac{\pi}{9}-\frac{\pi}{2}\right)\cos\left(\frac{\pi}{9}-\frac{\pi}{2}\right) =$

$\sin\left(2\cdot\left(\frac{\pi}{9}-\frac{\pi}{2}\right)\right) = \sin\left(\frac{2\pi}{9}-\pi\right) =$

$\sin\left(-\frac{7\pi}{9}\right) = -\sin\left(\frac{7\pi}{9}\right).$

35. $\tan\left(\frac{12°}{2}\right) = \tan 6°$

37. Rewrite the left side of the identity.

$$\begin{aligned}\cos^4 s - \sin^4 s &= \\ (\cos^2 s - \sin^2 s)(\cos^2 s + \sin^2 s) &= \\ \cos(2s)\cdot(1) &= \\ \cos(2s)\end{aligned}$$

39. Apply double angle identities to the left side of the identity.

$$\begin{aligned}\frac{\sin(4t)}{4} &= \\ \frac{2\sin(2t)\cos(2t)}{4} &= \\ \frac{2\cdot 2\sin t\cos t\cdot(\cos^2 t - \sin^2 t)}{4} &= \\ \sin t\cos t(\cos^2 t - \sin^2 t) &= \\ \cos^3 t\sin t - \sin^3 t\cos t\end{aligned}$$

41.

$$\begin{aligned}\frac{\cos(2x)+\cos(2y)}{\sin(x)+\cos(y)} &= \\ \frac{1-2\sin^2 x+2\cos^2 y-1}{\sin x+\cos y} &= \\ 2\frac{\cos^2 y-\sin^2 x}{\sin x+\cos y} &= \\ 2\frac{(\cos y-\sin x)(\cos y+\sin x)}{\sin x+\cos y} &= \\ 2\cos(y)-2\sin(x)\end{aligned}$$

43.

$$\begin{aligned}\frac{\cos(2x)}{\sin^2 x} &= \\ \frac{1-2\sin^2 x}{\sin^2 x} &= \\ \frac{1}{\sin^2 x}-2\cdot\frac{\sin^2 x}{\sin^2 x} &= \\ \csc^2 x-2\end{aligned}$$

45. Rewrite the right side of the identity.

$$\begin{aligned}&= \frac{\sin^2 u}{1+\cos u} \\ &= \frac{1-\cos^2 u}{1+\cos u} \\ &= \frac{(1-\cos u)(1+\cos u)}{1+\cos u} \\ &= (1-\cos u)\cdot\frac{2}{2} \\ &= 2\cdot\frac{1-\cos u}{2} \\ &2\sin^2(u/2)\end{aligned}$$

47. Multiply and divide by $\cos x$.

$$\begin{aligned}&= \frac{\sec x+\cos x-2}{\sec x-\cos x} \\ &= \frac{\sec x+\cos x-2}{\sec x-\cos x}\cdot\frac{\cos x}{\cos x} \\ &= \frac{1+\cos^2 x-2\cos x}{1-\cos^2 x} \\ &= \frac{\cos^2 x-2\cos x+1}{1-\cos^2 x} \\ &= \frac{(1-\cos x)^2}{(1+\cos x)(1-\cos x)} \\ &= \frac{1-\cos x}{1+\cos x} \\ &\tan^2(x/2)\end{aligned}$$

49.

$$\begin{aligned}\frac{1-\sin^2(x/2)}{1+\sin^2(x/2)} &= \\ \frac{1-\left(\dfrac{1-\cos x}{2}\right)}{1+\left(\dfrac{1-\cos x}{2}\right)}\cdot\frac{2}{2} &= \\ \frac{2-(1-\cos x)}{2+(1-\cos x)} &= \\ \frac{1+\cos x}{3-\cos x}\end{aligned}$$

51. Since $\cos(2\alpha)=2\cos^2\alpha-1$, we get

$$\begin{aligned}2\cos^2\alpha-1 &= \frac{3}{5} \\ 2\cos^2\alpha &= \frac{8}{5} \\ \cos^2\alpha &= \frac{4}{5} \\ \cos\alpha &= \pm\frac{2}{\sqrt{5}}.\end{aligned}$$

But $0° < \alpha < 45°$, so $\cos\alpha = \dfrac{2\sqrt{5}}{5}$ and $\sin\alpha = \sqrt{1-\left(\dfrac{2}{\sqrt{5}}\right)^2} = \sqrt{1-\dfrac{4}{5}} = \sqrt{\dfrac{1}{5}} = \dfrac{\sqrt{5}}{5}$. Furthermore, $\sec\alpha = \dfrac{\sqrt{5}}{2}$, $\csc\alpha = \sqrt{5}$, $\tan\alpha = \dfrac{1/\sqrt{5}}{2/\sqrt{5}} = \dfrac{1}{2}$, and $\cot\alpha = 2$.

53. Since $0 < \alpha < 90°$, we find

$$\cos(2\alpha) = \sqrt{1-\left(\frac{5}{13}\right)^2} = \sqrt{\frac{144}{169}} = \frac{12}{13}.$$

Then $\sin(\alpha) = \sqrt{\dfrac{1-\cos 2\alpha}{2}} = \sqrt{\dfrac{1-\dfrac{12}{13}}{2}} =$

$\sqrt{\frac{1}{26}} = \frac{\sqrt{26}}{26}$,

$\cos(\alpha) = \sqrt{\frac{1+\cos 2\alpha}{2}} = \sqrt{\frac{1+\frac{12}{13}}{2}} =$

$\sqrt{\frac{25}{26}} = \frac{5\sqrt{26}}{26}$,

$\tan\alpha = \frac{\sqrt{26}/26}{5\sqrt{26}26} = \frac{1}{5}$, $\csc\alpha = \sqrt{26}$,

$\sec\alpha = \sqrt{\frac{26}{25}} = \frac{\sqrt{26}}{5}$, and $\cot\alpha = 5$.

55. By a half-angle identity, we have

$$\begin{aligned} \cos(\alpha/2) = -\sqrt{\frac{1+\cos\alpha}{2}} &= -\frac{1}{4} \\ \frac{1+\cos\alpha}{2} &= \frac{1}{16} \\ 1+\cos\alpha &= \frac{1}{8} \\ \cos\alpha &= -\frac{7}{8}. \end{aligned}$$

Since $\pi \le \alpha \le 3\pi/2$, we get

$\sin\alpha = -\sqrt{1-\left(-\frac{7}{8}\right)^2} = -\sqrt{1-\frac{49}{64}} =$

$-\sqrt{\frac{15}{64}} = -\frac{\sqrt{15}}{8}$. Furthermore,

$\sec\alpha = -\frac{8}{7}$, $\csc\alpha = -\frac{8}{\sqrt{15}} = -\frac{8\sqrt{15}}{15}$,

$\tan\alpha = \frac{-\sqrt{15}/8}{-7/8} = \frac{\sqrt{15}}{7}$, and

$\cot\alpha = \frac{7}{\sqrt{15}} = \frac{7\sqrt{15}}{15}$.

57. By a half-angle identity, we find

$$\begin{aligned} \sin(\alpha/2) = \sqrt{\frac{1-\cos\alpha}{2}} &= \frac{4}{5} \\ \frac{1-\cos\alpha}{2} &= \frac{16}{25} \\ 1-\cos\alpha &= \frac{32}{25} \\ \cos\alpha &= -\frac{7}{25}. \end{aligned}$$

Since $(\pi/2+2k\pi) \le \alpha/2 \le (\pi+2k\pi)$ for some integer k, we get $(\pi+4k\pi) \le \alpha \le (2\pi+4k\pi)$. Then α is in quadrant III because $\cos\alpha < 0$.

Thus, $\sin\alpha = -\sqrt{1-\left(-\frac{7}{25}\right)^2} =$

$-\sqrt{1-\frac{49}{625}} = -\sqrt{\frac{576}{625}} = -\frac{24}{25}$.

Furthermore, $\sec\alpha = -\frac{25}{7}$, $\csc\alpha = -\frac{25}{24}$,

$\tan\alpha = \frac{-24/25}{-7/25} = \frac{24}{7}$, and $\cot\alpha = \frac{7}{24}$.

59. Since α lies in quadrant 2, we find

$$\cos\alpha = -\sqrt{1-\left(\frac{3}{5}\right)^2} = -\frac{4}{5}.$$

Then

$$\sin(2\alpha) = 2\sin\alpha\cos\alpha = 2\left(\frac{3}{5}\right)\left(\frac{-4}{5}\right) = -\frac{24}{25}.$$

61. Since $\sin\alpha = \frac{8}{17}$, we obtain

$$\cos 2\alpha = 1-2\sin^2\alpha = 1-2\left(\frac{8}{17}\right) = \frac{161}{289}.$$

63. Since $\tan\alpha = \frac{3}{5}$, we get $\sin\alpha = \frac{3}{\sqrt{34}}$ and

$$\cos\alpha = \frac{5}{\sqrt{34}}.$$

Applying a half-angle identity, we obtain

$$\begin{aligned} \tan\frac{\alpha}{2} &= \frac{\sin\alpha}{1+\cos\alpha} \\ &= \frac{\frac{3}{\sqrt{34}}}{1+\frac{5}{\sqrt{34}}} \\ &= \frac{3}{5+\sqrt{34}} \end{aligned}$$

and since $\tan\frac{\alpha}{2} = \frac{BD}{5}$ then

$$\begin{aligned} BD &= \frac{15}{5+\sqrt{34}} \\ &= \frac{15(5-\sqrt{34})}{25-34} \\ &= \frac{15(\sqrt{34}-5)}{9} \\ BD &= \frac{5\sqrt{34}-25}{3}. \end{aligned}$$

65. Since the base of the TV screen is $b = d\cos\alpha$ and its height is $h = d\sin\alpha$, then the area A is given by

$$\begin{aligned} A &= bh \\ &= (d\cos\alpha)(d\sin\alpha) \\ &= d^2\cos\alpha\sin\alpha \\ A &= \frac{d^2}{2}\sin(2\alpha). \end{aligned}$$

67. It is not an identity. If $x = \pi/4$, then

$$\sin(2\cdot\pi/4) = \sin(\pi/2) = 1$$

and

$$2\sin(\pi/4) = 2\cdot(\sqrt{2}/2) = \sqrt{2}.$$

69. It is not an identity. If $x = 2\pi/3$, then

$$\tan\left(\frac{2\pi/3}{2}\right) = \tan(\pi/3) = \sqrt{3}$$

and

$$\frac{1}{2}\cdot\tan(2\pi/3) = \frac{1}{2}\cdot(-\sqrt{3}).$$

71. It is not an identity. If $x = \pi/2$, then

$$\sin(2\cdot\pi/2)\sin\left(\frac{\pi/2}{2}\right) = \sin(\pi)\sin(\pi/4)$$

$$= 0\cdot\frac{\sqrt{2}}{2} = 0 \text{ and } \sin^2(\pi/2) = 1.$$

73. It is an identity. The proof below uses the double-angle identity for tangent.

$$\begin{aligned} \cot(x/2) - \tan(x/2) &= \\ \frac{1}{\tan(x/2)} - \tan(x/2) &= \\ \frac{1-\tan^2(x/2)}{\tan(x/2)} &= \\ 2\cdot\frac{1-\tan^2(x/2)}{2\cdot\tan(x/2)} &= \\ 2\cdot\frac{1}{\tan x} &= \\ 2\cdot\frac{\cos x}{\sin x}\cdot\frac{\sin x}{\sin x} &= \\ \frac{2\sin x\cos x}{\sin^2 x} &= \\ \frac{\sin(2x)}{\sin^2 x} & \end{aligned}$$

For Thought

1. True, $\sin 45^\circ\cos 15^\circ =$
$(1/2)\,[\sin(45^\circ+15^\circ)+\sin(45^\circ-15^\circ)] =$
$0.5\,[\sin 60^\circ + \sin 30^\circ]$.

2. False, $\cos(\pi/8)\sin(\pi/4) =$
$(1/2)\,[\sin(\pi/8+\pi/4)-\sin(\pi/8-\pi/4)] =$
$0.5\,[\sin(3\pi/8)-\sin(-\pi/8)] =$
$0.5\,[\sin(3\pi/8)+\sin(\pi/8)]$.

3. True, $2\cos(6^\circ)\cos(8^\circ) =$
$\cos(6^\circ-8^\circ)+\cos(6^\circ+8^\circ) =$
$\cos(-2^\circ)+\cos(14^\circ) = \cos(2^\circ)+\cos(14^\circ)$.

4. False, $\sin(5^\circ)-\sin(9^\circ) =$
$2\cos\left(\frac{5^\circ+9^\circ}{2}\right)\sin\left(\frac{5^\circ-9^\circ}{2}\right) =$
$2\cos(7^\circ)\sin(-2^\circ) = -2\cos(7^\circ)\sin(2^\circ)$.

5. True, $\cos(4)+\cos(12) =$
$2\cos\left(\frac{4+12}{2}\right)\cos\left(\frac{4-12}{2}\right) =$
$2\cos(8)\cos(-4) = 2\cos(8)\cos(4)$.

6. False, $\cos(\pi/3)-\cos(\pi/2) =$
$-2\sin\left(\frac{\pi/3+\pi/2}{2}\right)\sin\left(\frac{\pi/3-\pi/2}{2}\right) =$
$-2\sin(5\pi/12)\sin(-\pi/12) =$
$2\sin(5\pi/12)\sin(\pi/12)$.

7. True, $\sqrt{2}\sin(\pi/6+\pi/4) =$
$\sqrt{2}\,[\sin(\pi/6)\cos(\pi/4)+\cos(\pi/6)\sin(\pi/4)] =$
$\sqrt{2}\left[\sin(\pi/6)\cdot\frac{1}{\sqrt{2}}+\cos(\pi/6)\cdot\frac{1}{\sqrt{2}}\right] =$
$\sin(\pi/6)+\cos(\pi/6)$.

8. True, $\frac{1}{2}\sin(\pi/6)+\frac{\sqrt{3}}{2}\cos(\pi/6) =$
$\frac{1}{2}\cdot\frac{1}{2}+\frac{\sqrt{3}}{2}\cdot\frac{\sqrt{3}}{2} = \frac{1}{4}+\frac{3}{4} = 1 = \sin(\pi/2)$.

9. True, $y = \cos(\pi/3)\sin x + \sin(\pi/3)\cos x =$
$\sin(x+\pi/3)$.

10. True, since $y = \cos(\pi/4)\sin x + \sin(\pi/4)\cos x$ $= \sin(x+\pi/4)$ holds by a sum identity for sine.

3.6 Exercises

1.
$$\frac{1}{2}[\cos(13^\circ-9^\circ)-\cos(13^\circ+9^\circ)]=$$
$$0.5[\cos 4^\circ-\cos 22^\circ]$$

3.
$$\frac{1}{2}[\sin(16^\circ+20^\circ)+\sin(16^\circ-20^\circ)]=$$
$$0.5[\sin 36^\circ+\sin(-4^\circ)]=0.5[\sin 36^\circ-\sin 4^\circ]$$

5.
$$\frac{1}{2}\left[\cos\left(\frac{\pi}{6}-\frac{\pi}{5}\right)+\cos\left(\frac{\pi}{6}+\frac{\pi}{5}\right)\right]=$$
$$0.5\left[\cos\left(\frac{-\pi}{30}\right)+\cos\left(\frac{11\pi}{30}\right)\right]=$$
$$0.5\left[\cos\left(\frac{\pi}{30}\right)+\cos\left(\frac{11\pi}{30}\right)\right]$$

7.
$$\frac{1}{2}\left[\cos(5y^2-7y^2)+\cos(5y^2+7y^2)\right]=$$
$$0.5\left[\cos(-2y^2)+\cos(12y^2)\right]=$$
$$0.5\left[\cos(2y^2)+\cos(12y^2)\right]$$

9.
$$\frac{1}{2}\Big[\sin((2s-1)+(s+1))+$$
$$\sin((2s-1)-(s+1))\Big]=$$
$$0.5\Big[\sin(3s)+\sin(s-2)\Big]$$

11.
$$\frac{1}{2}[\cos(52.5^\circ-7.5^\circ)-\cos(52.5^\circ+7.5^\circ)]=$$
$$\frac{1}{2}[\cos 45^\circ-\cos 60^\circ]=$$
$$\frac{1}{2}\left[\frac{\sqrt{2}}{2}-\frac{1}{2}\right]=\frac{\sqrt{2}-1}{4}$$

13.
$$\frac{1}{2}\left[\sin\left(\frac{13\pi}{24}+\frac{5\pi}{24}\right)+\sin\left(\frac{13\pi}{24}-\frac{5\pi}{24}\right)\right]=$$
$$\frac{1}{2}[\sin(18\pi/24)+\sin(8\pi/24)]=$$
$$\frac{1}{2}[\sin(3\pi/4)+\sin(\pi/3)]=$$
$$\frac{1}{2}\left[\frac{\sqrt{2}}{2}+\frac{\sqrt{3}}{2}\right]=\frac{\sqrt{2}+\sqrt{3}}{4}$$

15.
$$2\cos\left(\frac{12^\circ+8^\circ}{2}\right)\sin\left(\frac{12^\circ-8^\circ}{2}\right)=$$
$$2\cos 10^\circ\sin 2^\circ$$

17.
$$-2\sin\left(\frac{80^\circ+87^\circ}{2}\right)\sin\left(\frac{80^\circ-87^\circ}{2}\right)=$$
$$-2\sin 83.5^\circ\sin(-3.5^\circ)=2\sin 83.5^\circ\sin 3.5^\circ$$

19.
$$2\cos\left(\frac{3.6+4.8}{2}\right)\sin\left(\frac{3.6-4.8}{2}\right)=$$
$$2\cos(4.2)\sin(-0.6)=-2\cos(4.2)\sin(0.6)$$

21.
$$-2\sin\left(\frac{(5y-3)+(3y+9)}{2}\right)\cdot$$
$$\sin\left(\frac{(5y-3)-(3y+9)}{2}\right)=$$
$$-2\sin(4y+3)\sin(y-6)$$

23.
$$2\cos\left(\frac{5\alpha+8\alpha}{2}\right)\sin\left(\frac{5\alpha-8\alpha}{2}\right)=$$
$$2\cos(6.5\alpha)\sin(-1.5\alpha)=$$
$$-2\cos(6.5\alpha)\sin(1.5\alpha)$$

25.
$$-2\sin\left(\frac{\pi/3+\pi/5}{2}\right)\sin\left(\frac{\pi/3-\pi/5}{2}\right)=$$
$$-2\sin\left(\frac{4\pi}{15}\right)\sin\left(\frac{\pi}{15}\right)$$

27.
$$2\sin\left(\frac{75^\circ+15^\circ}{2}\right)\cos\left(\frac{75^\circ-15^\circ}{2}\right)=$$
$$2\sin 45^\circ\cos(30^\circ)=2\cdot\frac{\sqrt{2}}{2}\frac{\sqrt{3}}{2}=\frac{\sqrt{6}}{2}$$

29.
$$-2\sin\left(\frac{\frac{-\pi}{24}+\frac{7\pi}{24}}{2}\right)\sin\left(\frac{\frac{-\pi}{24}-\frac{7\pi}{24}}{2}\right)=$$
$$-2\sin(3\pi/24)\sin(-4\pi/24)=$$
$$-2\sin(\pi/8)\sin(-\pi/6)=$$
$$-2\sin\left(\frac{\pi/4}{2}\right)\cdot\frac{-1}{2}=-2\sqrt{\frac{1-\cos(\pi/4)}{2}}\cdot\frac{-1}{2}$$
$$=\sqrt{\frac{1-\sqrt{2}/2}{2}\cdot\frac{2}{2}}=\sqrt{\frac{2-\sqrt{2}}{4}}=\frac{\sqrt{2-\sqrt{2}}}{2}$$

31. Since $a = 1$ and $b = -1$, we find

$$r = \sqrt{1^2 + (-1)^2} = \sqrt{2}.$$

If the terminal side of α passes through $(1, -1)$, then $\cos\alpha = a/r = 1/\sqrt{2}$ and $\sin\alpha = b/r = -1/\sqrt{2}$. Let $\alpha = -\pi/4$. Then $\sin x - \cos x = r\sin(x + \alpha) = \sqrt{2}\sin\left(x - \frac{\pi}{4}\right)$.

33. Since $a = -1/2$ and $b = \sqrt{3}/2$, we obtain

$$r = \sqrt{(-1/2)^2 + (\sqrt{3}/2)^2} = 1.$$

If the terminal side of α passes through $(-1/2, \sqrt{3}/2)$, then $\cos\alpha = a/r = a/1 = a = -1/2$ and $\sin\alpha = b/r = b/1 = b = \sqrt{3}/2$. Let $\alpha = 2\pi/3$. Thus, $-\frac{1}{2}\sin x + \frac{\sqrt{3}}{2}\cos x = r\sin(x + \alpha) = \sin\left(x + \frac{2\pi}{3}\right)$.

35. Since $a = \sqrt{3}/2$ and $b = -1/2$, we have

$$r = \sqrt{(\sqrt{3}/2)^2 + (-1/2)^2} = 1.$$

If the terminal side of α passes through $(\sqrt{3}/2, -1/2)$, then $\cos\alpha = a/r = a/1 = a = \sqrt{3}/2$ and $\sin\alpha = b/r = b/1 = b = -1/2$.

Let $\alpha = -\pi/6$. Thus, $\frac{\sqrt{3}}{2}\sin x - \frac{1}{2}\cos x = r\sin(x + \alpha) = \sin\left(x - \frac{\pi}{6}\right)$.

37. Since $a = -1$ and $b = 1$, we obtain $r = \sqrt{(-1)^2 + 1^2} = \sqrt{2}$. If the terminal side of α passes through $(-1, 1)$, then $\cos\alpha = a/r = -1/\sqrt{2}$ and $\sin\alpha = b/r = 1/\sqrt{2}$. Choose $\alpha = 3\pi/4$. Then $y = -\sin x + \cos x = r\sin(x + \alpha) = \sqrt{2}\sin(x + 3\pi/4)$. Amplitude is $\sqrt{2}$, period is 2π, and phase shift is $-3\pi/4$.

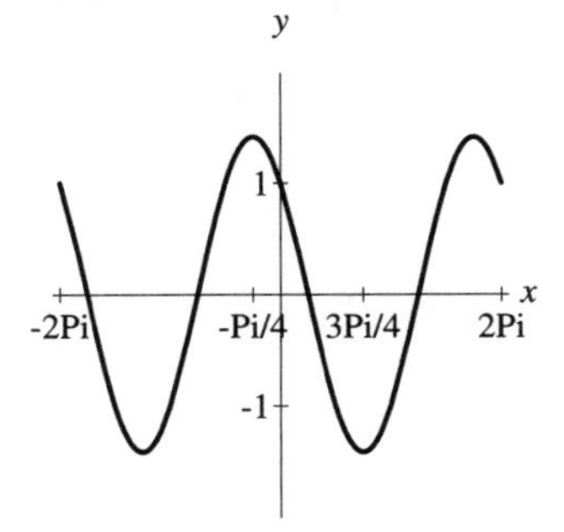

39. Since $a = \sqrt{2}$ and $b = -\sqrt{2}$, we obtain $r = \sqrt{\sqrt{2}^2 + (-\sqrt{2})^2} = 2$. If the terminal side of α passes through $(\sqrt{2}, -\sqrt{2})$, then $\cos\alpha = a/r = \sqrt{2}/2$ and $\sin\alpha = b/r = -\sqrt{2}/2$. Let $\alpha = -\pi/4$. Thus, $y = \sqrt{2}\sin x - \sqrt{2}\cos x = r\sin(x + \alpha) = 2\sin(x - \pi/4)$. Amplitude is 2, period is 2π, and phase shift is $\pi/4$.

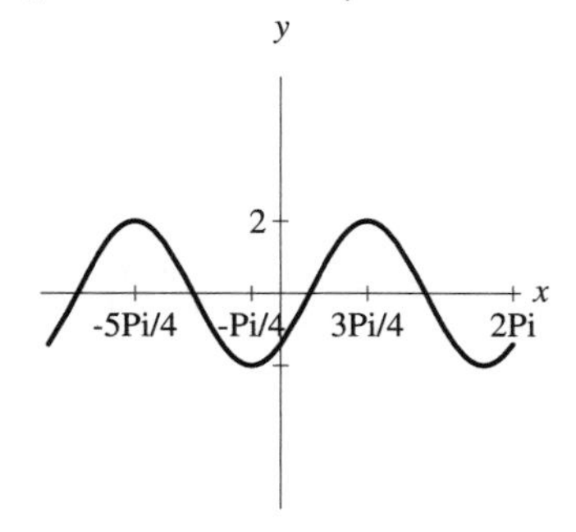

41. Since $a = -\sqrt{3}$ and $b = -1$, we find $r = \sqrt{(-\sqrt{3})^2 + (-1)^2} = 2$. If the terminal side of α passes through $(-\sqrt{3}, -1)$, then $\cos\alpha = a/r = -\sqrt{3}/2$ and $\sin\alpha = b/r = -1/2$. Choose $\alpha = 7\pi/6$. Then $y = -\sqrt{3}\sin x - \cos x = r\sin(x + \alpha) = 2\sin(x + 7\pi/6)$. Amplitude is 2, period is 2π, and phase shift is $-7\pi/6$.

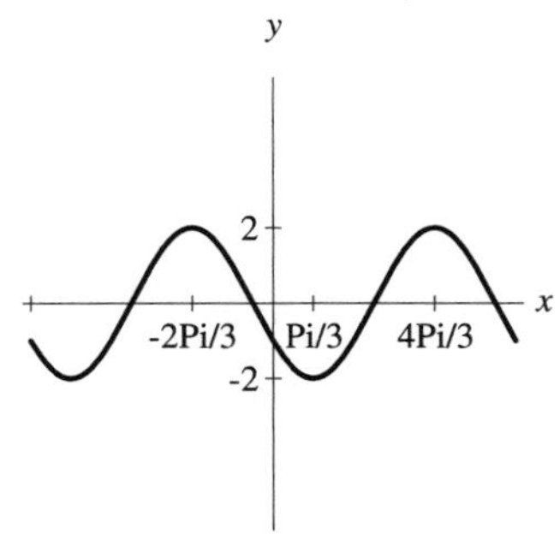

43. Since $a = 3$ and $b = 4$, the amplitude is $\sqrt{3^2 + 4^2} = \sqrt{25} = 5$. If the terminal side of α passes through $(3, 4)$, then $\tan\alpha = 3/4$ and $\alpha = \tan^{-1}(3/4) \approx 0.9$. Phase shift is -0.9.

45. Since $a = -6$ and $b = 1$, the amplitude is $\sqrt{(-6)^2 + 1^2} = \sqrt{37}$. If the terminal side of α passes through $(-6, 1)$, then $\tan\alpha = -1/6$. Using a calculator, one gets $\tan^{-1}(-1/6) \approx -0.165$ which is an angle in quadrant IV. Since $(-6, 1)$ is in quadrant II and π is the period of $\tan x$, we find $\alpha \approx -0.165 + \pi \approx 3.0$. Phase shift is -3.0.

47. Since $a = -3$ and $b = -5$, the amplitude is $\sqrt{(-3)^2 + (-5)^2} = \sqrt{34}$. If the terminal side of α passes through $(-3, -5)$, then $\tan\alpha = 5/3$. Using a calculator, one gets $\tan^{-1}(5/3) \approx 1.03$ which is an angle in quadrant I. Since $(-3, -5)$ is in quadrant III and π is the period of $\tan x$, we obtain $\alpha \approx 1.03 + \pi \approx 4.2$. The phase shift is -4.2.

49. By using a sum-to-product identity, we get

$$\frac{\sin(3t) - \sin(t)}{\cos(3t) + \cos(t)} =$$

$$\frac{2\cos\left(\frac{3t+t}{2}\right)\sin\left(\frac{3t-t}{2}\right)}{2\cos\left(\frac{3t+t}{2}\right)\cos\left(\frac{3t-t}{2}\right)} =$$

$$\frac{2\cos(2t)\sin t}{2\cos(2t)\cos t} =$$

$$\tan t$$

51. Using a sum-to-product identity, we find

$$\frac{\cos x - \cos(3x)}{\cos x + \cos(3x)} =$$

$$\frac{-2\sin\left(\frac{x+3x}{2}\right)\sin\left(\frac{x-3x}{2}\right)}{2\cos\left(\frac{x+3x}{2}\right)\cos\left(\frac{x-3x}{2}\right)} =$$

$$\frac{-2\sin(2x)\sin(-x)}{2\cos(2x)\cos(-x)} =$$

$$\frac{2\sin(2x)\sin x}{2\cos(2x)\cos x} =$$

$$\tan(2x)\tan(x)$$

53. By using a product-to-sum identity, we get

$$= -\sin(x+y)\sin(x-y)$$

$$= -\frac{1}{2}\Big[\cos\left((x+y)-(x-y)\right) - \cos\left((x+y)+(x-y)\right)\Big]$$

$$= -\frac{1}{2}\Big[\cos(2y) - \cos(2x)\Big]$$

$$= -\frac{1}{2}\Big[(2\cos^2 y - 1) - (2\cos^2 x - 1)\Big]$$

$$= -\frac{1}{2}\Big[2\cos^2 y - 2\cos^2 x\Big]$$

$$\cos^2 x - \cos^2 y.$$

55. Let $A = \frac{x+y}{2}$ and $B = \frac{x-y}{2}$. Note,

$$A + B = x \text{ and } A - B = y.$$

Expand the left-hand side and use product-to-sum identities.

$$(\sin A + \cos A)(\sin B + \cos B) =$$

$$\sin A\sin B + \sin A\cos B + \cos A\sin B + \cos A\cos B =$$

$$\frac{1}{2}\Big[\cos(A-B) - \cos(A+B)\Big] + \frac{1}{2}\Big[\sin(A+B) + \sin(A-B)\Big] + \frac{1}{2}\Big[\sin(A+B) - \sin(A-B)\Big] + \frac{1}{2}\Big[\cos(A-B) + \cos(A+B)\Big] =$$

$$\frac{1}{2}\Big[\cos y - \cos x\Big] + \frac{1}{2}\Big[\sin x + \sin y\Big] + \frac{1}{2}\Big[\sin x - \sin y\Big] + \frac{1}{2}\Big[\cos y + \cos x\Big] =$$

$$\frac{1}{2}\Big[2\cos y + 2\sin x\Big] =$$

$$\sin x + \cos y$$

57. Apply a sum-to-product identity in the 5th line.

$$\begin{aligned}
&= \sin^2(A+B) - \sin^2(A-B) \\
&= \sin(2A)\sin(2B) \quad \text{by Exercise 56} \\
&= (2\sin A\cos A)(2\sin B\cos B) \\
&= [2\cos A\cos B]\cdot[2\sin A\sin B] \\
&= \Big[\cos(A-B) + \cos(A+B)\Big]\cdot \Big[\cos(A-B) - \cos(A+B)\Big] \\
&= \cos^2(A-B) - \cos^2(A+B)
\end{aligned}$$

59. Note that x can be written in the form $x = a\sin(t+\alpha)$. The maximum displacement of $x = \sqrt{3}\sin t + \cos t$ is

$$a = \sqrt{\sqrt{3}^2 + 1^2} = 2.$$

Thus, 2 meters is the maximum distance between the block and its resting position.

Since the terminal side of α goes through $(\sqrt{3}, 1)$, we get $\tan\alpha = 1/\sqrt{3}$ and one can choose $\alpha = \pi/6$. Then $x = 2\sin(t + \pi/6)$.

Chapter 3 Review Exercises

1. $1 - \sin^2\alpha = \cos^2\alpha$

3. $(1 - \csc x)(1 + \csc x) = 1 - \csc^2 x = -\cot^2 x$

5.

$$\frac{1}{1+\sin\alpha} + \frac{\sin\alpha}{\cos^2\alpha} = \frac{1}{1+\sin\alpha} + \frac{\sin\alpha}{1-\sin^2\alpha} =$$
$$\frac{(1-\sin\alpha)+\sin\alpha}{1-\sin^2\alpha} = \frac{1}{\cos^2\alpha} = \sec^2\alpha$$

7. $\tan(4s)$, by the double angle idenity for tangent

9. $\sin(3\theta - 6\theta) = \sin(-3\theta) = -\sin(3\theta)$

11. $\tan\left(\frac{2z}{2}\right) = \tan z$, by a double-angle identity for tangent

13. Note, $\sin\alpha = \sqrt{1 - \left(\frac{-5}{13}\right)^2} = \sqrt{1 - \frac{25}{169}} =$ $\sqrt{\frac{144}{169}} = \frac{12}{13}$. Then $\tan\alpha = \frac{12/13}{-5/13} = -\frac{12}{5}$, $\cot\alpha = -\frac{5}{12}$, $\csc\alpha = \frac{13}{12}$, $\sec\alpha = -\frac{13}{5}$.

15. By using a cofunction identity, we get $\cos\alpha = \frac{-3}{5}$. Then $\sin\alpha = -\sqrt{1 - \left(\frac{-3}{5}\right)^2} =$ $-\sqrt{1 - \frac{9}{25}} = -\sqrt{\frac{16}{25}} = -\frac{4}{5}$, $\sec\alpha = -\frac{5}{3}$, $\csc\alpha = -\frac{5}{4}$, $\tan\alpha = \frac{-4/5}{-3/5} = \frac{4}{3}$, $\cot\alpha = \frac{3}{4}$.

17. By the half-angle identity for sine, we find

$$\begin{aligned}
\sqrt{\frac{1-\cos\alpha}{2}} &= \frac{3}{5} \\
\frac{1-\cos\alpha}{2} &= \frac{9}{25} \\
1-\cos\alpha &= \frac{18}{25} \\
\cos\alpha &= \frac{7}{25}.
\end{aligned}$$

Since $\frac{3\pi}{2} < \alpha < 2\pi$, α is in quadrant IV and $\sin\alpha = -\sqrt{1 - \left(\frac{7}{25}\right)^2} = -\sqrt{1 - \frac{49}{625}} =$ $-\sqrt{\frac{576}{625}} = -\frac{24}{25}$. Then $\tan\alpha = \frac{-24/25}{7/25} =$ $-\frac{24}{7}$, $\cot\alpha = -\frac{7}{24}$, $\sec\alpha = \frac{25}{7}$, $\csc\alpha = -\frac{25}{24}$.

19. It is an identity.

$$\begin{aligned}
(\sin x + \cos x)^2 &= \\
\sin^2 x + 2\sin x\cos x + cos^2 x &= \\
1 + 2\sin x\cos x &= \\
1 + \sin(2x)
\end{aligned}$$

21. It is not an identity since $\csc^2 x - \cot^2 x = 1$ and $\tan^2 - \sec^2 x = -1$.

23. Odd, since $f(-x) = \frac{\sin(-x) - \tan(-x)}{\cos(-x)} =$ $\frac{-\sin x + \tan x}{\cos x} = -\frac{\sin x - \tan x}{\cos x} = -f(x)$

25. It is neither even nor odd. Since $f(\pi/4) =$ $\frac{\cos(\pi/4) - \sin(\pi/4)}{\sec(\pi/4)} = \frac{\sqrt{2}/2 - \sqrt{2}/2}{\sqrt{2}} = 0$ and $f(-\pi/4) = \frac{\cos(-\pi/4) - \sin(-\pi/4)}{\sec(-\pi/4)} =$ $\frac{\sqrt{2}/2 + \sqrt{2}/2}{\sqrt{2}} = \frac{\sqrt{2}}{\sqrt{2}} = 1$, we find

$$f(\pi/4) \neq \pm f(-\pi/4).$$

27. Even, since $f(-x) = \frac{\sin(-x)\tan(-x)}{\cos(-x) + \sec(-x)} =$ $\frac{(-\sin x)(-\tan x)}{\cos x + \sec x} = \frac{\sin x \tan x}{\cos x + \sec x} = f(x)$

29.

$$\begin{aligned} &= \frac{1+\tan^2\theta}{1-\tan^2\theta} \\ &= \frac{\sec^2\theta}{1-\dfrac{\sin^2\theta}{\cos^2\theta}} \cdot \frac{\cos^2\theta}{\cos^2\theta} \\ &= \frac{1}{\cos^2\theta - \sin^2\theta} \\ &= \frac{1}{\cos 2\theta} \\ &= \sec 2\theta \end{aligned}$$

31.

$$\begin{aligned} &= \frac{\csc^2 x - \cot^2 x}{2\csc^2 x + 2\csc x \cot x} \\ &= \frac{1}{\dfrac{2}{\sin^2 x} + 2\cdot\dfrac{1}{\sin x}\cdot\dfrac{\cos x}{\sin x}} \\ &= \frac{1}{\dfrac{2}{\sin^2 x} + \dfrac{2\cos x}{\sin^2 x}} \cdot \frac{\sin^2 x}{\sin^2 x} \\ &= \frac{\sin^2 x}{2+2\cos x} \\ &= \frac{1-\cos^2 x}{2(1+\cos x)} \\ &= \frac{(1-\cos x)(1+\cos x)}{2(1+\cos x)} \\ &= \frac{1-\cos x}{2} \\ &= \sin^2\left(\frac{x}{2}\right) \end{aligned}$$

33.

$$\begin{aligned} \cot(\alpha - 45^\circ) &= \\ (\tan(\alpha - 45^\circ))^{-1} &= \\ \left(\frac{\tan\alpha - \tan 45^\circ}{1+\tan\alpha\tan 45^\circ}\right)^{-1} &= \\ \left(\frac{\tan\alpha - 1}{1+\tan\alpha}\right)^{-1} &= \\ \frac{1+\tan\alpha}{\tan\alpha - 1} &= \end{aligned}$$

35.

$$\begin{aligned} \frac{\sin 2\beta}{2\csc\beta} &= \\ \frac{2\sin\beta\cos\beta}{2/\sin\beta}\cdot\frac{\sin\beta}{\sin\beta} &= \\ \sin^2\beta\cos\beta &= \end{aligned}$$

37. Factor the numerator on the left-hand side as a difference of two cubes.
Note, $\cot y \tan y = 1$.

$$\begin{aligned} \frac{\cot^3 y - \tan^3 y}{\sec^2 y + \cot^2 y} &= \\ \frac{(\cot y - \tan y)(\cot^2 y + 1 + \tan^2 y)}{\sec^2 y + \cot^2 y} &= \\ \frac{(\cot y - \tan y)(\cot^2 y + \sec^2 y)}{\sec^2 y + \cot^2 y} &= \\ \cot y - \tan y &= \\ \frac{1}{\tan y} - \tan y &= \\ \frac{1-\tan^2 y}{\tan y} &= \\ 2\cdot\frac{1-\tan^2 y}{2\tan y} &= \\ 2\cdot(\tan 2y)^{-1} &= \\ 2\cot(2y) &= \end{aligned}$$

39. By using double-angle identities, we obtain

$$\begin{aligned} \cos(2\cdot 2x) &= \\ 1-2\sin^2(2x) &= \\ 1-2(2\sin x\cos x)^2 &= \\ 1-8\sin^2 x\cos^2 x &= \\ 1-8\sin^2 x(1-\sin^2 x) &= \\ 8\sin^4 x - 8\sin^2 x + 1 &= \end{aligned}$$

41. By the double-angle identity for sine, we get

$$\begin{aligned} \sin^4(2x) &= \\ (2\sin x\cos x)^4 &= \\ 16\sin^4 x\cos^4 x &= \\ 16\sin^4 x(1-\sin^2 x)^2 &= \\ 16\sin^4 x(1-2\sin^2 x+\sin^4 x) &= \\ 16\sin^4 x - 32\sin^6 x + 16\sin^8 x &= \end{aligned}$$

43. $\tan\left(\frac{-\pi/6}{2}\right) = \frac{1-\cos(-\pi/6)}{\sin(-\pi/6)} =$
$\frac{1-\sqrt{3}/2}{-1/2}\cdot\frac{2}{2} = \frac{2-\sqrt{3}}{-1} = \sqrt{3}-2$

45. $\sin\left(\frac{-150^\circ}{2}\right) = -\sqrt{\frac{1-\cos(-150^\circ)}{2}} =$
$-\sqrt{\frac{1-(-\sqrt{3}/2)}{2}\cdot\frac{2}{2}} = -\sqrt{\frac{2+\sqrt{3}}{4}} =$
$-\frac{\sqrt{2+\sqrt{3}}}{2}$ or alternatively by using a difference identity we find
$\sin(-30^\circ - 45^\circ) = -\frac{\sqrt{2}+\sqrt{6}}{4}$

47. Let $a = 4$, $b = 4$, and $r = \sqrt{4^2+4^2} = 4\sqrt{2}$. If the terminal side of α goes through $(4,4)$, then $\tan\alpha = 4/4 = 1$ and one can choose $\alpha = \pi/4$. So

$$y = 4\sqrt{2}\sin(x+\pi/4),$$

amplitude is $4\sqrt{2}$, period is 2π, and phase shift is $-\pi/4$.

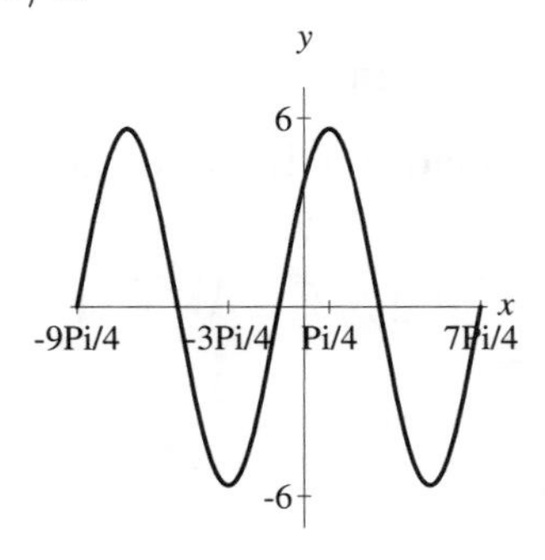

49. Let $a = -2$, $b = 1$, and

$$r = \sqrt{(-2)^2+1^2} = \sqrt{5}.$$

If the terminal side of α goes through $(-2,1)$, then $\tan\alpha = -1/2$. Since

$$\tan^{-1}(-1/2) \approx -0.46$$

and $(-2,1)$ is in quadrant II, one can choose

$$\alpha = \pi - 0.46 = 2.68.$$

Thus,

$$y = \sqrt{5}\sin(x+2.68)$$

and the amplitude is $\sqrt{5}$, the period is 2π, and the phase shift is -2.68.

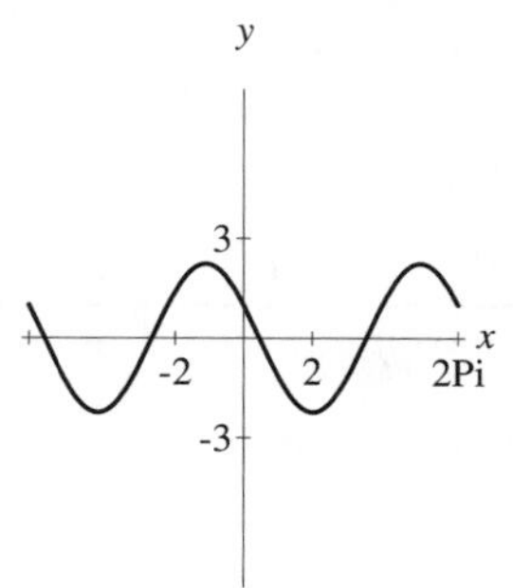

51. $\cos 15^\circ + \cos 19^\circ =$
$2\cos\left(\frac{15^\circ+19^\circ}{2}\right)\cos\left(\frac{15^\circ-19^\circ}{2}\right) =$
$2\cos 17^\circ\cos(-2^\circ) = 2\cos 17^\circ\cos 2^\circ$

53. $\sin(\pi/4) - \sin(-\pi/8) = \sin(\pi/4) + \sin(\pi/8) =$
$2\sin\left(\frac{\pi/4+\pi/8}{2}\right)\cos\left(\frac{\pi/4-\pi/8}{2}\right) =$
$2\sin\left(\frac{3\pi/8}{2}\right)\cos\left(\frac{\pi/8}{2}\right) =$
$2\sin\left(\frac{3\pi}{16}\right)\cos\left(\frac{\pi}{16}\right)$

55. $2\sin 11^\circ\cos 13^\circ =$
$\sin(11^\circ+13^\circ) + \sin(11^\circ-13^\circ) =$
$\sin 24^\circ + \sin(-2^\circ) = \sin 24^\circ - \sin 2^\circ$

57. $2\cos\frac{x}{4}\cos\frac{x}{3} =$
$\cos\left(\frac{x}{4}-\frac{x}{3}\right) + \cos\left(\frac{x}{4}+\frac{x}{3}\right) =$
$\cos\left(-\frac{x}{12}\right) + \cos\left(\frac{7x}{12}\right) =$
$\cos\left(\frac{x}{12}\right) + \cos\left(\frac{7x}{12}\right)$

59. Note, y can be written in the form $y = a\sin(t+\alpha)$ where

$$a = \sqrt{2^2+1^2} = \sqrt{5}$$

and $\alpha = \cos^{-1}\left(\frac{2}{\sqrt{5}}\right) \approx 0.46.$

Thus,

$$y = \sqrt{5}\sin(t+0.46)$$

and the maximum height above its normal position is $\sqrt{5}$ in.

Chapter 3 Test

1. $\dfrac{1}{\cos x} \cdot \dfrac{\cos x}{\sin x} \cdot 2 \sin x \cos x = 2 \cos x$

2. $\sin(2t + 5t) = \sin(7t)$

3. $\dfrac{1}{1-\cos y} + \dfrac{1}{1+\cos y} = \dfrac{1+\cos y + 1 - \cos y}{1 - cos^2 y} =$
$\dfrac{2}{\sin^2 y} = 2\csc^2 y$

4. $\tan\left(\dfrac{\pi}{5} + \dfrac{\pi}{10}\right) = \tan\left(\dfrac{3\pi}{10}\right)$

5.

$$\begin{aligned} \frac{\sin\beta\cos\beta}{\sin\beta/\cos\beta} &= \\ \sin\beta\cos\beta \cdot \frac{\cos\beta}{\sin\beta} &= \\ \cos^2\beta &= \\ 1 - \sin^2\beta &= \end{aligned}$$

6.

$$\begin{aligned} \frac{1}{\sec\theta - 1} - \frac{1}{\sec\theta + 1} &= \\ \frac{\sec\theta + 1 - (\sec\theta - 1)}{\sec^2\theta - 1} &= \\ \frac{2}{\tan^2\theta} &= \\ 2\cot^2\theta &= \end{aligned}$$

7. Using the cofunction identity for cosine, we obtain

$$\begin{aligned} \cos(\pi/2 - x)\cos(-x) &= \\ \sin x \cos x &= \\ \frac{2\sin x \cos x}{2} &= \\ \frac{\sin(2x)}{2} &= \end{aligned}$$

8. Factor the left-hand side and use a half-angle identity for tangent.

$$\begin{aligned} \tan(t/2)\cdot(\cos^2 t - 1) &= \\ \frac{1-\cos t}{\sin t}\cdot(-\sin^2 t) &= \\ (1-\cos t)\cdot(-\sin t) &= \\ (\cos t - 1)\sin t &= \\ \cos t \sin t - \sin t &= \\ \frac{\sin t}{\sec t} - \sin t &= \end{aligned}$$

9. Let $a = 1$, $b = -\sqrt{3}$, $r = \sqrt{1^2 + (-\sqrt{3})^2} = 2$. If the terminal side of α goes through $(1, -\sqrt{3})$, then $\tan\alpha = -\sqrt{3}$ and one can choose $\alpha = 5\pi/3$. Then

$$y = 2\sin(x + 5\pi/3),$$

the period is 2π, amplitude is 2, and the phase shift is $-5\pi/3$.

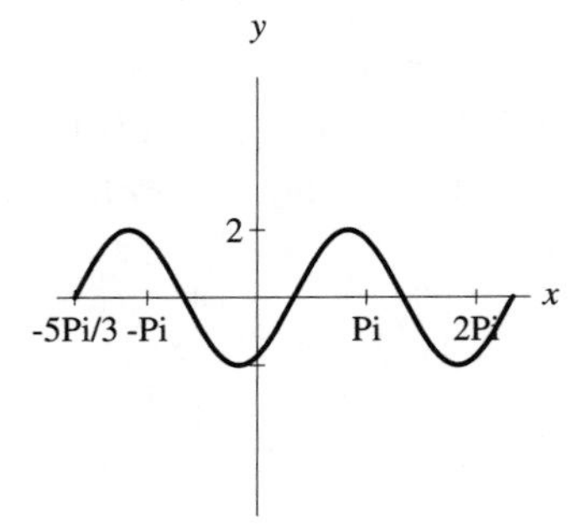

10. If $\csc\alpha = 2$, then $\sin\alpha = \dfrac{1}{2}$.
Since α is in quadrant II, we obtain
$\cos\alpha = -\sqrt{1 - \left(\dfrac{1}{2}\right)^2} = -\sqrt{1 - \dfrac{1}{4}} =$
$-\sqrt{\dfrac{3}{4}} = -\dfrac{\sqrt{3}}{2}$, $\sec\alpha = -\dfrac{2}{\sqrt{3}} = -\dfrac{2\sqrt{3}}{3}$,
$\tan\alpha = \dfrac{1/2}{-\sqrt{3}/2} = -\dfrac{1}{\sqrt{3}} = -\dfrac{\sqrt{3}}{3}$,
and $\cot\alpha = -\sqrt{3}$.

11. Even, $f(-x) = (-x)\sin(-x) =$ $(-x)(-\sin x) = x \sin x = f(x)$.

12. Using a half-angle identity, we obtain
$\sin\left(\dfrac{-\pi/6}{2}\right) = -\sqrt{\dfrac{1 - \cos(-\pi/6)}{2}} =$
$-\sqrt{\dfrac{1 - \sqrt{3}/2}{2}\cdot\dfrac{2}{2}} = -\sqrt{\dfrac{2-\sqrt{3}}{4}} =$
$-\dfrac{\sqrt{2-\sqrt{3}}}{2}$. Equivalently, using a

difference identity we find

$$\sin\left(-\frac{\pi}{12}\right) = \sin\left(\frac{\pi}{4} - \frac{\pi}{3}\right) = \frac{\sqrt{2}-\sqrt{6}}{4}.$$

13. If $x = y = \pi/6$, then we find that

$\tan x + \tan y = 2\tan(\pi/6) = 2 \cdot \frac{\sqrt{3}}{3}$ and

$\tan(x+y) = \tan(\pi/6+\pi/6) = \tan(\pi/3) = \sqrt{3}$. Thus, it is not an identity.

Tying It All Together

1. $\frac{\sqrt{2}}{2}$ **2.** $\frac{\sqrt{2}}{2}$

3. $\frac{1}{2}$ **4.** $\frac{1}{2}$

5. $\frac{1}{2}$ **6.** $\frac{1}{2}$

7. $\sin\left(\frac{55\pi}{6}\right) = \sin\left(\frac{7\pi}{6}\right) = -\frac{1}{2}$

8. $\sin\left(-\frac{23\pi}{2}\right) = \sin\left(\frac{\pi}{2}\right) = 1$

9. -1 **10.** 0

11. Odd, $f(-x) = (-x)^3 + \sin(-x) = -x^3 - \sin x = -f(x)$

12. Even, $f(-x) = (-x)^3 \sin(-x) = (-x^3)(-\sin x) = x^3 \sin x = f(x)$

13. Even, $f(-x) = \frac{\sin(-x)}{-x} = \frac{-\sin x}{-x} = \frac{\sin x}{x} = f(x)$

14. Even, $f(-x) = |\sin(-x)| = |-\sin x| = |\sin x| = f(x)$

15. Even, since $y = \cos^5 x$, $y = \cos^3 x$, $y = 2\cos x$ are even functions, and a sum and difference of even functions is again an even function.

16. Odd, since

$$\begin{aligned} f(-x) &= (-x)^3 \sin^4(-x) + (-x)\sin^2(-x) \\ &= -x^3 \sin^4(x) - x\sin^2(x) \\ f(-x) &= -f(x) \end{aligned}$$

17. Let $\alpha = \beta = \frac{\pi}{2}$. Then $\sin(\alpha+\beta) = \sin\pi = 0$ while $\sin\alpha + \sin\beta = 1 + 1 = 2$. Thus, $\sin(\alpha+\beta) = \sin\alpha + \sin\beta$ is not an identity.

18. Let $\alpha = \frac{\pi}{2}$ and $\beta = \frac{\pi}{4}$. Then

$$\cos(\alpha-\beta) = \cos\left(\frac{\pi}{2} - \frac{\pi}{4}\right) = \cos\frac{\pi}{4} = \frac{\sqrt{2}}{2}$$

while $\cos\alpha - \cos\beta = 0 - \frac{\sqrt{2}}{2} = -\frac{\sqrt{2}}{2}$. Thus, $\cos(\alpha-\beta) = \cos\alpha - \cos\beta$ is not an identity.

19. Note $\sin^{-1}(0.1) \approx 0.1$ and $\frac{1}{\sin 0.1} \approx 10.0$. Thus, $\sin^{-1} x = \frac{1}{\sin x}$ is not an identity.

20. Note $\sin^2(2) > 0$ and $\sin(2^2) = \sin(4) < 0$. Thus, $\sin^2 x = \sin(x^2)$ is not an identity.

21. Form a right triangle with $\alpha = 30°$, $a = 4$.

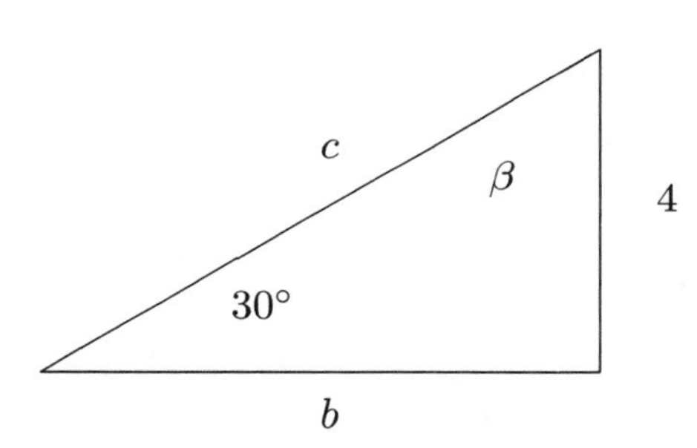

Since $\tan 30° = 4/b$ and $\sin 30° = 4/c$, we get

$$b = \frac{4}{\tan 30°} = 4\sqrt{3}$$

and

$$c = \frac{4}{\sin 30°} = 8.$$

Also, $\beta = 90° - 30° = 60°$.

22. Form a right triangle with $a = \sqrt{3}$, $b = 1$.

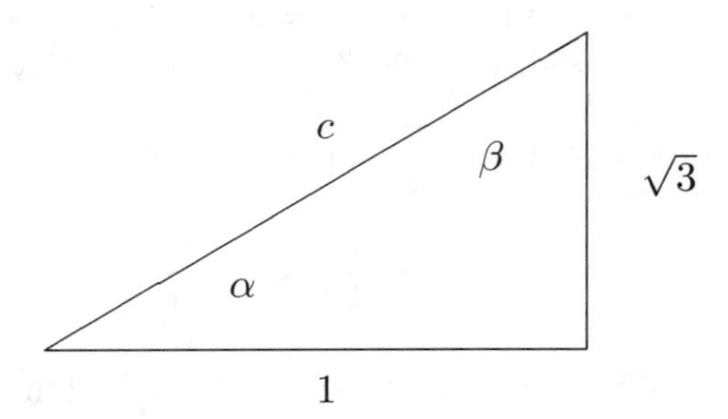

Then $c = \sqrt{\sqrt{3}^2 + 1^2} = 2$ by the Pythagorean Theorem. Since $\tan \alpha = \sqrt{3}/1 = \sqrt{3}$, we get $\alpha = 60^\circ$ and $\beta = 30^\circ$.

23. Form a right triangle with $b = 5$ as shown below.

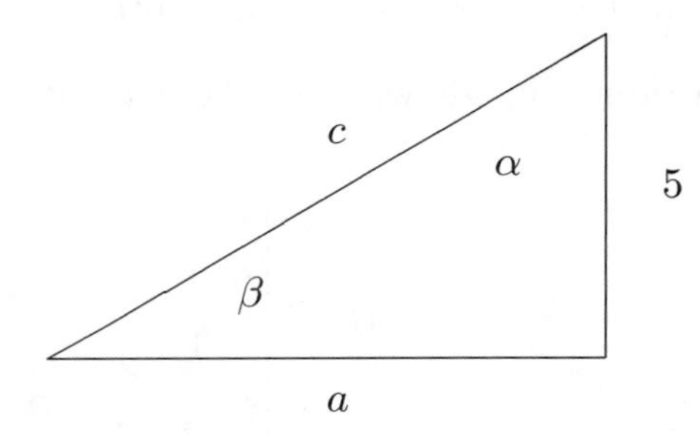

Since $\cos \beta = 0.3$, we find $\beta = \cos^{-1}(0.3) \approx 72.5^\circ$ and $\alpha = 17.5^\circ$. Since $\sin 72.5^\circ = 5/c$ and $\tan 72.5^\circ = 5/a$, we obtain $c = 5/\sin 72.5^\circ \approx 5.2$ and $a = 5/\tan 72.5^\circ \approx 1.6$.

24. Form a right triangle with $a = 2$.

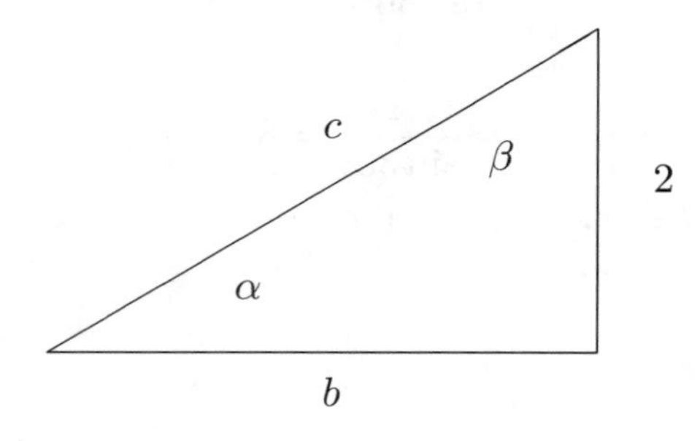

Since $\sin \alpha = 0.6$, we obtain $\alpha = \sin^{-1}(0.6) \approx 36.9^\circ$ and $\beta = 53.1^\circ$. Since $\sin 36.9^\circ = 2/c$ and $\tan 36.9^\circ = 2/b$, we get $c = 2/\sin 36.9^\circ \approx 3.3$ and $b = 2/\tan 36.9^\circ \approx 2.7$.

For Thought

1. True, $\sin^{-1}(0) = 0 = \sin(0)$.

2. True, since $\sin(3\pi/4) = \dfrac{\sqrt{2}}{2} = \dfrac{1}{\sqrt{2}}$.

3. False, $\cos^{-1}(0) = \pi/2$.

4. False, $\sin^{-1}(\sqrt{2}/2) = 45°$.

5. False, since it equals $\tan^{-1}(1/5)$.

6. True, since $1/5 = 0.2$.

7. True, $\sin(\cos^{-1}(\sqrt{2}/2)) = \sin(\pi/4) = 1/\sqrt{2}$.

8. True by definition of $y = \sec^{-1}(x)$.

9. False, since $f^{-1}(x) = \sin(x)$ where $-\pi/2 \le x \le \pi/2$.

10. False, since the secant and cotangent functions are not invertible functions.

4.1 Exercises

1. $-\pi/6$ **3.** $\pi/6$

5. $\pi/4$ **7.** $\pi/3$

9. $-90°$ **11.** $135°$

13. $30°$ **15.** $180°$

17. $-\pi/4$ **19.** $\pi/3$

21. $\pi/4$ **23.** $-\pi/6$

25. 0 **27.** $\pi/2$

29. $3\pi/4$ **31.** $2\pi/3$

33. 0.60 **35.** 3.02 **37.** -0.14

39. 1.87 **41.** 1.15 **43.** -0.36

45. 3.06 **47.** 0.06

49. $\dfrac{\pi}{6}$ **51.** $\dfrac{\pi}{4}$

53. $\dfrac{\pi}{4}$ **55.** $\dfrac{\pi}{3}$

57. $\dfrac{\pi}{6}$ **59.** 0

61. $\tan(\pi/3) = \sqrt{3}$

63. $\sin^{-1}(-1/2) = -\pi/6$

65. $\cot^{-1}(\sqrt{3}) = \pi/6$

67. $\arcsin(\sqrt{2}/2) = \pi/4$

69. $\tan(\pi/4) = 1$

71. $\cos^{-1}(0) = \pi/2$

73. $\sin^{-1}(\sin 3\pi/4) = \sin^{-1}(\sqrt{2}/2) = \pi/4$

75. $\cos(\cos^{-1}(-\sqrt{3}/2)) = \cos(5\pi/6) = -\sqrt{3}/2$

77. $\tan^{-1}(\tan\pi) = \tan^{-1}(0) = 0$

79. 0.8930

81. Undefined

83. -0.9802

85. -0.4082

87. 3.4583

89. 1.0183

91. Let $\theta = \arccos x$. Then $\cos\theta = x$ and θ lies in quadrant 1 or 2. Since $\sin^2\theta = 1 - \cos^2\theta = 1 - x^2$, we obtain $\sin(\arccos x) = \sin\theta = \pm\sqrt{1-x^2}$. Since sine is positive in both quadrants 1 and 2, we have $\sin(\arccos x) = \sqrt{1-x^2}$.

93. Note, $\cos(\theta) = \dfrac{1}{\sec\theta} = \dfrac{1}{\pm\sqrt{\tan^2\theta + 1}}$.

Since $\theta = \arctan x$ is an angle in quadrant 1 or 4, and cosine is positive in both quadrants 1 and 4, we get

$$\cos(\arctan x) = \frac{1}{\sqrt{\tan^2(\arctan x) + 1}} = \frac{1}{\sqrt{x^2+1}}.$$

95. Let $\theta = \text{arccot}\ x$ be an angle that lies in quadrant 1 or 2. Since cosine is positive in quadrants 1 and 2, we obtain

$$\begin{aligned}\cos(\text{arccot}\ x) &= \cos(\theta) \\ &= \sqrt{1-\sin^2\theta} \\ &= \sqrt{1-\frac{1}{\csc^2\theta}}\end{aligned}$$

$$\begin{aligned} &= \sqrt{1-\frac{1}{1+\cot^2\theta}} \\ &= \sqrt{1-\frac{1}{1+x^2}} \\ &= \sqrt{\frac{x^2}{1+x^2}} \\ &= \frac{\pm x}{\sqrt{1+x^2}} \end{aligned}$$

Note, $\cos(\text{arccot } x)$ is positive exactly when $x > 0$, and $\cos(\text{arccot } x)$ is negative exactly when $x < 0$. Thus,

$$\cos(\text{arccot } x) = \frac{x}{\sqrt{1+x^2}}.$$

97. Note, $\tan x = \pm\sqrt{\sec^2 x - 1}$ and $\cos(\arcsin x) = \sqrt{1-x^2}$. Then

$$\begin{aligned} \tan(\arcsin x) &= \pm\sqrt{\sec^2(\arcsin x) - 1} \\ &= \pm\sqrt{\left(\frac{1}{\sqrt{1-x^2}}\right)^2 - 1} \\ &= \pm\sqrt{\frac{1}{1-x^2} - 1} \\ &= \pm\sqrt{\frac{x^2}{1-x^2}} \\ &= \pm\frac{\sqrt{x^2}}{\sqrt{1-x^2}} \\ &= \pm\frac{\pm x}{\sqrt{1-x^2}} \\ &= \pm\frac{x}{\sqrt{1-x^2}}. \end{aligned}$$

Note, $\tan(\arcsin x)$ is positive exactly when $x > 0$, and $\tan(\arcsin x)$ is negative exactly when $x < 0$. Thus,

$$\tan(\arcsin x) = \frac{x}{\sqrt{1-x^2}}.$$

99. Note, $\arctan x$ is an angle in quadrant 1 or 4, and secant is positive in quadrants 1 and 4. Since $\sec(\theta) = \pm\sqrt{\tan^2(\theta) + 1}$, we have

$$\begin{aligned} \sec(\arctan x) &= \sqrt{\tan^2(\arctan x) + 1} \\ &= \sqrt{x^2+1}. \end{aligned}$$

101. 33.8°

103. 86.6°

105. 50.3°

107. 81.1°

109. Consider the right triangle with hypotenuse 2400, altitude 2000, and the angle between the hypotenuse and the altitude is $\theta/2$. Since

$$\cos\left(\frac{\theta}{2}\right) = \frac{2000}{2400},$$

we find

$$\begin{aligned} \theta &= 2\cos^{-1}\left(\frac{2000}{2400}\right) \\ &\approx 67.1^\circ \end{aligned}$$

Thus, the angle for which the airplane is within range of the gun is $\theta \approx 67.1^\circ$.

111. Note, the domain of $y = \sin\left(\sin^{-1} x\right)$ is $[-1, 1]$ and

$$y = \sin\left(\sin^{-1} x\right) = x.$$

Then the graph of $y = \sin\left(\sin^{-1} x\right)$ is a segment of the line $y = x$ with endpoints $(-1, -1)$ and $(1, 1)$.

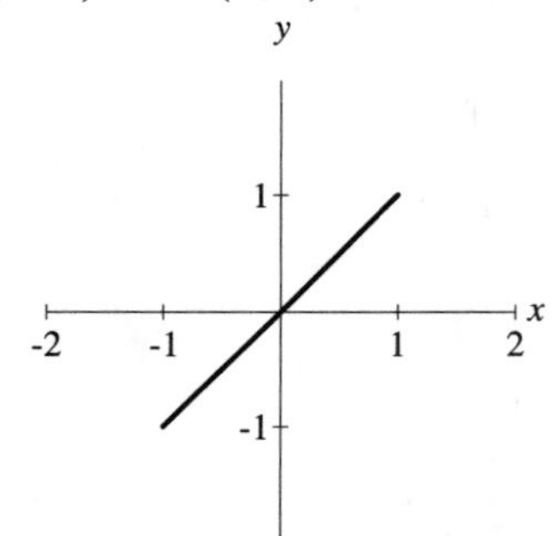

113. Since $\sin^{-1}(1/x) = \csc^{-1}(x)$ for $|x| \geq 1$, the graph of $y = \sin^{-1}(1/x)$ is the same as the graph of $y = \csc^{-1}(x)$.

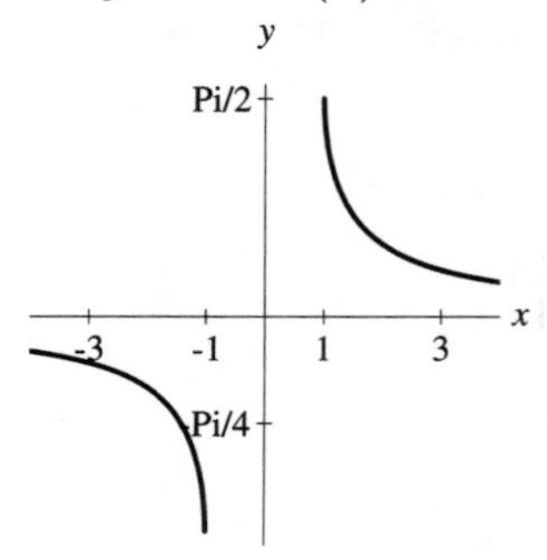

For Thought

1. False, since $\sin \pi = 0$. **2.** True

3. False, since $\sin(\pi/3) = \frac{\sqrt{3}}{2}$. **4.** True

5. True **6.** True

7. False, only solutions are 45° and 315°.

8. False, there is no solution in $[0, \pi)$.

9. True, since $-29°$ and $331°$ are coterminal angles.

10. True, since $\tan(3\pi/4) = \tan(7\pi/4) = -1$ and $\frac{3\pi}{4} = \frac{7\pi}{4} + (-1) \cdot \pi$.

4.2 Exercises

1. $\left\{x \mid x = \frac{\pi}{2} + k\pi, k \text{ an integer}\right\}$

3. $\{x \mid x = 2k\pi, k \text{ an integer}\}$

5. No solution, since $\sin x = -2$ and the range of $y = \sin x$ is $[-1, 1]$.

7. $\left\{x \mid x = \frac{3\pi}{2} + 2k\pi, k \text{ an integer}\right\}$

9. $\{x \mid x = k\pi, k \text{ an integer}\}$

11. Solution in $[0, \pi)$ is $x = \frac{3\pi}{4}$. The solution set is $\left\{x \mid x = \frac{3\pi}{4} + k\pi\right\}$.

13. Solutions in $[0, 2\pi)$ are $x = \frac{\pi}{3}, \frac{5\pi}{3}$. So solution set is $\left\{x \mid x = \frac{\pi}{3} + 2k\pi \text{ or } x = \frac{5\pi}{3} + 2k\pi\right\}$.

15. Solutions in $[0, 2\pi)$ are $x = \frac{\pi}{4}, \frac{3\pi}{4}$. So solution set is $\left\{x \mid x = \frac{\pi}{4} + 2k\pi \text{ or } x = \frac{3\pi}{4} + 2k\pi\right\}$.

17. Since $\tan x = \frac{1}{\sqrt{3}} = \frac{\sqrt{3}}{3}$, the solution in $[0, \pi)$ is $x = \frac{\pi}{6}$. Thus, the solution set is $\left\{x \mid x = \frac{\pi}{6} + k\pi\right\}$.

19. Solutions in $[0, 2\pi)$ are $x = \frac{5\pi}{6}, \frac{7\pi}{6}$. Then the solution set is $\left\{x \mid x = \frac{5\pi}{6} + 2k\pi \text{ or } x = \frac{7\pi}{6} + 2k\pi\right\}$.

21. Since $\sin x = -\frac{\sqrt{2}}{2}$, the solutions in $[0, 2\pi)$ are $x = \frac{5\pi}{4}, \frac{7\pi}{4}$. Thus, the solution set is $\left\{x \mid x = \frac{5\pi}{4} + 2k\pi \text{ or } x = \frac{7\pi}{4} + 2k\pi\right\}$.

23. Since $\tan x = \sqrt{3}$, the solution in $[0, \pi)$ is $x = \frac{\pi}{3}$. Thus, the solution set is $\left\{x \mid x = \frac{\pi}{3} + k\pi\right\}$.

25. Solutions in $[0, 360°)$ are $\alpha = 90°, 270°$. So solution set is $\{\alpha \mid \alpha = 90° + k \cdot 180°\}$.

27. Solution in $[0, 360°)$ is $\alpha = 90°$. So the solution set is $\{\alpha \mid \alpha = 90° + k \cdot 360°\}$.

29. Solution in $[0, 180°)$ is $\alpha = 0°$. The solution set is $\{\alpha \mid \alpha = k \cdot 180°\}$.

31. Since $\cos \alpha = \frac{\sqrt{2}}{2}$, the solutions in $[0, 360°)$ are $\alpha = 45°, 315°$. The solution set is $\{\alpha \mid \alpha = 45° + k \cdot 360°, \alpha = 315° + k \cdot 360°\}$.

33. Since $\sin \alpha = -\frac{1}{2}$, the solutions in $[0, 360°)$ are $\alpha = 210°, 330°$. The solution set is $\{\alpha \mid \alpha = 210° + k \cdot 360°, \alpha = 330° + k \cdot 360°\}$.

35. Since $\tan \alpha = 1$, the solution in $[0, 180°)$ is $\alpha = 45°$. The solution set is $\{\alpha \mid \alpha = 45° + k \cdot 180°\}$.

37. One solution is $\cos^{-1}(0.873) \approx 29.2°$. Another solution is $360° - 29.2° = 330.8°$. The solution set is $\{29.2°, 330.8°\}$.

39. One solution is $\sin^{-1}(-0.244) \approx -14.1°$. This is coterminal with $345.9°$. Another solution is $180° + 14.1° = 194.1°$. The solution set is $\{345.9°, 194.1°\}$.

41. One solution is $\tan^{-1}(5.42) \approx 79.5°$. Another solution is $180° + 79.5° = 259.5°$. The solution set is $\{79.5°, 259.5°\}$.

43. No solution, since $\cos\alpha = \sqrt{3} > 1$ and the range of $y = \cos\alpha$ is $[-1, 1]$.

45. Since $x = \cos^{-1}(0.66) \approx 0.85$ and $2\pi - \cos^{-1}(0.66) \approx 5.43$, the solution set is $\{0.85, 5.43\}$.

47. If $a = \sin^{-1}(-1/4) \approx -0.25$, then

$$x = 2\pi + a \approx 6.03$$

and

$$x = \pi - a \approx 3.39$$

are solutions to $\sin x = -1/4$. The solution set $\{3.39, 6.03\}$.

49. Since $\tan^{-1}(1/\sqrt{6}) \approx 0.39$ and $\pi + \tan^{-1}(1/\sqrt{6}) \approx 3.53$, the solution set is $\{0.39, 3.53\}$.

51. Let $a = \cos^{-1}(-2/\sqrt{5}) \approx 2.68$. Then $x = a$ and $x = 2\pi - a \approx 3.61$ are solutions to $\cos x = -2/\sqrt{5}$. The solution set is $\{2.68, 3.61\}$.

53. Since $\cos\alpha = 1$, the solution set is $\{-360°, 0°, 360°\}$.

55. Since $\sin\beta = -1/2$, the solution set is $\{210°, 330°\}$.

57. Multiplying by the LCD, we get

$$\begin{aligned} 25.9\sin\alpha &= 23.4\sin 67.2° \\ \sin\alpha &= \frac{23.4\sin 67.2°}{25.9} \\ \alpha &= \sin^{-1}\left(\frac{23.4\sin 67.2°}{25.9}\right) \\ \alpha &\approx 56.4°. \end{aligned}$$

The solution set is $\{56.4°\}$.

59. Isolate $\cos\alpha$ on one side.

$$\begin{aligned} 2(5.4)(8.2)\cos\alpha &= 5.4^2 + 8.2^2 - 3.6^2 \\ \cos\alpha &= \frac{5.4^2 + 8.2^2 - 3.6^2}{2(5.4)(8.2)} \end{aligned}$$

Then

$$\begin{aligned} \alpha &= \cos^{-1}\left(\frac{5.4^2 + 8.2^2 - 3.6^2}{2(5.4)(8.2)}\right) \\ \alpha &\approx 19.6°. \end{aligned}$$

The solution set is $\{19.6°\}$.

61. Since $\cos 3y = x/2$, we obtain

$$\begin{aligned} 3y &= \cos^{-1}(x/2) \\ y &= \frac{1}{3}\cos^{-1}(x/2). \end{aligned}$$

63. Since

$$\sin m = \frac{t-2}{-6} = \frac{2-t}{6}$$

we have

$$m = \sin^{-1}\left(\frac{2-t}{6}\right).$$

65. Since

$$\tan(a/3) = \frac{b+d}{7}$$

we find

$$\begin{aligned} a/3 &= \tan^{-1}\left(\frac{b+d}{7}\right) \\ a &= 3\tan^{-1}\left(\frac{b+d}{7}\right). \end{aligned}$$

67. Since

$$\sin(\pi b - \pi) = \frac{q}{3}$$

we get

$$\begin{aligned} \pi b - \pi &= \sin^{-1}\left(\frac{q}{3}\right) \\ \pi b &= \sin^{-1}\left(\frac{q}{3}\right) + \pi \\ b &= \frac{1}{\pi}\sin^{-1}\left(\frac{q}{3}\right) + 1. \end{aligned}$$

69. To find f^{-1}, interchange x and y, solve for y, and replace y by $f^{-1}(x)$:

$$\begin{aligned} x &= \sin 2y \\ \sin^{-1} x &= 2y \\ \frac{1}{2}\sin^{-1} x &= y \\ f^{-1}(x) &= \frac{1}{2}\sin^{-1} x. \end{aligned}$$

Since the domain of f is

$$-\frac{\pi}{4} \le x \le \frac{\pi}{4}$$

we obtain

$$\begin{aligned} -\frac{\pi}{2} \le\ & 2x \le \frac{\pi}{2} \\ -1 \le\ & \sin 2x \le 1. \end{aligned}$$

Then the range of f is $[-1, 1]$. Thus, the domain and range of f^{-1} are $[-1, 1]$ and $[-\pi/4, \pi/4]$, respectively.

71. To find f^{-1}, interchange x and y, solve for y, and replace y by $f^{-1}(x)$:

$$\begin{aligned} x &= 2\cos 3y \\ \cos^{-1}(x/2) &= 3y \\ \frac{1}{3}\cos^{-1}(x/2) &= y \\ f^{-1}(x) &= \frac{1}{3}\cos^{-1}(x/2). \end{aligned}$$

Since the domain of f is

$$0 \le x \le \frac{\pi}{3}$$

we find

$$\begin{aligned} 0 \le\ & 3x \le \pi \\ -1 \le\ & \cos 3x \le 1 \\ -2 \le\ & 2\cos 3x \le 2. \end{aligned}$$

Then the range of f is $[-2, 2]$. Thus, the domain and range of f^{-1} are $[-2, 2]$ and $[0, \pi/3]$, respectively.

73. To find f^{-1}, interchange x and y, solve for y, and replace y by $f^{-1}(x)$:

$$\begin{aligned} x &= 3 + \tan(\pi y) \\ \tan^{-1}(x-3) &= \pi y \\ \frac{1}{\pi}\tan^{-1}(x-3) &= y \\ f^{-1}(x) &= \frac{1}{\pi}\tan^{-1}(x-3). \end{aligned}$$

Since the domain of f is

$$-\frac{1}{2} < x < \frac{1}{2}$$

we find

$$\begin{aligned} -\frac{\pi}{2} <\ & \pi x < \frac{\pi}{2} \\ -\infty <\ & \tan \pi x < \infty \\ -\infty <\ & 3 + \tan \pi x < \infty. \end{aligned}$$

Then the range of f is $(-\infty, \infty)$. Thus, the domain and range of f^{-1} are $(-\infty, \infty)$ and $(-1/2, 1/2)$, respectively.

75. To find f^{-1}, interchange x and y, solve for y, and replace y by $f^{-1}(x)$:

$$\begin{aligned} x &= 2 - \sin(\pi y - \pi) \\ \pi y - \pi &= \sin^{-1}(2-x) \\ \pi y &= \sin^{-1}(2-x) + \pi \\ y &= \frac{1}{\pi}\sin^{-1}(2-x) + 1 \\ f^{-1}(x) &= \frac{1}{\pi}\sin^{-1}(2-x) + 1. \end{aligned}$$

Since the domain of f is

$$\frac{1}{2} \le x \le \frac{3}{2}$$

we obtain

$$\begin{aligned} \frac{\pi}{2} \le\ & \pi x \le \frac{3\pi}{2} \\ -\frac{\pi}{2} \le\ & \pi x - \pi \le \frac{\pi}{2} \\ -1 \le\ & \sin(\pi x - \pi) \le 1 \\ -1 \le\ & -\sin(\pi x - \pi) \le 1 \\ 1 \le\ & 2 - \sin(\pi x - \pi) \le 3. \end{aligned}$$

Then the range of f is $[1, 3]$. Thus, the domain and range of f^{-1} are $[1, 3]$ and $[1/2, 3/2]$, respectively.

77. To find f^{-1}, interchange x and y, solve for y, and replace y by $f^{-1}(x)$:

$$\begin{aligned} x &= \sin^{-1}(3y) \\ \sin x &= 3y \\ \frac{1}{3}\sin x &= y \\ f^{-1}(x) &= \frac{1}{3}\sin x. \end{aligned}$$

Since the domain of f is

$$-\frac{1}{3} \le x \le \frac{1}{3}$$

we obtain

$$\begin{aligned} -1 &\leq 3x \leq 1 \\ -\pi/2 &\leq \sin^{-1} 3x \leq \pi/2. \end{aligned}$$

Then the range of f is $[-\pi/2, \pi/2]$. Thus, the domain and range of f^{-1} are $[-\pi/2, \pi/2]$ and $[-1/3, 1/3]$, respectively.

79. To find f^{-1}, interchange x and y, solve for y, and replace y by $f^{-1}(x)$:

$$\begin{aligned} x &= \sin^{-1}(y/2) + 3 \\ \sin(x-3) &= y/2 \\ 2\sin(x-3) &= y \\ f^{-1}(x) &= 2\sin(x-3). \end{aligned}$$

Since the domain of f is

$$-2 \leq x \leq 2$$

we obtain

$$\begin{aligned} -1 &\leq x/2 \leq 1 \\ -\pi/2 &\leq \sin^{-1}(x/2) \leq \pi/2. \end{aligned}$$

Then the range of f is $[-\pi/2, \pi/2]$. Thus, the domain and range of f^{-1} are $[-\pi/2, \pi/2]$ and $[-2, 2]$, respectively.

81. Multiplying the equation by LCD, we get

$$\begin{aligned} 13.7 \sin 33.2^\circ &= a \cdot \sin 45.6^\circ \\ \frac{13.7 \sin 33.2^\circ}{\sin 45.6^\circ} &= a \\ 10.5 &\approx a. \end{aligned}$$

The solution set is $\{10.5^\circ\}$.

83. Since

$$\sin x = \frac{8.5 \sin(\pi/7)}{6.3}$$

we have

$$x = \sin^{-1}\left(\frac{8.5 \sin(\pi/7)}{6.3}\right) \approx 0.63 + 2k\pi$$

or

$$x = \pi - \sin^{-1}\left(\frac{8.5 \sin(\pi/7)}{6.3}\right) = 2.52 + 2k\pi$$

where k is an integer.

85. Since

$$\cos \alpha = \frac{5^2 + 6^2 - 7^2}{2(5)(6)}$$

we obtain

$$\alpha = \cos^{-1}\left(\frac{5^2 + 6^2 - 7^2}{2(5)(6)}\right) \approx 1.37 + 2k\pi$$

or

$$\alpha = 2\pi - \cos^{-1}\left(\frac{5^2 + 6^2 - 7^2}{2(5)(6)}\right) = 4.91 + 2k\pi$$

where k is an integer.

87. Since $c^2 = 19.34156...$, we get $c \approx \pm 4.40$.

89. Since

$$5 \sin x = 2$$

we find

$$x = \sin^{-1}\left(\frac{2}{5}\right) \approx 0.41 + 2k\pi$$

or

$$x = \pi - \sin^{-1}\left(\frac{2}{5}\right) = 2.73 + 2k\pi$$

where k is an integer.

91. We find

$$\begin{aligned} 6\cos^{-1} x &= 3 \\ \cos^{-1} x &= \frac{1}{2} \\ x &= \cos\left(\frac{1}{2}\right) \\ x &\approx 0.88 \end{aligned}$$

93. Since $\sin(\pm\pi) = \sin 0 = 0$, we find that $\sin x = 0$ has three solutions in $(-2\pi, 2\pi)$. Since $\sin(\pi/2) = 1$, the maximum value of $y = \sin x$ in $(-2\pi, 2\pi)$ is 1.

For Thought

1. False, rather $\alpha = \frac{3\pi}{2}$.

2. True

3. False, rather $x = \frac{\pi}{4} + \frac{k\pi}{2}$.

4. True, since $\cos(\pi/3) = \cos(5\pi/3) = 1/2$.

5. False, rather $5x = \frac{\pi}{4} + k\pi$.

6. True, since $\sin(\pi/4) = \sin(3\pi/4) = \sqrt{2}/2$.

7. True, since $\csc x = 1/\sin(x)$.

8. False, since $\cot^{-1} 3 = \tan^{-1}(1/3)$ and $\frac{1}{3} \neq 0.33$.

9. True, since $\cot^{-1}(3) = \tan^{-1}(1/3)$.

10. False, rather we have

$$\left\{x \mid 3x = \frac{\pi}{2} + 2k\pi\right\} = \left\{x \mid x = \frac{\pi}{6} + \frac{2k\pi}{3}\right\}.$$

4.3 Exercises

1. The values of $x/2$ in $[0, 2\pi)$ are $\pi/3$ and $5\pi/3$. Then we get

$$\frac{x}{2} = \frac{\pi}{3} + 2k\pi \text{ or } \frac{x}{2} = \frac{5\pi}{3} + 2k\pi$$

$$x = \frac{2\pi}{3} + 4k\pi \text{ or } x = \frac{10\pi}{3} + 4k\pi.$$

The solution set is

$$\left\{x \mid x = \frac{2\pi}{3} + 4k\pi \text{ or } x = \frac{10\pi}{3} + 4k\pi\right\}.$$

3. Value of $3x$ in $[0, 2\pi)$ is 0. Thus, $3x = 2k\pi$.

The solution set is $\left\{x \mid x = \frac{2k\pi}{3}\right\}$.

5. Since $\sin(x/2) = 1/2$, values of $x/2$ in $[0, 2\pi)$ are $\pi/6$ and $5\pi/6$. Then

$$\frac{x}{2} = \frac{\pi}{6} + 2k\pi \text{ or } \frac{x}{2} = \frac{5\pi}{6} + 2k\pi$$

$$x = \frac{\pi}{3} + 4k\pi \text{ or } x = \frac{5\pi}{3} + 4k\pi.$$

The solution set is

$$\left\{x \mid x = \frac{\pi}{3} + 4k\pi \text{ or } x = \frac{5\pi}{3} + 4k\pi\right\}.$$

7. Since $\sin(2x) = -\sqrt{2}/2$, values of $2x$ in $[0, 2\pi)$ are $5\pi/4$ and $7\pi/4$. Thus,

$$2x = \frac{5\pi}{4} + 2k\pi \text{ or } 2x = \frac{7\pi}{4} + 2k\pi$$

$$x = \frac{5\pi}{8} + k\pi \text{ or } x = \frac{7\pi}{8} + k\pi.$$

The solution set is

$$\left\{x \mid x = \frac{5\pi}{8} + k\pi \text{ or } x = \frac{7\pi}{8} + k\pi\right\}.$$

9. Value of $2x$ in $[0, \pi)$ is $\pi/3$. Then

$$2x = \frac{\pi}{3} + k\pi.$$

The solution set is $\left\{x \mid x = \frac{\pi}{6} + \frac{k\pi}{2}\right\}$.

11. Value of $4x$ in $[0, \pi)$ is 0. Then

$$4x = k\pi.$$

The solution set is $\left\{x \mid x = \frac{k\pi}{4}\right\}$.

13. The values of πx in $[0, 2\pi)$ are $\pi/6$ and $5\pi/6$. Then

$$\pi x = \frac{\pi}{6} + 2k\pi \text{ or } \pi x = \frac{5\pi}{6} + 2k\pi$$

$$x = \frac{1}{6} + 2k \text{ or } x = \frac{5}{6} + 2k.$$

The solution set is

$$\left\{x \mid x = \frac{1}{6} + 2k \text{ or } x = \frac{5}{6} + 2k\right\}.$$

15. Values of $2\pi x$ in $[0, 2\pi)$ are $\pi/2$ and $3\pi/2$. So

$$2\pi x = \frac{\pi}{2} + 2k\pi \text{ or } 2\pi x = \frac{3\pi}{2} + 2k\pi$$

$$x = \frac{1}{4} + k \text{ or } x = \frac{3}{4} + k.$$

The solution set is

$$\left\{x \mid x = \frac{1}{4} + \frac{k}{2}\right\}.$$

17. Since $\sin\alpha = -\sqrt{3}/2$, the solution set is $\{240°, 300°\}$.

19. Since $\sin 2x = -\sqrt{3}/2$, we obtain

$$2x = 240° + k \cdot 360°$$

or

$$2x = 300° + k \cdot 360°.$$

Then

$$x = 120° + k \cdot 180°$$

or

$$x = 150° + k \cdot 180°.$$

If $k = 0, 1$, then $x = 120°, 150°, 300°, 330°$.

21. Since $\cos 2x = -1/2$, we find

$$2x = 120^\circ + k \cdot 360^\circ$$

or

$$2x = 240^\circ + k \cdot 360^\circ.$$

Then

$$x = 60^\circ + k \cdot 180^\circ$$

or

$$x = 120^\circ + k \cdot 180^\circ.$$

If $k = 0, 1$, then $x = 60^\circ, 120^\circ, 240^\circ, 300^\circ$.

23. Since $\cos 2\alpha = 1/\sqrt{2}$, values of 2α in $[0, 360^\circ)$ are 45° and 315°. Thus,

$$2\alpha = 45^\circ + k \cdot 360^\circ \text{ or } 2\alpha = 315^\circ + k \cdot 360^\circ$$

$$\alpha = 22.5^\circ + k \cdot 180^\circ \text{ or } \alpha = 157.5^\circ + k \cdot 180^\circ.$$

Then let $k = 0, 1$. The solution set is

$$\{22.5^\circ, 157.5^\circ, 202.5^\circ, 337.5^\circ\}.$$

25. Values of 3α in $[0, 360^\circ)$ are 135° and 225°. Then

$$3\alpha = 135^\circ + k \cdot 360^\circ \text{ or } 3\alpha = 225^\circ + k \cdot 360^\circ$$

$$\alpha = 45^\circ + k \cdot 120^\circ \text{ or } \alpha = 75^\circ + k \cdot 120^\circ.$$

By choosing $k = 0, 1, 2$, one obtains the solution set $\{45^\circ, 75^\circ, 165^\circ, 195^\circ, 285^\circ, 315^\circ\}$.

27. The value of $\alpha/2$ in $[0, 180^\circ)$ is 30°. Then

$$\frac{\alpha}{2} = 30^\circ + k \cdot 180^\circ$$

$$\alpha = 60^\circ + k \cdot 360^\circ.$$

By choosing $k = 0$, the solution set is $\{60^\circ\}$.

29. Since $\csc(4\alpha) = \sqrt{2}$, the values of 4α in $[0, 360^\circ)$ are 45° and 135°. Then

$$4\alpha = 45^\circ + k \cdot 360^\circ \text{ or } 4\alpha = 135^\circ + k \cdot 360^\circ$$

$$\alpha = 11.25^\circ + k \cdot 90^\circ \text{ or } \alpha = 33.75^\circ + k \cdot 90^\circ.$$

Choosing $k = 0, 1, 2, 3$, the solution set is $\{11.25^\circ, 33.75^\circ, 101.25^\circ, 123.75^\circ, 191.25^\circ, 213.75^\circ, 281.25^\circ, 303.75^\circ\}$.

31. Since $\sin 2x = \sqrt{3}/2$, we obtain

$$2x = \frac{\pi}{3} + k \cdot 2\pi$$

or

$$2x = \frac{2\pi}{3} + k \cdot 2\pi$$

where k is an integer. Then

$$x = \frac{\pi}{6} + k \cdot \pi$$

or

$$x = \frac{\pi}{3} + k \cdot \pi$$

If $k = 0, 1$, then $x = \frac{\pi}{6}, \frac{\pi}{3}, \frac{7\pi}{6}, \frac{4\pi}{3}$.

33. Since $\cos 2x = 1/2$, we find

$$2x = \frac{\pi}{3} + k \cdot 2\pi$$

or

$$2x = \frac{5\pi}{3} + k \cdot 2\pi$$

where k is an integer. Then

$$x = \frac{\pi}{6} + k \cdot \pi$$

or

$$x = \frac{5\pi}{6} + k \cdot \pi$$

If $k = 0, 1$, then

$$x = \frac{\pi}{6}, \frac{5\pi}{6}, \frac{7\pi}{6}, \frac{11\pi}{6}.$$

35. Since $\tan 3x = 1$, we find

$$3x = \frac{\pi}{4} + k \cdot \pi.$$

where k is an integer. Then

$$x = \frac{\pi}{12} + \frac{k\pi}{3}.$$

If $k = 0, 1, ..., 5$, then

$$x = \frac{\pi}{12}, \frac{5\pi}{12}, \frac{9\pi}{12}, \frac{13\pi}{12}, \frac{17\pi}{12}, \frac{21\pi}{12}.$$

37. Since $\sin(x/3) = 1/\sqrt{2}$, we obtain

$$\frac{x}{3} = \frac{\pi}{4} + k \cdot 2\pi$$

or

$$\frac{x}{3} = \frac{3\pi}{4} + k \cdot 2\pi$$

where k is an integer. Then

$$x = \frac{3\pi}{4} + k \cdot 6\pi$$

or

$$x = \frac{9\pi}{4} + k \cdot 6\pi$$

Thus, $x = \dfrac{3\pi}{4}$.

39. Since $\tan(x/2) = 1/\sqrt{3}$, we obtain

$$\frac{x}{2} = \frac{\pi}{6} + k \cdot \pi$$

where k is an integer. Then

$$x = \frac{\pi}{3} + k \cdot 2\pi.$$

If $k = 0$, then $x = \dfrac{\pi}{3}$.

41. A solution is $3\alpha = \sin^{-1}(0.34) \approx 19.88°$. Another solution is $3\alpha = 180° - 19.88° = 160.12°$. Then

$$3\alpha = 19.88° + k \cdot 360° \text{ or } 3\alpha = 160.12° + k \cdot 360°$$

$$\alpha \approx 6.6° + k \cdot 120° \text{ or } \alpha \approx 53.4° + k \cdot 120°.$$

Solution set is

$$\{\alpha \mid \alpha = 6.6° + k \cdot 120° \text{ or } \alpha = 53.4° + k \cdot 120°\}.$$

43. A solution is $3\alpha = \sin^{-1}(-0.6) \approx -36.87°$. This is coterminal with $323.13°$. Another solution is $3\alpha = 180° + 36.87° = 216.87°$. Then

$$3\alpha = 323.13° + k \cdot 360° \text{ or } 3\alpha = 216.87° + k \cdot 360°$$

$$\alpha \approx 107.7° + k \cdot 120° \text{ or } \alpha \approx 72.3° + k \cdot 120°.$$

The solution set is

$$\{\alpha \mid \alpha = 72.3° + k120° \text{ or } \alpha = 107.7° + k120°\}.$$

45. A solution is $2\alpha = \cos^{-1}(1/4.5) \approx 77.16°$. Another solution is $2\alpha = 360° - 77.16° = 282.84°$. Thus,

$$2\alpha = 77.16° + k \cdot 360° \text{ or } 2\alpha = 282.84° + k \cdot 360°$$

$$\alpha \approx 38.6° + k \cdot 180° \text{ or } \alpha \approx 141.4° + k \cdot 180°.$$

The solution set is

$$\{\alpha \mid \alpha = 38.6° + k180° \text{ or } \alpha = 141.4° + k180°\}.$$

47. A solution is $\alpha/2 = \sin^{-1}(-1/2.3) \approx -25.77°$. This is coterminal with $334.23°$. Another solution is $\alpha/2 = 180° + 25.77° = 205.77°$. Thus,

$$\frac{\alpha}{2} = 334.23° + k \cdot 360° \text{ or } \frac{\alpha}{2} = 205.77° + k \cdot 360°$$

$$\alpha \approx 668.5° + k \cdot 720° \text{ or } \alpha \approx 411.5° + k \cdot 720°.$$

The solution set is

$$\{\alpha \mid \alpha = 668.5° + k720° \text{ or } \alpha = 411.5° + k720°\}.$$

49. Note, $5x = \sin^{-1}(1/3) + 2k\pi$ or $5x = \pi - \sin^{-1}(1/3) + 2k\pi$. Solving for x, we get $x = \dfrac{\sin^{-1}(1/3)}{5} + \dfrac{2k\pi}{5}$ or $x = \dfrac{\pi - \sin^{-1}(1/3)}{5} + \dfrac{2k\pi}{5}$. Since $\dfrac{\sin^{-1}(1/3)}{5} \approx 0.07$ and $\dfrac{\pi - \sin^{-1}(1/3)}{5} \approx 0.56$, the solution set is

$$\left\{x \mid x = 0.07 + \frac{2k\pi}{5}, x = 0.56 + \frac{2k\pi}{5}\right\}.$$

51. Note, $x/2 = \sin^{-1}(-6/10) + 2\pi + 2k\pi$ or $x/2 = \pi + \sin^{-1}(6/10) + 2k\pi$. Solving for x, we get $x = 2\left(\sin^{-1}(-6/10) + 2\pi\right) + 4k\pi$ or $x = 2\left(\pi + \sin^{-1}(6/10)\right) + 4k\pi$. Since $2\left(\sin^{-1}(-6/10) + 2\pi\right) \approx 11.28$ and $2\left(\pi + \sin^{-1}(6/10)\right) \approx 7.57$, the solution set is

$$\{x \mid x = 7.57 + 4k\pi \text{ or } x = 11.28 + 4k\pi\}.$$

53. Note, $\pi x = \cos^{-1}(2/9) + 2k\pi$ or $\pi x = 2\pi - \cos^{-1}(2/9) + 2k\pi$. Solving for x, we find $x = \dfrac{\cos^{-1}(2/9)}{\pi} + 2k$ or $x = \dfrac{2\pi - \cos^{-1}(2/9)}{\pi} + 2k$.

Since $\dfrac{\cos^{-1}(2/9)}{\pi} \approx 0.43$ and $\dfrac{2\pi - \cos^{-1}(2/9)}{\pi} \approx 1.57$, the solution set is

$$\{x \mid x = 0.43 + 2k \text{ or } x = 1.57 + 2k\}.$$

55. Note, $\pi x - 1 = \tan^{-1}(3) + k\pi$. Solving for x, we obtain $x = \dfrac{1 + \tan^{-1}(3)}{\pi} + k$.

Since $\dfrac{1 + \tan^{-1}(3)}{\pi} \approx 0.72$, the solution set is

$$\{x \mid x = 0.72 + k\}.$$

57. Given below is the graph of $y = \sin(x/2) - \cos(3x)$.
The intercepts or solutions on $[0, 2\pi)$ are approximately $\{0.4, 1.9, 2.2, 4.0, 4.4, 5.8\}$.

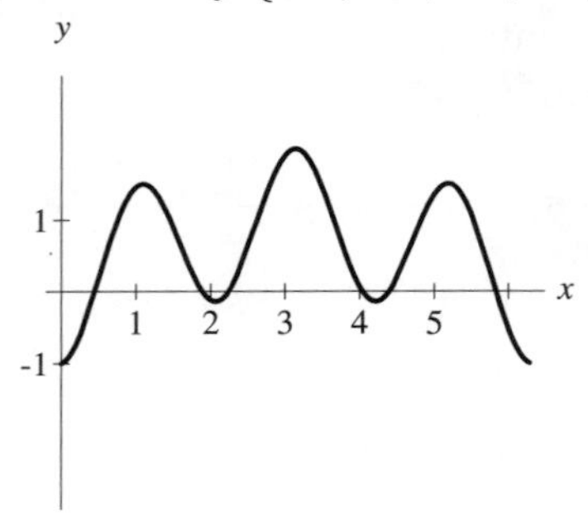

59. The graph of $y = \dfrac{x}{2} - \dfrac{\pi}{6} + \dfrac{\sqrt{3}}{2} - \sin x$ is shown. The solution set is $\{\pi/3\}$.

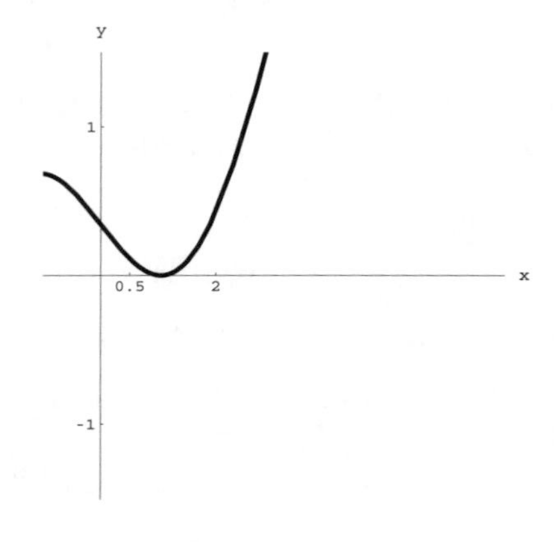

61. Since $v_o = 325$ and $d = 3300$, we have

$$\begin{aligned} 325^2 \sin 2\theta &= 32(3300) \\ \sin 2\theta &= \frac{32(3300)}{325^2} \\ 2\theta &= \sin^{-1}\left(\frac{32(3300)}{325^2}\right) \\ 2\theta &\approx 88.74^\circ \\ \theta &\approx 44.4^\circ. \end{aligned}$$

Another angle is given by $2\theta = 180^\circ - 88.74^\circ = 91.26^\circ$ or $\theta = 91.26^\circ/2 \approx 45.6^\circ$.
The muzzle was aimed at 44.4° or 45.6°.

63. Note, 90 mph$= 90 \cdot \dfrac{5280}{3600}$ ft/sec$=$ 132 ft/sec.
In $v_o^2 \sin 2\theta = 32d$, let $v_o = 132$ and $d = 230$.

$$\begin{aligned} 132^2 \sin 2\theta &= 32(230) \\ 2\theta &= \sin^{-1}\left(\frac{32(230)}{132^2}\right) \\ 2\theta &\approx 25.0^\circ \end{aligned}$$

Another possibility is $2\theta \approx 180^\circ - 25^\circ = 155^\circ$. The two possible angles are

$$\theta \approx 12.5^\circ, 77.5^\circ.$$

The time it takes the ball to reach home plate can be found by using $x = v_o t \cos\theta$. (See Example 5). For the angle 12.5°, it takes

$$t = \frac{230}{132 \cos 12.5^\circ} \approx 1.78 \text{ sec}$$

while for 77.5° it takes

$$t = \frac{230}{132 \cos 77.5^\circ} \approx 8.05 \text{ sec.}$$

The difference in time is $8.05 - 1.78 \approx 6.3$ sec.

65. First, find the values of t when $x = \sqrt{3}$.

$$\begin{aligned} 2\sin\left(\frac{\pi t}{3}\right) &= \sqrt{3} \\ \sin\left(\frac{\pi t}{3}\right) &= \frac{\sqrt{3}}{2} \end{aligned}$$

$$\frac{\pi t}{3} = \frac{\pi}{3} + 2k\pi \quad \text{or} \quad \frac{\pi t}{3} = \frac{2\pi}{3} + 2k\pi$$

$$\pi t = \pi + 6k\pi \quad \text{or} \quad \pi t = 2\pi + 6k\pi$$

$$t = 1 + 6k \quad \text{or} \quad t = 2 + 6k$$

Then the ball is $\sqrt{3}$ ft above sea level for the values of t satisfying

$$1 + 6k < t < 2 + 6k$$

where k is a nonnegative integer.

67. Infinitely many solutions

For Thought

1. False, only solutions are 60° and 300°.

2. False, since there is no solution is $[0, \pi)$.

3. True, since -30° and 330° are coterminal angles.

4. True **5.** True, since the right-side is a factorization of the left-side.

6. False, $x = 0$ is a solution to the first equation and not to the second equation.

7. False, $\cos^{-1} 2$ is undefined. **8.** True

9. False, $x = 3\pi/4$ is not a solution to the first equation but is a solution to the second equation.

10. False, since $x = 3\pi/2$ does not satisfy both equations.

4.4 Exercises

1. Since $-\sin x = \sin x$, we get $0 = 2\sin x$ or $\sin x = 0$. The solution set is $\{0, \pi\}$.

3. Since $-\tan x = \tan x$, we get $0 = 2\tan x$ or $\tan x = 0$. The solution set is $\{0, \pi\}$.

5. Set the right-hand side to zero and factor.

$$\begin{aligned} 3\sin^2 x - \sin x &= 0 \\ \sin x(3\sin x - 1) &= 0 \end{aligned}$$

Set each factor to zero.

$$\begin{array}{lcl} \sin x = 0 & \text{or} & \sin x = 1/3 \\ x = 0, \pi & \text{or} & x = \sin^{-1}(1/3) \approx 0.3 \end{array}$$

Another solution to $\sin x = 1/3$ is $x = \pi - 0.3 \approx 2.8$.
The solution set is $\{0, 0.3, 2.8, \pi\}$.

7. Since

$$\cos x(2\cos x - 3) = 0$$

we obtain

$$\cos x = 0$$

or

$$\cos x = \frac{3}{2}.$$

The latter equation has no solution. The solutions of $\cos x = 0$ are

$$x = \frac{\pi}{2} + k \cdot \pi.$$

Thus,

$$x = \frac{\pi}{2}, \frac{3\pi}{2}.$$

9. Set the right-hand side to zero and factor.

$$\begin{aligned} 2\cos^2 x + 3\cos x + 1 &= 0 \\ (2\cos x + 1)(\cos x + 1) &= 0 \end{aligned}$$

Set the factors to zero.

$$\begin{array}{rcl} \cos x = -1/2 & \text{or} & \cos x = -1 \\ x = 2\pi/3, 4\pi/3 & \text{or} & x = \pi \end{array}$$

The solution set is $\{\pi, 2\pi/3, 4\pi/3\}$.

11. Squaring both sides of the equation, we obtain

$$\begin{aligned} \tan^2 x &= \sec^2 x - 2\sqrt{3}\sec x + 3 \\ \sec^2 x - 1 &= \sec^2 x - 2\sqrt{3}\sec x + 3 \\ -4 &= -2\sqrt{3}\sec x \\ \sec x &= 2/\sqrt{3} \\ x &= \pi/6, 11\pi/6. \end{aligned}$$

Checking $x = \pi/6$, one gets $\tan(\pi/6) = 1/\sqrt{3}$ and $\sec(\pi/6) - \sqrt{3} = 2/\sqrt{3} - \sqrt{3} = -1/\sqrt{3}$. Then $x = \pi/6$ is an extraneous root and the solution set is $\{11\pi/6\}$.

13. Square both sides of the equation.

$$\begin{aligned} \sin^2 x + 2\sqrt{3}\sin x + 3 &= 27\cos^2 x \\ \sin^2 x + 2\sqrt{3}\sin x + 3 &= 27(1 - \sin^2 x) \\ 28\sin^2 x + 2\sqrt{3}\sin x - 24 &= 0 \\ 14\sin^2 x + \sqrt{3}\sin x - 12 &= 0 \end{aligned}$$

By the quadratic formula, we get

$$\begin{aligned}\sin x &= \frac{-\sqrt{3} \pm \sqrt{675}}{28} \\ \sin x &= \frac{-\sqrt{3} \pm 15\sqrt{3}}{28} \\ \sin x &= \frac{\sqrt{3}}{2}, \frac{-4\sqrt{3}}{7}.\end{aligned}$$

Thus,

$$x = \frac{\pi}{3}, \frac{2\pi}{3} \quad \text{or} \quad x = \sin^{-1}\left(\frac{-4\sqrt{3}}{7}\right)$$

$$x = \frac{\pi}{3}, \frac{2\pi}{3} \quad \text{or} \quad x \approx -1.427.$$

Checking $x = 2\pi/3$, one finds $\sin(2\pi/3) + \sqrt{3} = \sqrt{3}/2 + \sqrt{3}$ and $3\sqrt{3}\cos(2\pi/3)$ is a negative number. Then $x = 2\pi/3$ is an extraneous root.

An angle coterminal with -1.427 is $2\pi - 1.427 \approx 4.9$. In a similar way, one checks that $\pi + 1.427 \approx 4.568$ is an extraneous root. Thus, the solution set is $\{\pi/3, 4.9\}$.

15. Substitute $\cos^2 x = 1 - \sin^2 x$.

$$\begin{aligned}5\sin^2 x - 2\sin x &= 1 - \sin^2 x \\ 6\sin^2 x - 2\sin x - 1 &= 0\end{aligned}$$

Apply the quadratic formula.

$$\begin{aligned}\sin x &= \frac{2 \pm \sqrt{28}}{12} \\ \sin x &= \frac{1 \pm \sqrt{7}}{6}\end{aligned}$$

Then

$$x = \sin^{-1}\left(\frac{1+\sqrt{7}}{6}\right) \quad \text{or} \quad x = \sin^{-1}\left(\frac{1-\sqrt{7}}{6}\right)$$

$$x \approx 0.653 \quad \text{or} \quad x \approx -0.278.$$

Another solution is $\pi - 0.653 \approx 2.5$. An angle coterminal with -0.278 is $2\pi - 0.278 \approx 6.0$. Another solution is $\pi + 0.278 \approx 3.4$. The solution set is $\{0.7, 2.5, 3.4, 6.0\}$.

17. Express the equation in terms of $\sin x$ and $\cos x$.

$$\begin{aligned}\frac{\sin x}{\cos x} \cdot 2\sin x \cos x &= 0 \\ 2\sin^2 x &= 0 \\ \sin x &= 0\end{aligned}$$

The solution set is $\{0, \pi\}$.

19. Substitute the double-angle identity for $\sin x$.

$$\begin{aligned}2\sin x\cos x - \sin x \cos x &= \cos x \\ \sin x \cos x - \cos x &= 0 \\ \cos x(\sin x - 1) &= 0 \\ \cos x = 0 \quad &\text{or} \quad \sin x = 1 \\ x = \pi/2, 3\pi/2 \quad &\text{or} \quad x = \pi/2\end{aligned}$$

The solution set is $\{\pi/2, 3\pi/2\}$.

21. Use the sum identity for sine.

$$\sin(x + \pi/4) = 1/2$$

$$x + \frac{\pi}{4} = \frac{\pi}{6} + 2k\pi \quad \text{or} \quad x + \frac{\pi}{4} = \frac{5\pi}{6} + 2k\pi$$

$$x = \frac{-\pi}{12} + 2k\pi \quad \text{or} \quad x = \frac{7\pi}{12} + 2k\pi$$

By choosing $k = 1$ in the first case and $k = 0$ in the second case, one finds the solution set is $\{7\pi/12, 23\pi/12\}$.

23. Apply the difference identity for sine.

$$\begin{aligned}\sin(2x - x) &= -1/2 \\ \sin x &= -1/2\end{aligned}$$

The solution set is $\{7\pi/6, 11\pi/6\}$.

25. Use a half-angle identity for cosine and express the equation in terms of $\cos\theta$.

$$\begin{aligned}\frac{1+\cos\theta}{2} &= \frac{1}{\cos\theta} \\ \cos\theta + \cos^2\theta &= 2 \\ \cos^2\theta + \cos\theta - 2 &= 0 \\ (\cos\theta + 2)(\cos\theta - 1) &= 0 \\ \cos\theta = -2 \quad &\text{or} \quad \cos\theta = 1 \\ \text{no solution} \quad &\text{or} \quad \theta = 0^\circ\end{aligned}$$

Solution set is $\{0^\circ\}$.

27. Divide both sides of the equation by $2\cos\theta$.

$$\begin{aligned}\frac{\sin\theta}{\cos\theta} &= \frac{1}{2} \\ \tan\theta &= 0.5 \\ \theta &= \tan^{-1}(0.5) \approx 26.6^\circ\end{aligned}$$

Another solution is $180^\circ + 26.6^\circ = 206.6^\circ$. Solution set is $\{26.6^\circ, 206.6^\circ\}$.

29. Since
$$2\sin\theta\cos\theta = 3\cos\theta$$
we have
$$\sin\theta(2\cos\theta - 3) = 0.$$
Then
$$\sin\theta = 0 \quad \text{or} \quad \cos\theta = \frac{3}{2}.$$
The latter equation has no solution. The solutions of $\sin\theta = 0$ are
$$x = k \cdot 180^\circ.$$
Thus,
$$x = 0^\circ, 180^\circ.$$

31. Express the equation in terms of $\sin 3\theta$.
$$\begin{aligned}\sin 3\theta &= \frac{1}{\sin 3\theta}\\ \sin^2 3\theta &= 1\\ \sin 3\theta &= \pm 1\end{aligned}$$
Then
$$\begin{aligned}3\theta = 90^\circ + k\cdot 360^\circ \quad &\text{or} \quad 3\theta = 270^\circ + k\cdot 360^\circ\\ \theta = 30^\circ + k\cdot 120^\circ \quad &\text{or} \quad \theta = 90^\circ + k\cdot 120^\circ.\end{aligned}$$
By choosing $k = 0, 1, 2$, one finds that the solution set is
$$\{30^\circ, 90^\circ, 150^\circ, 210^\circ, 270^\circ, 330^\circ\}.$$

33. By the method of completing the square, we get
$$\begin{aligned}\tan^2\theta - 2\tan\theta &= 1\\ \tan^2\theta - 2\tan\theta + 1 &= 2\\ (\tan\theta - 1)^2 &= 2\\ \tan\theta - 1 &= \pm\sqrt{2}\\ \theta = \tan^{-1}(1+\sqrt{2}) \quad &\text{or} \quad \theta = \tan^{-1}(1-\sqrt{2})\\ \theta \approx 67.5^\circ \quad &\text{or} \quad \theta = -22.5^\circ.\end{aligned}$$
Other solutions are $180^\circ + 67.5^\circ = 247.5^\circ$, $180^\circ - 22.5^\circ = 157.5^\circ$, and $180^\circ + 157.5^\circ = 337.5^\circ$. The solution set is
$$\{67.5^\circ, 157.5^\circ, 247.5^\circ, 337.5^\circ\}.$$

35. Factor as a perfect square.
$$\begin{aligned}(3\sin\theta + 2)^2 &= 0\\ \sin\theta &= -2/3\\ \theta &= \sin^{-1}(-2/3) \approx -41.8^\circ\end{aligned}$$
An angle coterminal with -41.8° is $360^\circ - 41.8^\circ = 318.2^\circ$. Another solution is $180^\circ + 41.8^\circ = 221.8^\circ$. The solution set is $\{221.8^\circ, 318.2^\circ\}$.

37. Factoring, we find
$$\begin{aligned}3\sin x - \sin x\cos x + 6 - 2\cos x &= 0\\ \sin x(3 - \cos x) + 2(3 - \cos x) &= 0\\ (\sin x + 2)(3 - \cos x) &= 0\\ \sin x = -2 \quad \text{or} \quad & \cos x = 3.\end{aligned}$$
Since
$$-1 \le \sin x, \cos x \le 1$$
we see that there are no solutions. The solution set is the empty set $\emptyset$.

39. By using the sum identity for tangent, we get
$$\begin{aligned}\tan(3\theta - \theta) &= \sqrt{3}\\ 2\theta &= 60^\circ + k\cdot 180^\circ\\ \theta &= 30^\circ + k\cdot 90^\circ.\end{aligned}$$
By choosing $k = 1, 3$, one obtains that the solution set is $\{120^\circ, 300^\circ\}$. Note, 30° and 210° are not solutions.

41. Factoring, we get
$$(4\cos^2\theta - 3)(2\cos^2\theta - 1) = 0.$$
Then
$$\begin{aligned}\cos^2\theta = 3/4 \quad &\text{or} \quad \cos^2\theta = 1/2\\ \cos\theta = \pm\sqrt{3}/2 \quad &\text{or} \quad \cos\theta = \pm 1/\sqrt{2}.\end{aligned}$$
The solution set is
$$\{30^\circ, 45^\circ, 135^\circ, 150^\circ, 210^\circ, 225^\circ, 315^\circ, 330^\circ\}.$$

43. Factoring, we obtain
$$\begin{aligned}(\sec^2\theta - 1)(\sec^2\theta - 4) &= 0\\ \sec^2\theta = 1 \quad &\text{or} \quad \sec^2\theta = 4\\ \sec\theta = \pm 1 \quad &\text{or} \quad \sec\theta = \pm 2.\end{aligned}$$
Solution set is $\{0^\circ, 60^\circ, 120^\circ, 180^\circ, 240^\circ, 300^\circ\}$.

45. Since $\cot x = \dfrac{\cos x}{\sin x}$, we get

$$\begin{aligned} \frac{\cos x}{\sin x} &= \sin x \\ \sin^2 x &= \cos x \\ 1 - \cos^2 &= \cos x \\ \cos^2 + \cos x - 1 &= 0. \end{aligned}$$

By the quadratic formula, we get $\cos x = \dfrac{-1 \pm \sqrt{5}}{2}$. Note, $\left|\dfrac{-1-\sqrt{5}}{2}\right| > 1$ and let $a = \cos^{-1}\left(\dfrac{-1+\sqrt{5}}{2}\right)$. Since $a \approx 0.905$ and $2\pi - a \approx 5.379$, the solution set is $\{x \mid x = 0.905 + 2k\pi, x = 5.379 + 2k\pi\}$.

47. Since $a = \sqrt{3}$ and $b = 1$, we obtain $r = \sqrt{\sqrt{3}^2 + 1^2} = 2$. If the terminal side of α goes through $(\sqrt{3}, 1)$, then $\tan\alpha = 1/\sqrt{3}$. Then one can choose $\alpha = \pi/6$ and $x = 2\sin(2t + \pi/6)$. The times when $x = 0$ are given by

$$\begin{aligned} \sin\left(2t + \frac{\pi}{6}\right) &= 0 \\ 2t + \frac{\pi}{6} &= k \cdot \pi \\ 2t &= -\frac{\pi}{6} + k \cdot \pi \\ t &= -\frac{\pi}{12} + \frac{k \cdot \pi}{2} \\ t &= -\frac{\pi}{12} + \frac{\pi}{2} + \frac{k \cdot \pi}{2} \\ t &= \frac{5\pi}{12} + \frac{k \cdot \pi}{2} \end{aligned}$$

where k is a nonnegative integer.

49. Note, $\dfrac{1 - \cos x}{x} = 0$ implies $\cos x = 1$ and $x \neq 0$. Thus, $x = \pm 2\pi, \pm 4\pi, \pm 6\pi$.
In the interval $(-2\pi, 2\pi)$, using a calculator, we get that the maximum value of $\dfrac{1 - \cos x}{x}$ is about 0.7246.

Review Exercises

1. $-\pi/6$

3. $-\pi/4$

5. $\pi/4$

7. $\cos(\arcsin(1/2) = \cos(\pi/6) = \sqrt{3}/2$

9. $\tan(\arcsin(\sqrt{2}/2) = \tan(\pi/4) = 1$

11. $\sin^{-1}(\sin(-\pi/4)) = \sin^{-1}(-\sqrt{2}/2) = -\pi/4$

13. $\sin^{-1}(\sin(3\pi/4)) = \sin^{-1}(\sqrt{2}/2) = \pi/4$

15. $\cos^{-1}(\cos(-\pi/6)) = \cos^{-1}(\sqrt{3}/2) = \pi/6$

17. $\csc^{-1}(\sec(\pi/3)) = \csc^{-1}(2) = \pi/6$

19. $90°$

21. $135°$

23. $30°$

25. $90°$

27. Since $\cos x = -1$, the solution set is $\{\pi\}$.

29. Since $\sin x = 1/2$, the solution set is $\{\pi/6, 5\pi/6\}$.

31. Since $\tan x = -1$, the solution set is $\{3\pi/4, 7\pi/4\}$.

33. Since $-2\sin(x) + 1 = 0$, we get $\sin x = 1/2$. The solution set is $\{30°, 150°\}$.

35. Since $2\cos(x) = \sqrt{2}$, we get $\cos x = \sqrt{2}/2$. The solution set is $\{45°, 315°\}$.

37. Since $-\sqrt{3}\tan(x) - 1 = 0$, we get $\tan x = -1/\sqrt{3}$. The solution set is $\{150°, 330°\}$.

39. Note, $a = \cos^{-1}(-3/5) \approx 2.21$ and $2\pi - a \approx 4.07$. The solution set is $\{2.21, 4.07\}$.

41. Since $2x = \sin^{-1}(2/9) + 2k\pi$ or $2x = \pi - \sin^{-1}(2/9) + 2k\pi$, we obtain

$$x = \frac{\sin^{-1}(2/9)}{2} + k\pi \text{ or}$$

$$x = \frac{\pi - \sin^{-1}(2/9)}{2} + k\pi.$$

By letting $k = 0, 1$, we find that the solution set is $\{0.11, 1.46, 3.25, 4.60\}$.

43. Note,

$$a = \cot^{-1}(\sqrt{2}) = \tan^{-1}(1/\sqrt{2}) \approx 0.62$$

and

$$\pi + a \approx 3.78.$$

The solution set is $\{0.62, 3.76\}$.

45. Solving for t, we find

$$\begin{aligned} \sin t &= \frac{x}{3} \\ t &= \sin^{-1}\left(\frac{x}{3}\right) \end{aligned}$$

47. Solving for y, we obtain

$$\begin{aligned} \sin(2y) &= \frac{a}{3} \\ 2y &= \sin^{-1}\left(\frac{a}{3}\right) \\ y &= \frac{1}{2}\sin^{-1}\left(\frac{a}{3}\right) \end{aligned}$$

49. Solving for h, we get

$$\begin{aligned} -2\cos h &= q-1 \\ \cos h &= \frac{q-1}{-2} \\ \cos h &= \frac{1-q}{2} \\ h &= \cos^{-1}\left(\frac{1-q}{2}\right) \end{aligned}$$

51. Solving for x, we find

$$\begin{aligned} 5\tan(\pi x) &= b+3 \\ \tan(\pi x) &= \frac{b+3}{5} \\ \pi x &= \tan^{-1}\left(\frac{b+3}{5}\right) \\ x &= \frac{1}{\pi}\tan^{-1}\left(\frac{b+3}{5}\right) \end{aligned}$$

53. Solving for y, we obtain

$$\begin{aligned} \cos^{-1}(y-2) &= \frac{x}{a} \\ y-2 &= \cos\left(\frac{x}{a}\right) \\ y &= \cos\left(\frac{x}{a}\right)+2 \end{aligned}$$

55. To find f^{-1}, interchange x and y, solve for y, and replace y by $f^{-1}(x)$:

$$\begin{aligned} x &= \sin y \\ \sin^{-1} x &= y \\ f^{-1}(x) &= \sin^{-1} x. \end{aligned}$$

Since the domain of f is

$$-\pi/2 \le x \le \pi/2$$

we obtain

$$-1 \le \sin x \le 1.$$

Then the range of f is $[-1, 1]$. Thus, the domain and range of f^{-1} are $[-1, 1]$ and $[-\pi/2, \pi/2]$, respectively.

57. To find f^{-1}, interchange x and y, solve for y, and replace y by $f^{-1}(x)$:

$$\begin{aligned} x &= \sin^{-1} y \\ \sin x &= y \\ f^{-1}(x) &= \sin x. \end{aligned}$$

Since the domain of f is

$$-1 \le x \le 1$$

we find

$$-\pi/2 \le \sin^{-1} x \le \pi/2.$$

Then the range of f is $[-\pi/2, \pi/2]$. Thus, the domain and range of f^{-1} are $[-\pi/2, \pi/2]$ and $[-1, 1]$, respectively.

59. Interchange x and y, solve for y, and replace y by $f^{-1}(x)$:

$$\begin{aligned} x &= \sin 3y \\ \sin^{-1} x &= 3y \\ \frac{1}{3}\sin^{-1} x &= y \\ f^{-1}(x) &= \frac{1}{3}\sin^{-1} x. \end{aligned}$$

Since the domain of f is

$$-\pi/6 \le x \le \pi/6$$

we obtain

$$\begin{aligned} -\pi/2 \le\ & 3x \le \pi/2 \\ -1 \le\ & \sin 3x \le 1. \end{aligned}$$

Then the range of f is $[-1,1]$. Thus, the domain and range of f^{-1} are $[-1,1]$ and $[-\pi/6, \pi/6]$, respectively.

61. Interchange x and y, solve for y, and replace y by $f^{-1}(x)$:

$$\begin{aligned} x &= 6\cos 4y \\ \frac{x}{6} &= \cos 4y \\ \cos^{-1}\left(\frac{x}{6}\right) &= 4y \\ \frac{1}{4}\cos^{-1}\left(\frac{x}{6}\right) &= y \\ f^{-1}(x) &= \frac{1}{4}\cos^{-1}\left(\frac{x}{6}\right). \end{aligned}$$

Since the domain of f is

$$0 \le x \le \pi/4$$

we obtain

$$\begin{aligned} 0 \le\ & 4x \le \pi \\ -1 \le\ & \cos 4x \le 1 \\ -6 \le\ & 6\cos 4x \le 6. \end{aligned}$$

Then the range of f is $[-6,6]$. Thus, the domain and range of f^{-1} are $[-6,6]$ and $[0,\pi/4]$, respectively.

63. Interchange x and y, solve for y, and replace y by $f^{-1}(x)$:

$$\begin{aligned} x &= 4+\tan\left(\frac{\pi y}{2}\right) \\ x-4 &= \tan\left(\frac{\pi y}{2}\right) \\ \tan^{-1}(x-4) &= \frac{\pi y}{2} \\ \frac{2}{\pi}\tan^{-1}(x-4) &= y \\ f^{-1}(x) &= \frac{2}{\pi}\tan^{-1}(x-4). \end{aligned}$$

Since the domain of f is

$$-1 < x < 1$$

we obtain

$$\begin{aligned} -\frac{\pi}{2} <\ & \frac{\pi x}{2} < \frac{\pi}{2} \\ -\infty <\ & \tan\left(\frac{\pi x}{2}\right) < \infty \\ -\infty <\ & 4+\tan\left(\frac{\pi x}{2}\right) < \infty. \end{aligned}$$

Then the range of f is $(-\infty,\infty)$. Thus, the domain and range of f^{-1} are $(-\infty,\infty)$ and $(-1,1)$, respectively.

65. $-\pi/6$

67. None, since cosine is negative in quadrants 2 and 3.

69. $-\pi/4$

71. $\pm\pi/4$

73. $-\pi/6$

75. Isolate $\cos 2x$ on one side.

$$\begin{aligned} 2\cos 2x &= -1 \\ \cos 2x &= -\frac{1}{2} \end{aligned}$$

$$2x = \frac{2\pi}{3} + 2k\pi \quad \text{or} \quad 2x = \frac{4\pi}{3} + 2k\pi$$

$$x = \frac{\pi}{3} + k\pi \quad \text{or} \quad x = \frac{2\pi}{3} + k\pi$$

The solution set is

$$\left\{x \mid x = \frac{\pi}{3} + k\pi \text{ or } x = \frac{2\pi}{3} + k\pi\right\}.$$

77. Set each factor to zero.

$$(\sqrt{3}\csc x - 2)(\csc x - 2) = 0$$

$$\csc x = \frac{2}{\sqrt{3}} \quad \text{or} \quad \csc x = 2$$

Thus, $x = \frac{\pi}{3}, \frac{2\pi}{3}, \frac{\pi}{6}, \frac{5\pi}{6}$ plus multiples of 2π.

The solution set is

$$\left\{x \mid x = \frac{\pi}{3} + 2k\pi, \frac{2\pi}{3} + 2k\pi, \frac{\pi}{6} + 2k\pi, \frac{5\pi}{6} + 2k\pi\right\}$$

79. Set the right-hand side to zero and factor.

$$\begin{aligned} 2\sin^2 x - 3\sin x + 1 &= 0 \\ (2\sin x - 1)(\sin x - 1) &= 0 \\ \sin x = \frac{1}{2} \quad &\text{or } \sin x = 1 \end{aligned}$$

The $x = \frac{\pi}{6}, \frac{5\pi}{6}, \frac{\pi}{2}$ plus multiples of 2π. The solution set is

$$\left\{x \mid x = \frac{\pi}{6} + 2k\pi, \frac{5\pi}{6} + 2k\pi, \frac{\pi}{2} + 2k\pi\right\}.$$

81. Isolate $\sin \frac{x}{2}$ on one side.

$$\begin{aligned} \sin\frac{x}{2} &= \frac{12}{8\sqrt{3}} \\ \sin\frac{x}{2} &= \frac{3}{2\sqrt{3}} \\ \sin\frac{x}{2} &= \frac{\sqrt{3}}{2} \end{aligned}$$

$$\frac{x}{2} = \frac{\pi}{3} + 2k\pi \quad \text{or} \quad \frac{x}{2} = \frac{2\pi}{3} + 2k\pi$$

$$x = \frac{2\pi}{3} + 4k\pi \quad \text{or} \quad x = \frac{4\pi}{3} + 4k\pi$$

The solution set is

$$\left\{x \mid x = \frac{2\pi}{3} + 4k\pi \text{ or } x = \frac{4\pi}{3} + 4k\pi\right\}.$$

83. By using the double-angle identity for sine, we get

$$\begin{aligned} \cos\frac{x}{2} - \sin\left(2\cdot\frac{x}{2}\right) &= 0 \\ \cos\frac{x}{2} - 2\sin\frac{x}{2}\cos\frac{x}{2} &= 0 \\ \cos\frac{x}{2}(1 - 2\sin\frac{x}{2}) &= 0 \end{aligned}$$

$$\cos\frac{x}{2} = 0 \quad \text{or} \quad \sin\frac{x}{2} = \frac{1}{2}.$$

Then $\frac{x}{2} = \frac{\pi}{2} + k\pi$, or $x = \frac{\pi}{6}, \frac{5\pi}{6}$ plus mutiples of 2π. Thus, $x = \pi + 2k\pi$ or $x = \frac{\pi}{3}, \frac{5\pi}{3}$ plus multiples of 4π. The solution set is

$$\left\{x \mid x = \pi + 2k\pi, \frac{\pi}{3} + 4k\pi, \frac{5\pi}{3} + 4k\pi\right\}.$$

85. By the double-angle identity for cosine, we find

$$\begin{aligned} \cos 2x + \sin^2 x &= 0 \\ \cos^2 x - \sin^2 x + \sin^2 x &= 0 \\ \cos^2 x &= 0 \\ x &= \frac{\pi}{2} + k\pi. \end{aligned}$$

The solution set is $\left\{x \mid x = \frac{\pi}{2} + k\pi\right\}$.

87. By factoring, we obtain

$$\begin{aligned} \sin x(\cos x + 1) + (\cos x + 1) &= 0 \\ (\sin x + 1)(\cos x + 1) &= 0. \end{aligned}$$

Then

$$\sin x = -1 \quad \text{or} \quad \cos x = -1$$

$$x = \frac{3\pi}{2} + 2k\pi \quad \text{or} \quad x = \pi + 2k\pi.$$

The solution set is

$$\left\{x \mid x = \frac{3\pi}{2} + 2k\pi \text{ or } x = \pi + 2k\pi\right\}.$$

89. By multiplying the equation by 2, we obtain

$$\begin{aligned} 2\sin\alpha\cos\alpha &= 1 \\ \sin 2\alpha &= 1 \\ 2\alpha &= 90^\circ + k360^\circ \\ \alpha &= 45^\circ + k180^\circ. \end{aligned}$$

If $k = 0, 1$, then the solution set is

$$\{45^\circ, 225^\circ\}.$$

91. Suppose $1 + \cos\alpha \neq 0$. Dividing the equation by $1 + \cos\alpha$, we get

$$\begin{aligned} \frac{\sin\alpha}{1 + \cos\alpha} &= 1 \\ \tan\frac{\alpha}{2} &= 1 \\ \frac{\alpha}{2} &= 45^\circ + k180^\circ \\ \alpha &= 90^\circ + k360^\circ. \end{aligned}$$

One solution is 90°. On the other hand if $1 + \cos\alpha = 0$, then $\cos\alpha = -1$ and $\alpha = 180^\circ$. Note $\alpha = 180^\circ$ satisfies the given equation. The solution set is $\{90^\circ, 180^\circ\}$.

93. No solution since the left-hand side is equal to 1 by an identity. The solution set is $\emptyset$.

95. Isolate $\sin 2\alpha$ on one side.

$$\begin{aligned} \sin^4 2\alpha &= \frac{1}{4} \\ \sin 2\alpha &= \pm\sqrt[4]{\frac{1}{4}} \\ \sin 2\alpha &= \pm\frac{1}{\sqrt{2}} \\ 2\alpha &= 45^\circ + k90^\circ \\ \alpha &= 22.5^\circ + k45^\circ \end{aligned}$$

By choosing $k = 0, 1, ..., 7$, one gets the solution set $\{22.5^\circ, 67.5^\circ, 112.5^\circ, 157.5^\circ, 202.5^\circ, 247.5^\circ, 292.5^\circ, 337.5^\circ\}$.

97. Suppose $\tan\alpha \neq 0$. Divide the equation by $\tan\alpha$.

$$\begin{aligned} \frac{2\tan\alpha}{1-\tan^2\alpha} &= \tan\alpha \\ \frac{2}{1-\tan^2\alpha} &= 1 \\ 2 &= 1-\tan^2\alpha \\ \tan^2\alpha &= -1 \end{aligned}$$

The last equation is inconsistent since $\tan^2\alpha$ is nonnegative. But if $\tan\alpha = 0$, then $\alpha = 0^\circ, 180^\circ$ and these two values of α satisfy the given equation.
The solution set is $\{0^\circ, 180^\circ\}$.

99. Using the sum identity for sine, we obtain

$$\begin{aligned} \sin(2\alpha+\alpha) &= \cos 3\alpha \\ \sin 3\alpha &= \cos 3\alpha \\ \tan 3\alpha &= 1 \\ 3\alpha &= 45^\circ + k180^\circ \\ \alpha &= 15^\circ + k60^\circ. \end{aligned}$$

By choosing $k = 0, 1, ..., 5$, one gets the solution set $\{15^\circ, 75^\circ, 135^\circ, 195^\circ, 255^\circ, 315^\circ\}$.

101. If $\sin x = \cos x$, then $\tan x = 1$ and

$x = \dfrac{\pi}{4} + k\pi$ where k is an integer.

The points of intersection are

$$\left(\frac{\pi}{4} + 2k\pi, \frac{\sqrt{2}}{2}\right)$$

and

$$\left(\frac{5\pi}{4} + 2k\pi, -\frac{\sqrt{2}}{2}\right).$$

103. Let $a = 0.6$, $b = 0.4$, and

$$r = \sqrt{0.6^2 + 0.4^2} \approx 0.72.$$

If the terminal side of α goes through $(0.6, 0.4)$, then

$$\tan\alpha = 0.4/0.6$$

and we can choose

$$\alpha = \tan^{-1}(2/3) \approx 0.588.$$

Thus,

$$x = 0.72\sin(2t + 0.588).$$

The values of t when $x = 0$ are given by

$$\begin{aligned} \sin(2t+0.588) &= 0 \\ 2t + 0.588 &= k\pi \\ 2t &= -0.588 + k\pi \\ t &= -0.294 + \frac{k\pi}{2} \end{aligned}$$

When $k = 1, 2$, we get

$$t \approx 1.28 \text{ sec}, 2.85 \text{ sec}.$$

Chapter 4 Test

1. $-\pi/6$ **2.** $2\pi/3$

3. $-\pi/4$ **4.** $\pi/6$

5. $-\pi/4$ **6.** $\sqrt{1-\left(-\frac{1}{3}\right)^2} = \dfrac{2\sqrt{2}}{3}$

7. Since $-\sin\theta = 1$, we get $\sin\theta = -1$ and the solution set is $\left\{\theta \mid \theta = \dfrac{3\pi}{2} + 2k\pi\right\}$.

8. Since $\cos 3s = \frac{1}{2}$, we obtain

$$3s = \frac{\pi}{3} + 2k\pi \quad \text{or} \quad 3s = \frac{5\pi}{3} + 2k\pi$$
$$s = \frac{\pi}{9} + \frac{2k\pi}{3} \quad \text{or} \quad s = \frac{5\pi}{9} + \frac{2k\pi}{3}.$$

The solution set is

$$\left\{ s \mid s = \frac{\pi}{9} + \frac{2k\pi}{3} \text{ or } s = \frac{5\pi}{9} + \frac{2k\pi}{3} \right\}.$$

9. Since $\tan 2t = -\sqrt{3}$, we have

$$\begin{aligned} 2t &= \frac{2\pi}{3} + k\pi \\ t &= \frac{\pi}{3} + \frac{k\pi}{2}. \end{aligned}$$

The solution set is $\left\{ t \mid t = \frac{\pi}{3} + \frac{k\pi}{2} \right\}$.

10.

$$\begin{aligned} 2\sin\theta\cos\theta &= \cos\theta \\ \cos\theta(2\sin\theta - 1) &= 0 \\ \cos\theta = 0 \quad &\text{or} \quad \sin\theta = 1/2 \end{aligned}$$

The solution set is

$$\left\{ \theta \mid \theta = \frac{\pi}{2} + k\pi, \frac{\pi}{6} + 2k\pi, \frac{5\pi}{6} + 2k\pi \right\}.$$

11. Since $\csc\alpha = \frac{8}{4}$, we get $\sin\alpha = \frac{1}{2}$.

The solution set is $\{30°, 150°\}$.

12. Since $\cot(\alpha/2) = -1$, we get $\frac{\alpha}{2} = 135° + k \cdot 180°$ or $\alpha = 270° + k \cdot 360°$ The solution set is $\{270°\}$.

13. Since $\sec\alpha = \frac{2}{\sqrt{3}}$, we obtain $\cos\alpha = \frac{\sqrt{3}}{2}$.

The solution set is $\{30°, 330°\}$.

14. Since $\sin\alpha = 0$ or $\cos\alpha = 0$, the solution set is $\{0°, 90°, 180°, 270°\}$.

15. Since $\sin(2\alpha) = \pm 1$, we find $2\alpha = 90° + k \cdot 180°$ or $\alpha = 45° + k \cdot 90°$. Thus, the solution set is $\{45°, 135°, 225°, 315°\}$.

16. By factoring, we obtain

$$\begin{aligned} (3\sin\alpha - 1)(\sin\alpha - 1) &= 0 \\ \sin\alpha = 1/3 \quad &\text{or} \quad \sin\alpha = 1 \\ \alpha = \sin^{-1}(1/3) \approx 19.5° \quad &\text{or} \quad \alpha = 90°. \end{aligned}$$

Another solution is $\alpha = 180° - 19.5° = 160.5°$. The solution set is $\{19.5°, 90°, 160.5°\}$.

17.

$$\begin{aligned} \tan(2\alpha - 7\alpha) &= 1 \\ \tan(-5\alpha) &= 1 \\ -\tan 5\alpha &= 1 \\ \tan 5\alpha &= -1 \\ 5\alpha &= 135° + k180° \\ \alpha &= 27° + k36° \end{aligned}$$

The solution set is

$$\{27°, 63°, 99°, 171°, 207°, 243°, 279°, 351°\}.$$

Note, 135° and 315° are not solutions.

18. Solving for t, we get

$$\begin{aligned} \sin 2t &= \frac{a}{5} \\ 2t &= \sin^{-1}\left(\frac{a}{5}\right) \\ t &= \frac{1}{2}\sin^{-1}\left(\frac{a}{5}\right) \end{aligned}$$

19. Interchange x and y, solve for y, and replace y by $f^{-1}(x)$:

$$\begin{aligned} x &= \frac{1}{5}\cos\left(\frac{y}{4}\right) + \frac{1}{5} \\ 5x &= \cos\left(\frac{y}{4}\right) + 1 \\ 5x - 1 &= \cos\left(\frac{y}{4}\right) \\ \cos^{-1}(5x - 1) &= \frac{y}{4} \\ 4\cos^{-1}(5x - 1) &= y \\ f^{-1}(x) &= 4\cos^{-1}(5x - 1). \end{aligned}$$

Since the domain of f is

$$0 \le x \le 4\pi$$

we obtain

$$\begin{aligned} 0 \le & \frac{x}{4} & \le \pi \\ -1 \le & \cos\left(\frac{x}{4}\right) & \le 1 \\ -\frac{1}{5} \le & \frac{1}{5}\cos\left(\frac{x}{4}\right) & \le \frac{1}{5} \\ 0 \le & \frac{1}{5}\cos\left(\frac{x}{4}\right)+\frac{1}{5} & \le \frac{2}{5}. \end{aligned}$$

Then the range of f is $[0, 2/5]$. Thus, the domain and range of f^{-1} are $[0, 2/5]$ and $[0, 4\pi]$, respectively.

20. If $\cos x = \dfrac{1}{\cos x}$, then $\cos^2 x = 1$.
Thus, $x = k\pi$ where k is an integer. The points of intersection are $(2k\pi, 1)$ and $(\pi + 2k\pi, -1)$.

21. Let $a = 2$, $b = -4$, and

$$r = \sqrt{2^2 + (-4)^2} = \sqrt{20}.$$

If the terminal side of α goes through $(2, -4)$, then one can choose

$$\alpha = \tan^{-1}(-4/2) \approx -1.107.$$

Then

$$d = \sqrt{20}\sin(3t - 1.107).$$

The values of t when $d = 0$ are given by

$$\begin{aligned} \sin(3t - 1.107) &= 0 \\ 3t - 1.107 &= k\pi \\ 3t &= 1.107 + k\pi \\ t &\approx 0.4 + \frac{k\pi}{3}. \end{aligned}$$

By choosing $k = 0, 1, 2, 3$, we obtain the values of t in $[0, 4]$, namely, 0.4 sec, 1.4 sec, 2.5 sec, and 3.5 sec.

Tying It All Together

1. $\sqrt{2}/2$ **2.** $-1/2$

3. 1 **4.** $1/2$

5. $-\pi/6$ **6.** $2\pi/3$

7. $-\pi/4$ **8.** 0

9. $\sin(5\pi/3) = -\sqrt{3}/2$ **10.** $\cos(11\pi/6) = \sqrt{3}/2$

11. 1 **12.** 1

13. Period 2π, amplitude 1, phase shift $\pi/6$, and since the solutions of $\sin(x - \pi/6) = 0$ are $x = \pi/6 + k\pi$ it follows that the x-intercepts are $(\pi/6 + k\pi, 0)$ where k is an integer.

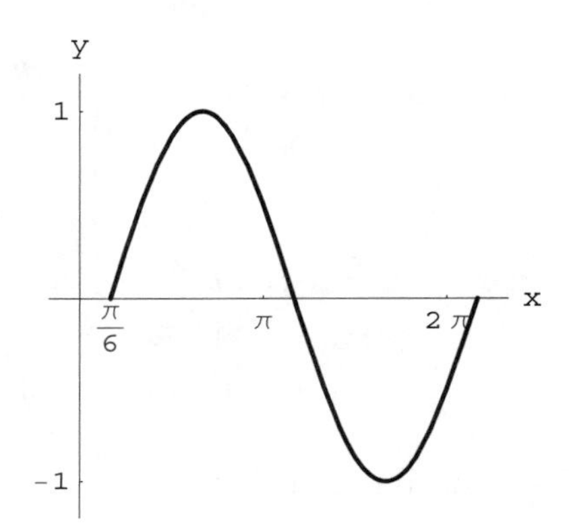

14. Period π, amplitude 1, phase shift 0, and since the solutions of $\sin(2x) = 0$ are $x = \dfrac{k\pi}{2}$ it follows that the x-intercepts are

$$\left(\frac{k\pi}{2}, 0\right)$$

where k is an integer.

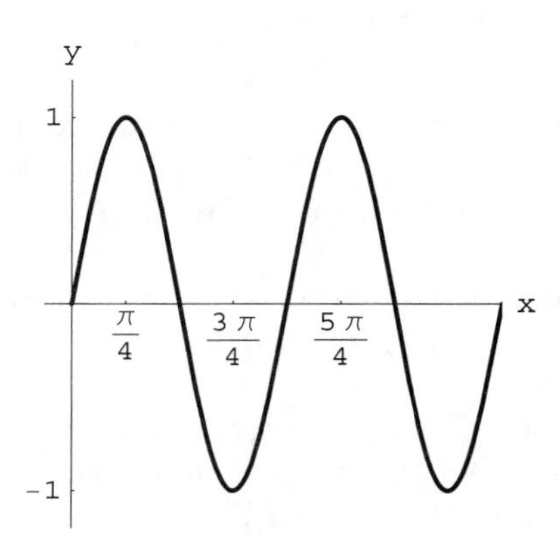

15. Period $2\pi/3$, amplitude 1, phase shift 0, and since the solutions of $\cos(3x) = 0$ are $x = \frac{\pi}{6} + \frac{k\pi}{3}$ it follows that the x-intercepts are

$$\left(\frac{\pi}{6} + \frac{k\pi}{3}, 0\right)$$

where k is an integer.

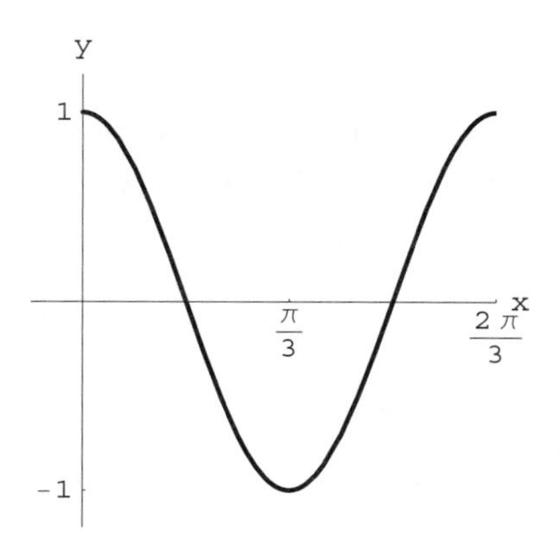

16. Note,

$$y = \cos\left(2\left(x - \frac{\pi}{2}\right)\right).$$

The period is π, amplitude is 1, and the phase shift is $\pi/2$. Since the solutions of

$$\cos(2x - \pi) = 0$$

are $x = \frac{\pi}{4} + \frac{k\pi}{2}$, the x-intercepts are

$$\left(\frac{\pi}{4} + \frac{k\pi}{2}, 0\right) \quad \text{or} \quad \left(\frac{3\pi}{4} + \frac{k\pi}{2}, 0\right)$$

where k is an integer.

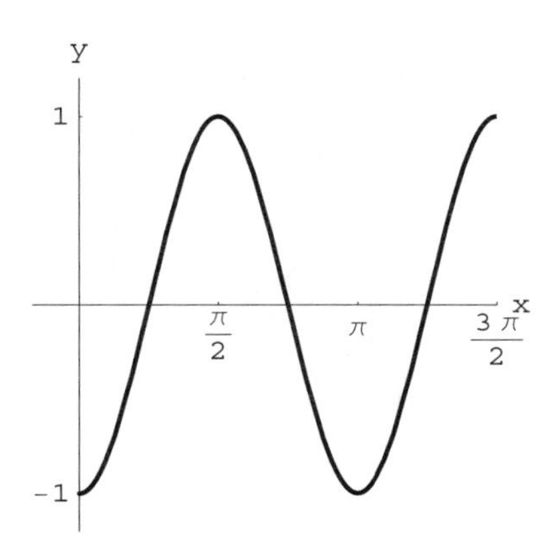

17. Note,

$$y = 3\sin\left(\frac{1}{2}(x - \pi)\right).$$

The period is 4π, amplitude is 3, and the phase shift is π. Since the solutions of

$$\sin\left(\frac{x}{2} - \frac{\pi}{2}\right) = 0$$

are $x = \pi + 2k\pi$, the x-intercepts are

$$(\pi + 2k\pi, 0)$$

where k is an integer.

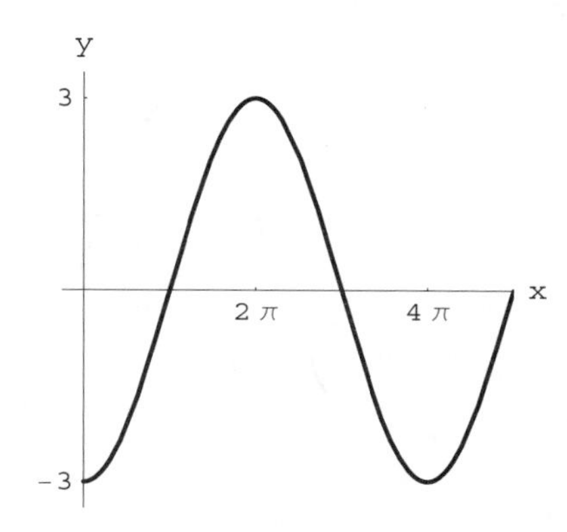

18. Period 2π, amplitude 4, phase shift $-\pi/3$, and since the solutions of

$$\cos(x + \pi/3) = 0$$

are $x = \pi/6 + k\pi$, the x-intercepts are

$$(\pi/6 + k\pi, 0)$$

where k is an integer.

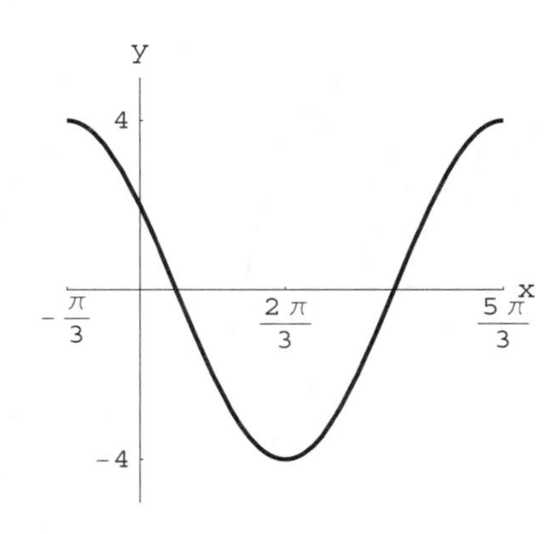

19. Note,

$$y = -4\cos\left(\frac{\pi}{3}x\right).$$

Then we obtain the following: period $\frac{2\pi}{\pi/3}$ or 6, amplitude 4, phase shift 0. Since the solutions of

$$\cos(\pi x/3) = 0$$

are $x = \frac{3}{2} + 3k$, the x-intercepts are

$$\left(\frac{3}{2} + 3k, 0\right)$$

where k is an integer.

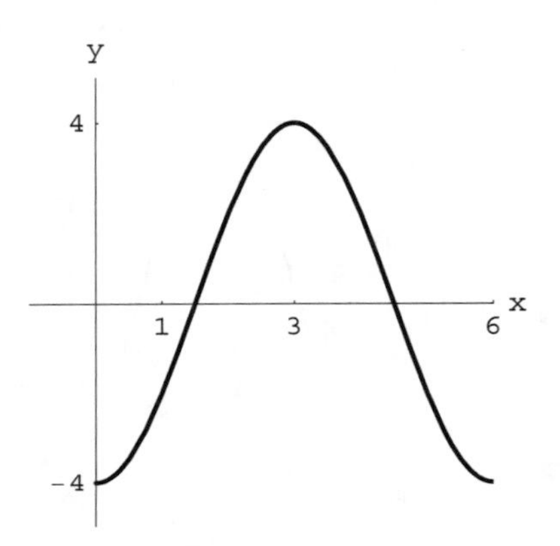

20. Note,

$$y = -2\sin(\pi(x-1)).$$

The period is 2, amplitude is 2, and the phase shift is 1. Since the solutions of

$$\sin(\pi x - \pi) = 0$$

are $x = k$, the x-intercepts are

$$(k, 0)$$

where k is an integer.

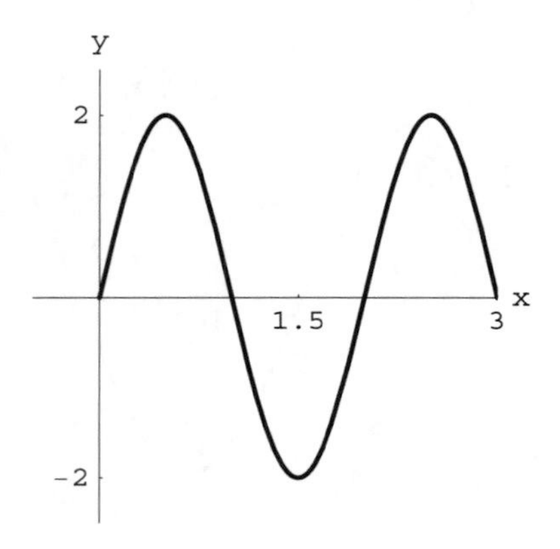

21. Since α is in Quadrant III, we get

$$\cos\alpha = -\sqrt{1 - \left(-\frac{1}{3}\right)^2} = -\frac{2\sqrt{2}}{3},$$

$$\csc\alpha = -3,\ \sec\alpha = -\frac{3}{2\sqrt{2}} = -\frac{3\sqrt{2}}{4},$$

$$\tan\alpha = \frac{-1/3}{-2\sqrt{2}/3} = \frac{\sqrt{2}}{4},$$

and $\cot\alpha = \frac{4}{\sqrt{2}} = 2\sqrt{2}$.

22. Since α is in Quadrant II, we get

$$\sin\alpha = \sqrt{1 - \left(-\frac{1}{4}\right)^2} = \frac{\sqrt{15}}{4},$$

$$\sec\alpha = -4,\ \csc\alpha = \frac{4}{\sqrt{15}} = \frac{4\sqrt{15}}{15},$$

$$\tan\alpha = \frac{\sqrt{15}/4}{-1/4} = -\sqrt{15},$$

and $\cot\alpha = -\frac{1}{\sqrt{15}} = -\frac{\sqrt{15}}{15}$.

23. Since α is in Quadrant I, we get

$$\sec\alpha = \sqrt{1 + \left(\frac{3}{5}\right)^2} = \frac{\sqrt{34}}{5},$$

$$\cot\alpha = \frac{5}{3},\ \cos\alpha = \frac{5}{\sqrt{34}} = \frac{5\sqrt{34}}{34},$$

$$\csc\alpha = \sqrt{1 + \left(\frac{5}{3}\right)^2} = \frac{\sqrt{34}}{3},$$

and $\sin\alpha = \frac{3}{\sqrt{34}} = \frac{3\sqrt{34}}{34}$.

24. Since α is in Quadrant IV, we get

$$\cos\alpha = \sqrt{1 - \left(-\frac{4}{5}\right)^2} = \frac{3}{5},$$

$$\csc\alpha = -\frac{5}{4},\ \sec\alpha = \frac{5}{3},$$

$\tan\alpha = \frac{-4/5}{3/5} = -\frac{4}{3}$, and

$$\cot\alpha = -\frac{3}{4}.$$

For Thought

1. True, the sum of the measurements of the three angles is 180°.

2. False, since similar triangles have the same corresponding angles but their corresponding sides are not necessarily equal.

3. True, since three angles do not uniquely determine a triangle.

4. False, $a \sin 17^\circ = 88 \sin 9^\circ$ and $a = \dfrac{88 \sin 9^\circ}{\sin 17^\circ}$.

5. False, since $\alpha = \sin^{-1}\left(\dfrac{5 \sin 44^\circ}{18}\right) \approx 11^\circ$ and $\alpha = 180 - 11^\circ = 169^\circ$.

6. True, since $\sin \beta = \dfrac{2.3 \sin 39^\circ}{1.6}$.

7. True, since $\dfrac{\sin 60^\circ}{\sqrt{3}} = \dfrac{\sqrt{3}/2}{\sqrt{3}} = \dfrac{1}{2}$ and $\dfrac{\sin 30^\circ}{1} = \sin 30^\circ = \dfrac{1}{2}$.

8. False, a triangle exists since $a = 500$ is bigger than $h = 10 \sin 60^\circ \approx 8.7$.

9. True, since the triangle that exists is a right triangle.

10. False, there exists only one triangle and it is an obtuse triangle.

5.1 Exercises

1. Note $\gamma = 180^\circ - (64^\circ + 72^\circ) = 44^\circ$.

By the sine law $\dfrac{b}{\sin 72^\circ} = \dfrac{13.6}{\sin 64^\circ}$ and $\dfrac{c}{\sin 44^\circ} = \dfrac{13.6}{\sin 64^\circ}$. Then

$$b = \frac{13.6}{\sin 64^\circ} \cdot \sin 72^\circ \approx 14.4$$

and

$$c = \frac{13.6}{\sin 64^\circ} \cdot \sin 44^\circ \approx 10.5.$$

3. Note $\beta = 180^\circ - (12.2^\circ + 33.6^\circ) = 134.2^\circ$.

By the sine law $\dfrac{a}{\sin 12.2^\circ} = \dfrac{17.6}{\sin 134.2^\circ}$ and $\dfrac{c}{\sin 33.6^\circ} = \dfrac{17.6}{\sin 134.2^\circ}$. Then

$$a = \frac{17.6}{\sin 134.2^\circ} \cdot \sin 12.2^\circ \approx 5.2$$

and

$$c = \frac{17.6}{\sin 134.2^\circ} \cdot \sin 33.6^\circ \approx 13.6.$$

5. Note $\beta = 180^\circ - (10.3^\circ + 143.7^\circ) = 26^\circ$.

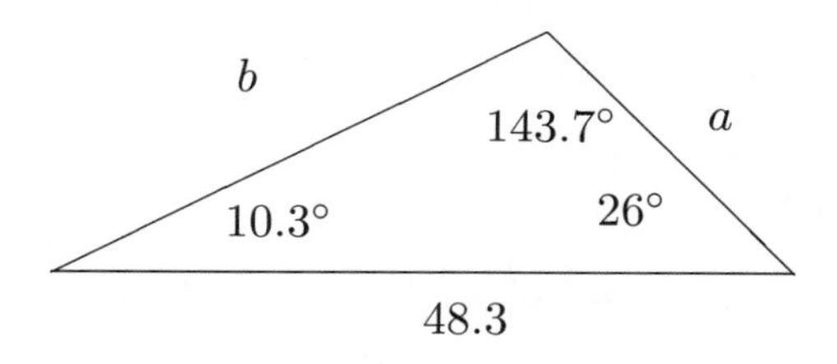

Since

$$\frac{a}{\sin 10.3^\circ} = \frac{48.3}{\sin 143.7^\circ}$$

and

$$\frac{b}{\sin 26^\circ} = \frac{48.3}{\sin 143.7^\circ}$$

we have

$$a = \frac{48.3}{\sin 143.7^\circ} \cdot \sin 10.3^\circ \approx 14.6$$

and

$$b = \frac{48.3}{\sin 143.7^\circ} \cdot \sin 26^\circ \approx 35.8$$

7. Note $\alpha = 180^\circ - (120.7^\circ + 13.6^\circ) = 45.7^\circ$.

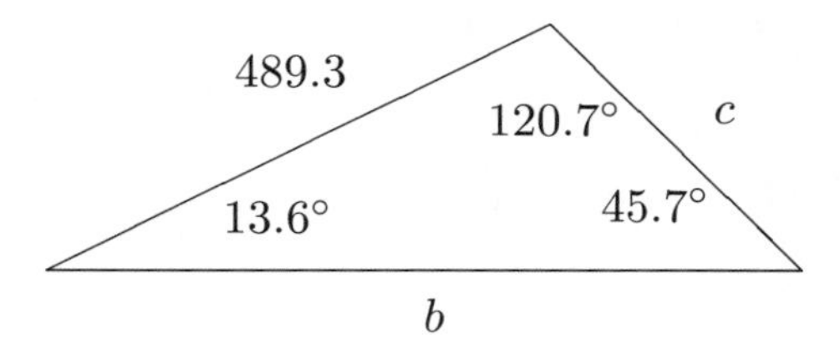

Since

$$\frac{c}{\sin 13.6^\circ} = \frac{489.3}{\sin 45.7^\circ}$$

and

$$\frac{b}{\sin 120.7^\circ} = \frac{489.3}{\sin 45.7^\circ}$$

we have

$c = \dfrac{489.3}{\sin 45.7^\circ} \cdot \sin 13.6^\circ \approx 160.8$ and

$b = \dfrac{489.3}{\sin 45.7^\circ} \cdot \sin 120.7^\circ \approx 587.9$

9. Draw angle $\alpha = 39.6°$ and let h be the height.

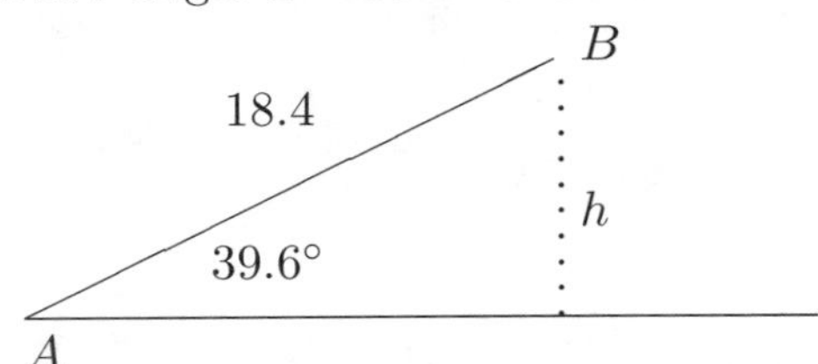

Since $\sin 39.6° = \dfrac{h}{18.4°}$, we have

$$h = 18.4 \sin 39.6° \approx 11.7.$$

There is no triangle since $a = 3.7$ is smaller than $h \approx 11.7$.

11. Draw angle $\gamma = 60°$ and let h be the height.

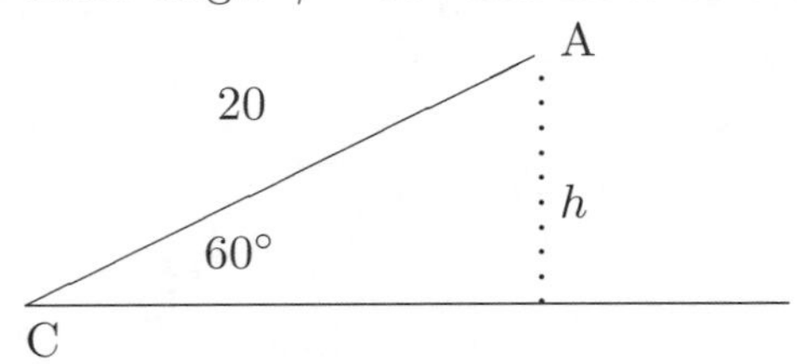

Since

$$h = 20 \sin 60° = 10\sqrt{3}$$

and $c = h$, there is exactly one triangle and it is a right triangle. Then $\beta = 90°$ and $\alpha = 30°$. By the Pythagorean Theorem,

$$a = \sqrt{20^2 - (10\sqrt{3})^2} = \sqrt{400 - 300} = 10.$$

13. Draw angle $\beta = 138.1°$.

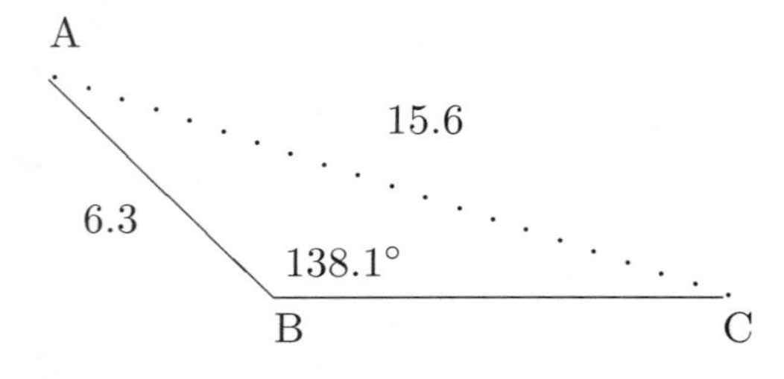

Apply the sine law.

$$\begin{aligned} \frac{15.6}{\sin 138.1°} &= \frac{6.3}{\sin \gamma} \\ \sin \gamma &= \frac{6.3 \sin 138.1°}{15.6} \\ \sin \gamma &\approx 0.2697 \\ \gamma = \sin^{-1}(0.2697) &\approx 15.6° \end{aligned}$$

So $\alpha = 180° - (15.6° + 138.1°) = 26.3°$.

Using the sine law, we obtain

$$a = \frac{15.6}{\sin 138.1°} \sin 26.3° \approx 10.3.$$

15. Draw angle $\beta = 32.7°$ and let h be the height.

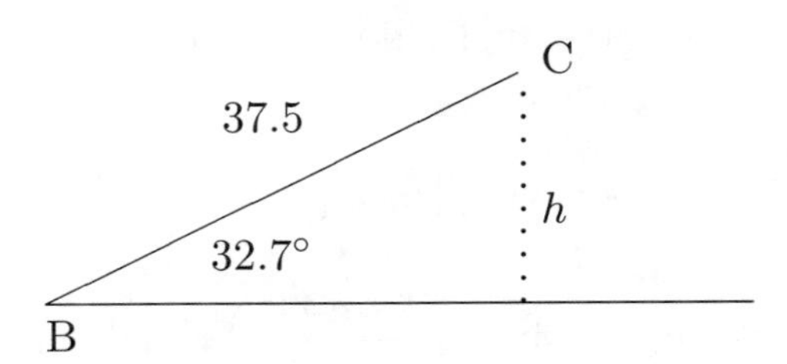

Since $h = 37.5 \sin 32.7° \approx 20.3$ and $20.3 < b < 37.5$, there are two triangles and they are given by

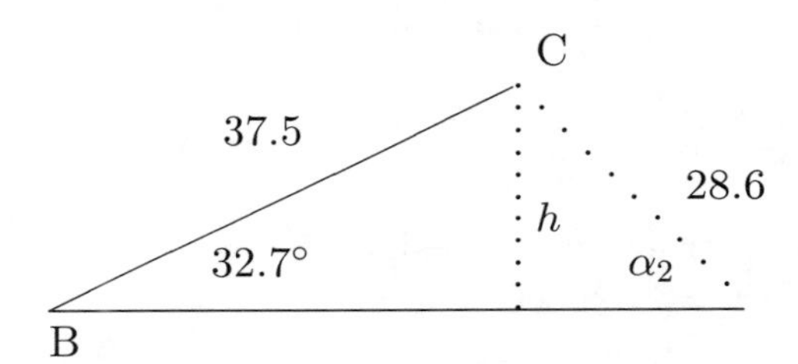

and

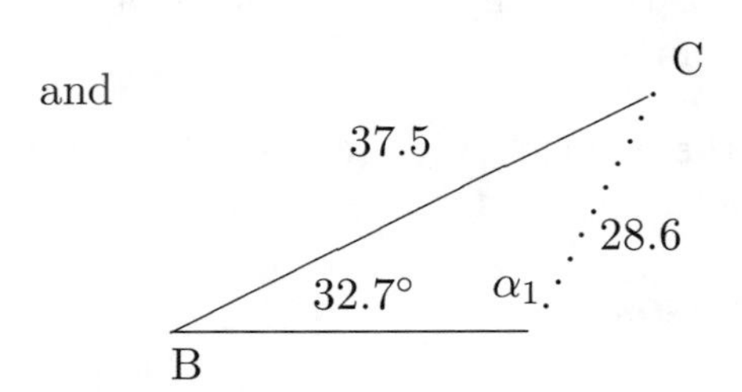

Apply the sine law to the acute triangle.

$$\begin{aligned} \frac{28.6}{\sin 32.7°} &= \frac{37.5}{\sin \alpha_2} \\ \sin \alpha_2 &= \frac{37.5 \sin 32.7°}{28.6} \\ \sin \alpha_2 &\approx 0.708 \\ \alpha_2 = \sin^{-1}(0.708) &\approx 45.1° \end{aligned}$$

So $\gamma_2 = 180° - (45.1° + 32.7°) = 102.2°$.

By the sine law,

$$c_2 = \frac{28.6}{\sin 32.7°} \sin 102.2° \approx 51.7.$$

On the obtuse triangle, we find
$\alpha_1 = 180° - \alpha_2 = 134.9°$ and
$\gamma_1 = 180° - (134.9° + 32.7°) = 12.4°$.

By the sine law,

$$c_1 = \frac{28.6}{\sin 32.7°} \sin 12.4° \approx 11.4.$$

17. Draw angle $\gamma = 99.6°$. Note, there is exactly one triangle since $12.4 > 10.3$.

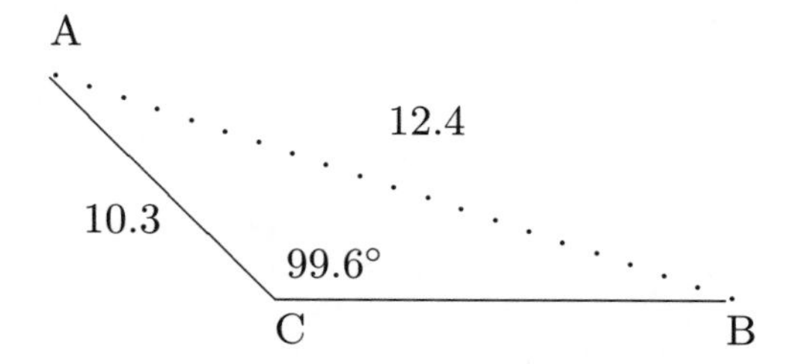

By the sine law, we obtain

$$\begin{aligned} \frac{12.4}{\sin 99.6°} &= \frac{10.3}{\sin \beta} \\ \sin \beta &= \frac{10.3 \sin 99.6°}{12.4} \\ \sin \beta &\approx 0.819 \\ \beta = \sin^{-1}(0.819) &\approx 55.0°. \end{aligned}$$

So $\alpha = 180° - (55.0° + 99.6°) = 25.4°$.

Using the sine law, we find

$$a = \frac{12.4}{\sin 99.6°} \sin 25.4° \approx 5.4.$$

19. Let x be the number of miles flown along I-20.

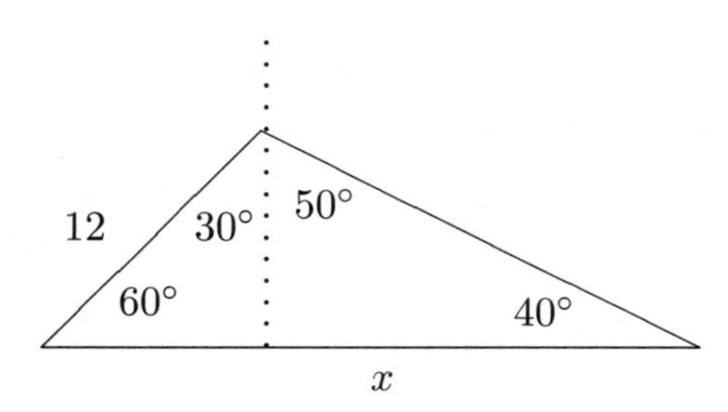

Applying the sine law, we obtain

$$\frac{x}{\sin 80°} = \frac{12}{\sin 40°}.$$

Then $x \approx 18.4$ miles.

21. Let x and y be the lengths of the missing sides.

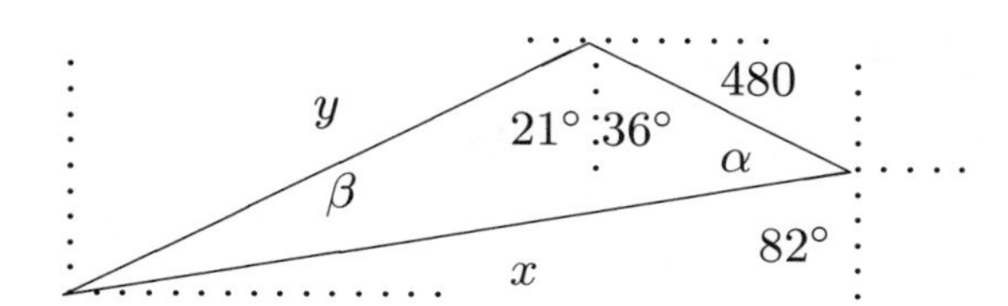

There is a $21°$ angle because of the $S21°W$ direction. There are $36°$ and $82°$ angles because opposite angles are equal and because of the directions $N36°W$ and $N82°E$.
Note $\alpha = 180° - (82° + 36°) = 62°$ and
$\beta = 180 - (21° + 36° + 62°) = 61°$.

By the sine law, we find

$$x = \frac{480}{\sin 61°} \sin 57° \approx 460.27$$

and

$$y = \frac{480}{\sin 61°} \sin 62° \approx 484.57.$$

The perimeter is $x + y + 480 \approx 1425$ ft.

23. Applying the sine law, we find

$$x = \frac{19.2 \sin 82°}{\sin 30°} \approx 38.0 \text{ ft.}$$

25. Let h be the height of the tower.

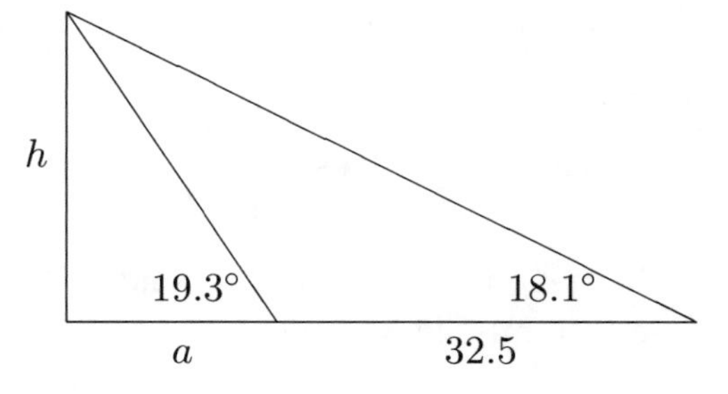

Using right triangle trigonometry, we get

$$\tan 19.3° = \frac{h}{a}$$

or

$$a = \frac{h}{\tan 19.3°}.$$

Similarly, we have

$$\tan 18.1^\circ = \frac{h}{a+32.5}.$$

Then

$$\begin{aligned} \tan 18.1^\circ(a+32.5) &= h \\ a\tan 18.1^\circ + 32.5\tan 18.1^\circ &= h \\ \frac{h}{\tan 19.3^\circ}\cdot \tan 18.1^\circ + 32.5\tan 18.1^\circ &= h \\ h\cdot\frac{\tan 18.1^\circ}{\tan 19.3^\circ} + 32.5\tan 18.1^\circ &= h. \end{aligned}$$

Solving for h, we find that the height of the tower is

$$h \approx 159.4 \text{ ft}.$$

27. Note, $\tan\gamma = 6/12$ and $\gamma = \tan^{-1}(0.5) \approx 26.565^\circ$. Also, $\tan\alpha = 3/12$ and $\alpha = \tan^{-1}(0.25) \approx 14.036^\circ$.

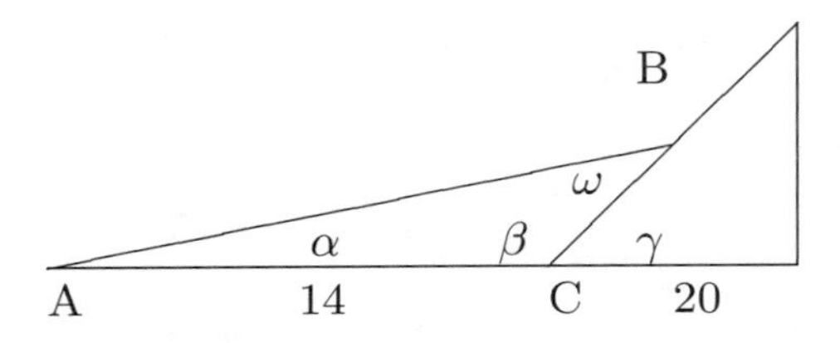

The remaining angles are $\beta = 153.435^\circ$ and $\omega = 12.529^\circ$. By the sine law, we obtain

$$\frac{AB}{\sin 153.435^\circ} = \frac{14}{\sin 12.529^\circ}$$

and

$$\frac{BC}{\sin 14.036^\circ} = \frac{14}{\sin 12.529^\circ}.$$

Then $AB \approx 28.9$ ft and $BC \approx 15.7$ ft.

29. By the sine law, we get

$$x = \frac{24\sin 47^\circ}{\sin 104^\circ} \approx 18.1 \text{ in}.$$

31. Let t be the number of seconds since the cruise missile was spotted.

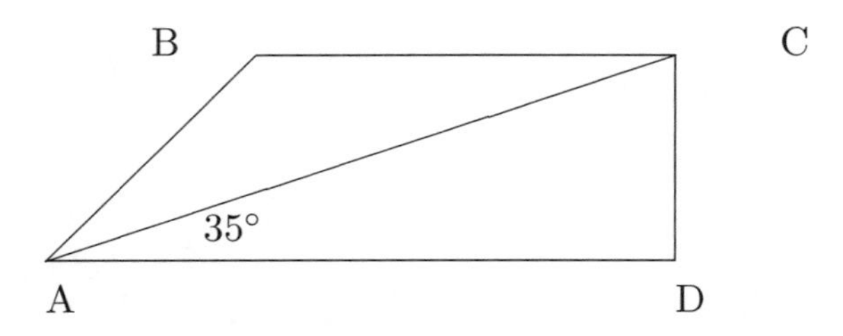

Let β be the angle at B. The angle formed by BAC is $180^\circ - 35^\circ - \beta$. After t seconds, the cruise missile would have traveled $548\frac{t}{3600}$ miles and the projectile $688\frac{t}{3600}$ miles. Using the law of sines, we have

$$\begin{aligned} \frac{\frac{548t}{3600}}{\sin(145^\circ-\beta)} &= \frac{\frac{688t}{3600}}{\sin 35^\circ} \\ \frac{548}{\sin(145^\circ-\beta)} &= \frac{688}{\sin 35^\circ} \\ \beta &= 145^\circ - \sin^{-1}\left(\frac{548\sin 35^\circ}{688}\right) \\ \beta &\approx 117.8^\circ. \end{aligned}$$

Then angle BAC is 27.2°. The angle of elevation of the projectile must be angle DAB which is 62.2° $(= 35^\circ + 27.2^\circ)$.

For Thought

1. True, since $\cos 90^\circ = 0$ in the law of cosines.

2. False, $a = \sqrt{c^2+b^2-2bc\cos\alpha}$.

3. False, $c^2 = a^2+b^2-2ab\cos\gamma$.

4. True, this follows from the sine law.

5. False, it has only one solution in $[0^\circ, 180^\circ]$.

6. True, since the sum of the angles is 180°.

7. True, since $\beta = \sin^{-1}(0.1235)$ or $\beta = 180^\circ - \sin^{-1}(0.1235)$.

8. True, since the law of cosines will be used and cosine is a one-to-one function in $[0^\circ, 180^\circ]$.

9. True, since

$$\cos\gamma = \frac{3.4^2+4.2^2-8.1^2}{2(3.4)(4.2)} \approx -1.27$$

has no real solution γ.

10. False, there is exactly one triangle.

5.2 Exercises

1. By the cosine law, we obtain

$c = \sqrt{3.1^2 + 2.9^2 - 2(3.1)(2.9)\cos 121.3^\circ}$

$\approx 5.23 \approx 5.2$. By the sine law, we find

$$\begin{aligned} \frac{3.1}{\sin\alpha} &= \frac{5.23}{\sin 121.3^\circ} \\ \sin\alpha &= \frac{3.1\sin 121.3^\circ}{5.23} \\ \sin\alpha &\approx 0.50647 \\ \alpha \approx \sin^{-1}(0.50647) &\approx 30.4^\circ. \end{aligned}$$

Then $\beta = 180^\circ - (30.4^\circ + 121.3^\circ) = 28.3^\circ$.

3. By the cosine law, we find

$$\cos\beta = \frac{6.1^2 + 5.2^2 - 10.3^2}{2(6.1)(5.2)} \approx -0.6595$$

and so

$$\beta \approx \cos^{-1}(-0.6595) \approx 131.3^\circ.$$

By the sine law,

$$\begin{aligned} \frac{6.1}{\sin\alpha} &= \frac{10.3}{\sin 131.3^\circ} \\ \sin\alpha &= \frac{6.1\sin 131.3^\circ}{10.3} \\ \sin\alpha &\approx 0.4449 \\ \alpha \approx \sin^{-1}(0.4449) &\approx 26.4^\circ. \end{aligned}$$

So $\gamma = 180^\circ - (26.4^\circ + 131.3^\circ) = 22.3^\circ$.

5. By the cosine law,

$b = \sqrt{2.4^2 + 6.8^2 - 2(2.4)(6.8)\cos 10.5^\circ}$

$\approx 4.46167 \approx 4.5$ and

$$\cos\alpha = \frac{2.4^2 + 4.46167^2 - 6.8^2}{2(2.4)(4.46167)} \approx -0.96066.$$

So $\alpha = \cos^{-1}(-0.96066) \approx 163.9^\circ$ and
$\gamma = 180^\circ - (163.9^\circ + 10.5^\circ) = 5.6^\circ$

7. By the cosine law,

$$\cos\alpha = \frac{12.2^2 + 8.1^2 - 18.5^2}{2(12.2)(8.1)} \approx -0.6466.$$

Then $\alpha = \cos^{-1}(-0.6466) \approx 130.3^\circ$.
By the sine law,

$$\begin{aligned} \frac{12.2}{\sin\beta} &= \frac{18.5}{\sin 130.3^\circ} \\ \sin\beta &= \frac{12.2\sin 130.3^\circ}{18.5} \\ \sin\beta &\approx 0.5029 \\ \beta \approx \sin^{-1}(0.5029) &\approx 30.2^\circ \end{aligned}$$

So $\gamma = 180^\circ - (30.2^\circ + 130.3^\circ) = 19.5^\circ$

9. By the cosine law, we obtain

$a = \sqrt{9.3^2 + 12.2^2 - 2(9.3)(12.2)\cos 30^\circ}$

$\approx 6.23 \approx 6.2$ and

$$\cos\gamma = \frac{6.23^2 + 9.3^2 - 12.2^2}{2(6.23)(9.3)} \approx -0.203.$$

So $\gamma = \cos^{-1}(-0.203) \approx 101.7^\circ$ and
$\beta = 180^\circ - (101.7^\circ + 30^\circ) = 48.3^\circ$.

11. By the cosine law,

$$\cos\beta = \frac{6.3^2 + 6.8^2 - 7.1^2}{2(6.3)(6.8)} \approx 0.4146.$$

So $\beta = \cos^{-1}(0.4146) \approx 65.5^\circ$.
By the sine law, we have

$$\begin{aligned} \frac{6.8}{\sin\gamma} &= \frac{7.1}{\sin 65.5^\circ} \\ \sin\gamma &= \frac{6.8\sin 65.5^\circ}{7.1} \\ \sin\gamma &\approx 0.8715 \\ \gamma \approx \sin^{-1}(0.8715) &\approx 60.6^\circ. \end{aligned}$$

So $\alpha = 180^\circ - (60.6^\circ + 65.5^\circ) = 53.9^\circ$.

13. Note, $\alpha = 180^\circ - 25^\circ - 35^\circ = 120^\circ$.
Then by the sine law, we obtain

$$\frac{7.2}{\sin 120^\circ} = \frac{b}{\sin 25^\circ} = \frac{c}{\sin 35^\circ}$$

from which we have

$$b = \frac{7.2\sin 25^\circ}{\sin 120^\circ} \approx 3.5$$

and

$$c = \frac{7.2\sin 35^\circ}{\sin 120^\circ} \approx 4.8.$$

15. There is no such triangle. Note, $a+b=c$ and in a triangle the sum of the lengths of two sides is greater than the length of the third side.

17. One triangle exists. The angles are uniquely determined by the law of cosines.

19. There is no such triangle since the sum of the angles in a triangle is 180°.

21. Exactly one triangle exists. This is seen by constructing a 179°-angle with two sides that have lengths 1 and 10. The third side is constructed by joining the endpoints of the first two sides.

23. Consider the figure below.

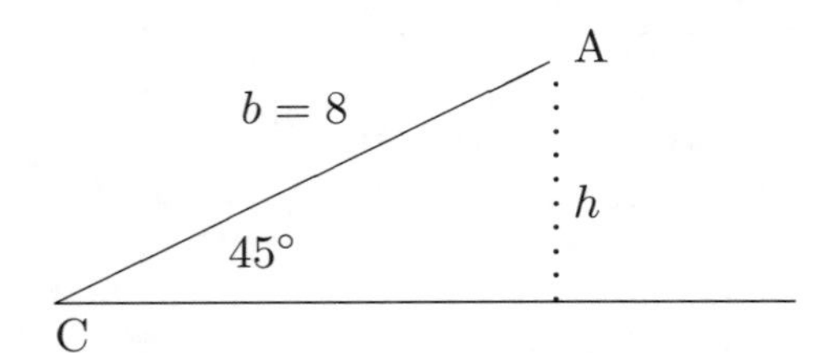

Note, $h = 8\sin 45^\circ = 4\sqrt{2}$. So the minimum value of c so that we will be able to make a triangle is $4\sqrt{2}$. Since $c = 2$, no such triangle is possible.

25. Recall, a central angle α in a circle of radius r intercepts a chord of length $r\sqrt{2-2\cos\alpha}$. Since $r = 30$ and $\alpha = 19^\circ$, the length is

$$30\sqrt{2-2\cos 19^\circ} \approx 9.90 \text{ ft.}$$

27. Note, a central angle α in a circle of radius r intercepts a chord of length $r\sqrt{2-2\cos\alpha}$. Since

$$921 = r\sqrt{2-2\cos 72^\circ}$$

(where $360 \div 5 = 72$), we obtain

$$r = \frac{921}{\sqrt{2-2\cos 72^\circ}} \approx 783.45 \text{ ft.}$$

29. After 6 hours, Jan hiked a distance of 24 miles and Dean hiked 30 miles. Let x be the distance between them after 6 hrs.

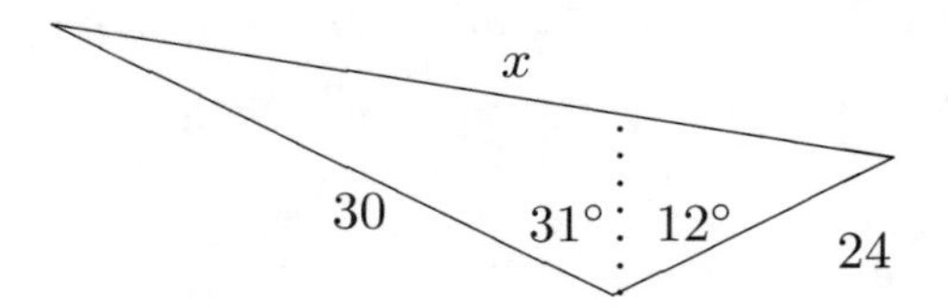

By the cosine law, we find $x = \sqrt{30^2+24^2-2(30)(24)\cos 43^\circ} = \sqrt{1476-1440\cos 43^\circ} \approx 20.6$ miles.

31. By the cosine law, we find

$$\begin{aligned} \cos\alpha &= \frac{1.2^2+1.2^2-0.4^2}{2(1.2)(1.2)} \\ \cos\alpha &\approx 0.9444 \\ \alpha &\approx \cos^{-1}(0.9444) \\ \alpha &\approx 19.2^\circ. \end{aligned}$$

33. Let α, β, and γ be the angles at pipes A, B, and C. The length of the sides of the triangle are 5, 6, and 7. By the cosine law,

$$\begin{aligned} \cos\alpha &= \frac{5^2+6^2-7^2}{2(5)(6)} \\ \cos\alpha &= 0.2 \\ \alpha &= \cos^{-1}(0.2) \\ \alpha &\approx 78.5^\circ. \end{aligned}$$

By the sine law,

$$\begin{aligned} \frac{6}{\sin\beta} &= \frac{7}{\sin 78.5^\circ} \\ \sin\beta &\approx 0.8399 \\ \beta &\approx \sin^{-1}(0.8399) \\ \beta &\approx 57.1^\circ. \end{aligned}$$

Then $\gamma = 180^\circ - (57.1^\circ + 78.5^\circ) = 44.4^\circ$.

35. By the cosine law,
$AB = \sqrt{5.3^2 + 7.6^2 - 2(5.3)(7.6)\cos 28^\circ} =$
$\sqrt{85.85 - 80.56\cos 28^\circ} \approx 3.8$ miles.
By using the exact value of AB, we get

$$\begin{aligned}\cos(\angle CBA) &= \frac{AB^2 + 5.3^2 - 7.6^2}{2(AB)(5.3)} \\ \angle CBA &= \cos^{-1}\left(\frac{AB^2 + 5.3^2 - 7.6^2}{2(AB)(5.3)}\right) \\ \angle CBA &\approx 111.6^\circ\end{aligned}$$

and $\angle CAB = 180^\circ - (111.6^\circ + 28^\circ) = 40.4^\circ$.

37. The pentagon consists of 5 chords each of which intercepts a $\frac{360^\circ}{5} = 72^\circ$ angle.
By the cosine law, the length of a chord is given by

$$\sqrt{10^2 + 10^2 - 2(10)(10)\cos 72^\circ} =$$

$$\sqrt{200 - 200\cos 72^\circ} \approx 11.76 \text{ m.}$$

39. The lower-left corner is the origin $(0, 0)$.

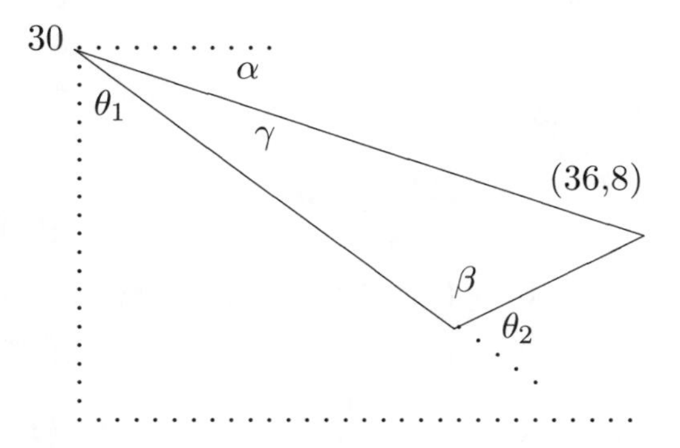

Note $\tan\alpha = 22/36$ and

$$\alpha = \tan^{-1}(22/36) \approx 31.4^\circ.$$

The distance between $(36, 8)$ and $(0, 30)$ is approximately 42.19. By the cosine law,

$$\begin{aligned}\beta &= \cos^{-1}\left(\frac{30^2 + 30^2 - 42.19^2}{2(30)(30)}\right) \\ \beta &\approx 89.4^\circ.\end{aligned}$$

So $\theta_2 = 180^\circ - 89.4^\circ = 90.6^\circ$.

By the sine law, we find

$$\frac{30}{\sin\gamma} = \frac{42.19}{\sin 89.4^\circ}$$

$$\gamma = \sin^{-1}\left(\frac{30\sin 89.4^\circ}{42.19}\right) \approx 45.3^\circ.$$

Then $\theta_1 = 90^\circ - (45.3^\circ + 31.4^\circ) = 13.3^\circ$.

41. **a)** Let α_m and α_M be the minimum and maximum values of α, respectively. By the law of cosines, we get

$$865,000^2 = 2(91,400,000)^2 - 2(91,400,000)^2\cos\alpha_M.$$

Then

$$\begin{aligned}\alpha_M &= \cos^{-1}\left(\frac{2(91400000)^2 - 865000^2}{2(91400000)^2}\right) \\ \alpha_M &\approx 0.54^\circ.\end{aligned}$$

Likewise,

$$\begin{aligned}\alpha_m &= \cos^{-1}\left(\frac{2(94500000)^2 - 865000^2}{2(94500000)^2}\right) \\ \alpha_m &\approx 0.52^\circ.\end{aligned}$$

b) Let β_m and β_M be the minimum and maximum values of β, respectively. By the law of cosines, one obtains

$$2163^2 = 2(225,800)^2 - 2(225,800)^2\cos\beta_M.$$

Then

$$\begin{aligned}\beta_M &= \cos^{-1}\left(\frac{2(225800)^2 - 2163^2}{2(225800)^2}\right) \\ \beta_M &\approx 0.55^\circ.\end{aligned}$$

Likewise,

$$\begin{aligned}\beta_m &= \cos^{-1}\left(\frac{2(252000)^2 - 2163^2}{2(252000)^2}\right) \\ \beta_m &\approx 0.49^\circ.\end{aligned}$$

c) Yes, even in perfect alignment a total eclipse may not occur, for instance when $\beta = 0.49^\circ$ and $\alpha = 0.52^\circ$.

43. Let d_b and d_h be the distance from the bear and hiker, respectively, to the base of the tower. Then $d_b = 150\tan 80°$ and $d_h = 150\tan 75°$.

Since the line segments joining the base of the tower to the bear and hiker form a 45° angle, by the cosine law the distance, d, between the bear and the hiker is

$$\begin{aligned} d &= \sqrt{d_b^2 + d_h^2 - 2(d_b)(d_h)\cos 45°} \\ &\approx \left((850.69)^2 + (559.81)^2 - \right. \\ &\qquad \left. 2(850.69)(559.81)\cos 45°\right)^{1/2} \\ &\approx 603 \text{ feet.} \end{aligned}$$

45. Using the cosine law, we obtain

$a = \sqrt{2r^2 - 2r^2\cos(\theta)} = \sqrt{4r^2\dfrac{1-\cos\theta}{2}} =$

$2r\sin(\theta/2)$.

47. If $\alpha = 90°$ in $a^2 = b^2 + c^2 - 2bc\cos\alpha$, then $a^2 = b^2 + c^2$ since $\cos 90° = 0°$. Thus the Pythagorean Theorem is a special case (i.e., when the angle is 90°) of the law of cosines.

For Thought

1. False, rather in a right triangle the area is one-half the product of its legs.

2. True

3. False, rather the area is one-half the product of two lengths of two sides and the sine of the included angle.

4. True

5. True, since one can use Heron's formula.

5.3 Exercises

1. Since two sides and an included angle are given, the area is

$A = \dfrac{1}{2}(12.9)(6.4)\sin 13.7° \approx 9.8.$

3. Draw angle $\alpha = 39.4°$.

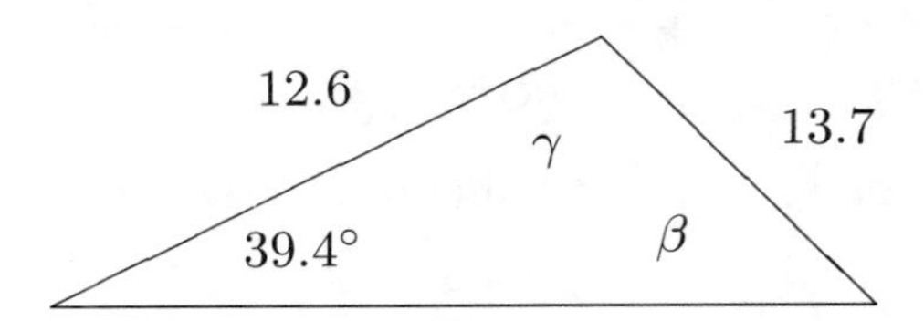

By the sine law, we obtain

$$\begin{aligned} \frac{12.6}{\sin\beta} &= \frac{13.7}{\sin 39.4°} \\ \sin\beta &= \frac{12.6\sin 39.4°}{13.7} \\ \beta &= \sin^{-1}\left(\frac{12.6\sin 39.4°}{13.7}\right) \\ \beta &\approx 35.7°. \end{aligned}$$

Then $\gamma = 180° - (35.7° + 39.4°) = 104.9°$.

The area is $A = \dfrac{1}{2}\cdot ab\sin\gamma =$

$\dfrac{1}{2}\cdot(13.7)(12.6)\sin 104.9° \approx 83.4.$

5. Draw angle $\alpha = 42.3°$.

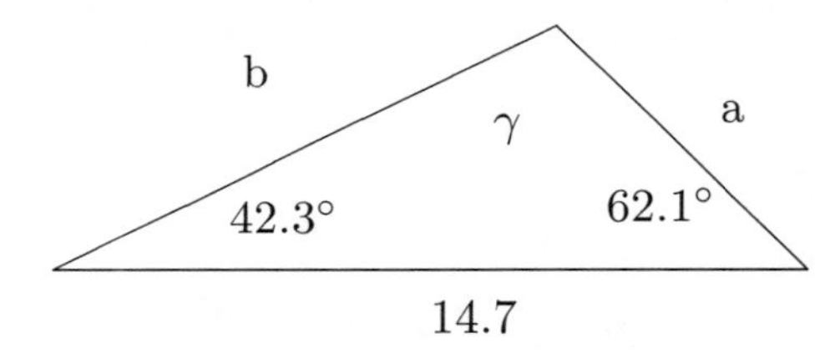

Note $\gamma = 180° - (42.3° + 62.1°) = 75.6°$. By the sine law,

$$\begin{aligned} \frac{b}{\sin 62.1°} &= \frac{14.7}{\sin 75.6°} \\ b &= \frac{14.7}{\sin 75.6°}\cdot\sin 62.1° \\ b &\approx 13.41. \end{aligned}$$

The area is $A = \dfrac{1}{2}bc\sin\alpha =$

$\dfrac{1}{2}(13.41)(14.7)\sin 42.3° \approx 66.3.$

7. Draw angle $\alpha = 56.3°$.

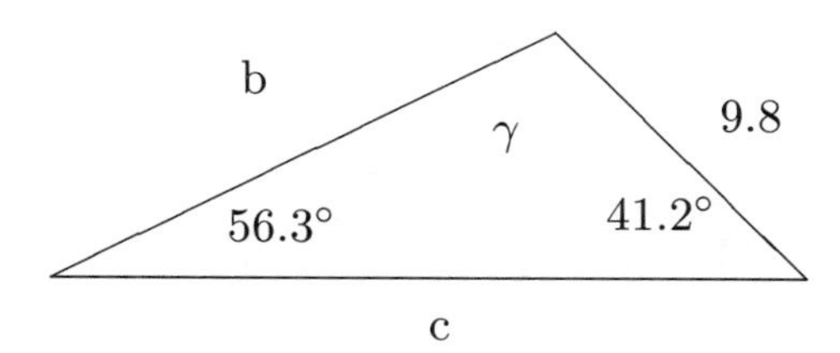

Note $\gamma = 180° - (56.3° + 41.2°) = 82.5°$. By the sine law, we obtain

$$\frac{c}{\sin 82.5°} = \frac{9.8}{\sin 56.3°}$$
$$c = \frac{9.8}{\sin 56.3°} \sin 82.5°$$
$$c \approx 11.679.$$

The area is $A = \frac{1}{2}ac\sin\beta =$

$\frac{1}{2}(9.8)(11.679)\sin 41.2° \approx 37.7.$

9. Note, the area of the triangle below is

$$\frac{1}{2} \cdot (1.5)(1.5\sqrt{3}) = 1.125\sqrt{3}.$$

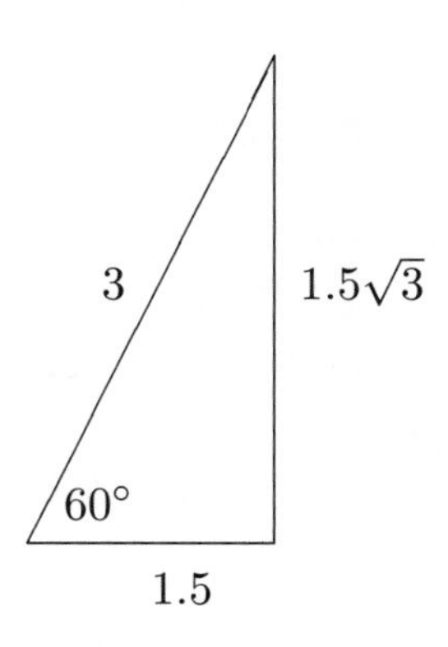

The trapezoid in the problem can be divided into two triangles and a rectangle with dimensions 2.7 by $1.5\sqrt{3}$. Thus, the area of the trapezoid is the area of the rectangle plus twice the area of the triangle shown above. That is, the area of the trapezoid is

$$2.7(1.5\sqrt{3}) + 2(1.125\sqrt{3}) = 6.3\sqrt{3} \approx 11 \text{ ft}^2.$$

11. Divide the given 4-sided polygon into two triangles by drawing the diagonal that connects the 60° angle to the 135° angle. On each triangle two sides and an included angle are given. The area of the polygon is equal to the sum of the areas of the two triangles. Namely,

$$\frac{1}{2}(4)(10)\sin 120° + \frac{1}{2}(12+2\sqrt{3})(2\sqrt{6})\sin 45° =$$
$$20(\sqrt{3}/2) + \frac{1}{2}(24\sqrt{6} + 4\sqrt{18})(\sqrt{2}/2) =$$
$$10\sqrt{3} + \frac{1}{2}(12\sqrt{12} + 2\sqrt{36}) =$$
$$10\sqrt{3} + 6\sqrt{12} + \sqrt{36} = 10\sqrt{3} + 12\sqrt{3} + 6 \approx$$
44 square inches.

13. Note,

$$S = \frac{16 + 9 + 10}{2} = 17.5.$$

The area is

$A = \sqrt{17.5(17.5-16)(17.5-9)(17.5-10)}$
$= \sqrt{17.5(1.5)(8.5)(7.5)} \approx 40.9.$

15. Note,

$$S = \frac{3.6 + 9.8 + 8.1}{2} = 10.75.$$

The area is

$\sqrt{10.75(10.75-3.6)(10.75-9.8)(10.75-8.1)}$
$= \sqrt{10.75(7.15)(0.95)(2.65)} \approx 13.9.$

17. Note,

$$S = \frac{346 + 234 + 422}{2} = 501.$$

The area is

$\sqrt{501(501-346)(501-234)(501-422)} =$
$\sqrt{501(155)(267)(79)} \approx 40,471.9.$

19. Since the base is 20 and the height is 10, the area is $\frac{1}{2}bh = \frac{1}{2}(20)(10) = 100.$

21. Since two sides and an included angle are given, the area is

$$\frac{1}{2}(6)(8)\sin 60° \approx 20.8.$$

23. Note, $S = \dfrac{9+5+12}{2} = 13.$

The area is $\sqrt{13(13-9)(13-5)(13-12)} =$

$$\sqrt{13(4)(8)(1)} \approx 20.4.$$

25. The kite consists of two equal triangles. The area of the kite is twice the area of the triangle.

Then the area of the kite is

$$2\left(\frac{1}{2}\right)(24)(18)\sin 40^\circ \approx 277.7 \text{ in.}^2.$$

27. The largest angle γ is opposite the 13-inch side. By the cosine law, we find

$$\gamma = \cos^{-1}\left(\frac{8^2+9^2-13^2}{2(8)(9)}\right) \approx 99.6^\circ.$$

Thus, the area is

$$\frac{1}{2}(8)(9)\sin(99.6^\circ) \approx 35.5 \quad \text{in.}^2.$$

29. Let x be the length of the third side.

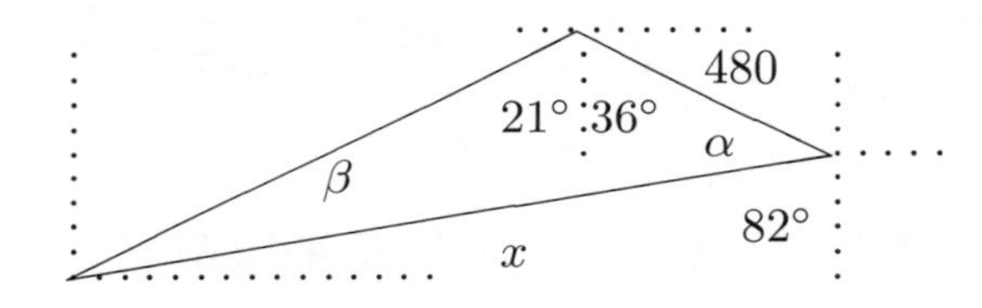

There is a 21° angle because of the $S21^\circ W$ direction. There are 36° and 82° angles because opposite angles are equal and because of the directions $N36^\circ W$ and $N82^\circ E$.
Note,

$$\alpha = 180^\circ - (82^\circ + 36^\circ) = 62^\circ$$

and $\beta = 180 - (21^\circ + 36^\circ + 62^\circ) = 61^\circ$. By the sine law, we obtain

$$x = \frac{480}{\sin 61^\circ}\sin 57^\circ.$$

The area is

$$\frac{1}{2}(480x)\sin 62^\circ \approx 97,534.8 \text{ sq ft}.$$

31. Note the angles in the quadrilateral property.

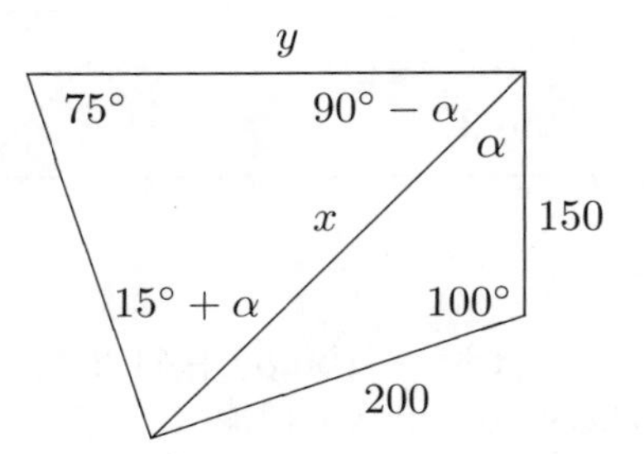

By the cosine law, we obtain

$$x = \sqrt{200^2 + 150^2 - 2(200)(150)\cos 100^\circ}.$$

Then the area of the triangle on the right

$$A_R = \frac{1}{2}200(150)\sin 100^\circ \approx 14,772.1163.$$

By the sine law, we find that in degrees

$$\alpha = \sin^{-1}\left(\frac{200\sin 100^\circ}{x}\right) \approx 46.8355^\circ.$$

Similarly, by the sine law, we get

$$y = \frac{x\sin(15^\circ + \alpha)}{\sin 75^\circ} \approx 246.4597$$

and the area of the triangle on the left is

$$A_L = \frac{1}{2}(xy)\sin(90^\circ - \alpha) \approx 22,764.2076.$$

Thus, the area of the property is

$$A_R + A_L \approx 37,536.3 \text{ ft}^2.$$

35. Note,

$$S = \frac{37+48+86}{2} = 85.5.$$

By Heron's formula, the area of the triangle is suppose to be

$$\sqrt{85.5(85.5-37)(85.5-48)(85.5-86)}.$$

But this area is undefined since the area is a complex imaginary number. Thus, no triangle exists with sides 37, 48, and 86.

37. Let $a = 6$, $b = 9$,and $c = 13$. Then

$$4b^2c^2 = 54,756$$

and

$$(b^2 + c^2 - a^2)^2 = 45,796.$$

The area is given by
$\frac{1}{4}\sqrt{4b^2c^2 - (b^2 + c^2 - a^2)^2} =$
$\frac{1}{4}\sqrt{54,756 - 45,796} = \frac{1}{4}\sqrt{8960} =$
$4\sqrt{35} \approx 23.7 \text{ ft}^2$.

Next, we verify that that

$$A_1 = \frac{1}{4}\sqrt{4b^2c^2 - (b^2 + c^2 - a^2)^2}$$

or equivalently

$$\sqrt{\frac{4b^2c^2 - (b^2 + c^2 - a^2)^2}{16}}$$

gives the area of a triangle. To do this, we will use Heron's formula. Let $s = \frac{a+b+c}{2}$. Since it can be shown that

$$s(s-a)(s-b)(s-c) = \frac{4b^2c^2 - (b^2 + c^2 - a^2)^2}{16}$$

it then follows that formula A_1 gives the area of a triangle.

For Thought

1. True, since if $v = \langle x, y\rangle$ then $2v = \langle 2x, 2y\rangle$ and $\tan^{-1}\left(\frac{2y}{2x}\right) = \tan^{-1}\left(\frac{y}{x}\right)$ and
$|2v| = \sqrt{4x^2 + 4y^2} = 2\sqrt{x^2 + y^2} = 2|v|$.

2. False, if $\boldsymbol{A} = \langle 1, 0\rangle$ and $\boldsymbol{B} = \langle 0, 1\rangle$ then $|\boldsymbol{A} + \boldsymbol{B}| = |\langle 1, 1\rangle| = \sqrt{2}$ and $|\boldsymbol{A}| + |\boldsymbol{B}| = 2$.

3. True, since if $A = \langle x, y\rangle$ then $-A = \langle -x, -y\rangle$ and $|-A| = \sqrt{(-x)^2 + (-y)^2} = \sqrt{x^2 + y^2} = |A|$.

4. True, since $\langle x, y\rangle + \langle -x, -y\rangle = \langle 0, 0\rangle$.

5. False, rather the parallelogram law says that the magnitude of $\boldsymbol{A} + \boldsymbol{B}$ is the length of a diagonal of the parallelogram formed by $\boldsymbol{A}$ and $\boldsymbol{B}$.

6. False, the direction angle is formed with the positive x-axis.

7. True, this follows from the fact that the horizontal component makes a $0°$-angle with the positive x-axis and $\cos\theta = \text{adjacent}/\text{hypotenuse}$.

8. True, since $|\langle 3, -4\rangle| = \sqrt{3^2 + (-4)^2} = \sqrt{9 + 16} = 5$.

9. True, the direction angle of a vector is unchanged when it is multipied by a positive scalar.

10. True, since $r = \sqrt{(-2)^2 + 2^2} = \sqrt{8}$ and $\cos\theta = \frac{x}{r} = \frac{-2}{\sqrt{8}}$.

5.4 Exercises

1. $\boldsymbol{A} + \boldsymbol{B} = 5\boldsymbol{j} + 4\boldsymbol{i} = 4\boldsymbol{i} + 5\boldsymbol{j}$

and

$$\boldsymbol{A} - \boldsymbol{B} = 5\boldsymbol{j} - 4\boldsymbol{i} = -4\boldsymbol{i} + 5\boldsymbol{j}$$

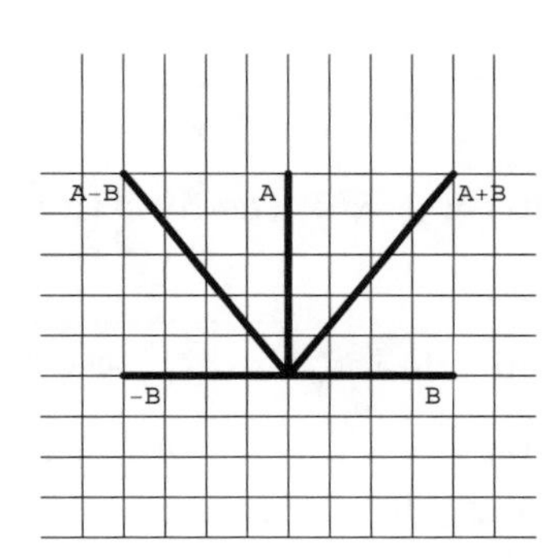

3. $\boldsymbol{A} + \boldsymbol{B} = (\boldsymbol{i} + 3\boldsymbol{j}) + (4\boldsymbol{i} + \boldsymbol{j}) = 5\boldsymbol{i} + 4\boldsymbol{j}$

and

$$\boldsymbol{A} - \boldsymbol{B} = (\boldsymbol{i} + 3\boldsymbol{j}) - (4\boldsymbol{i} + \boldsymbol{j}) = -3\boldsymbol{i} + 2\boldsymbol{j}$$

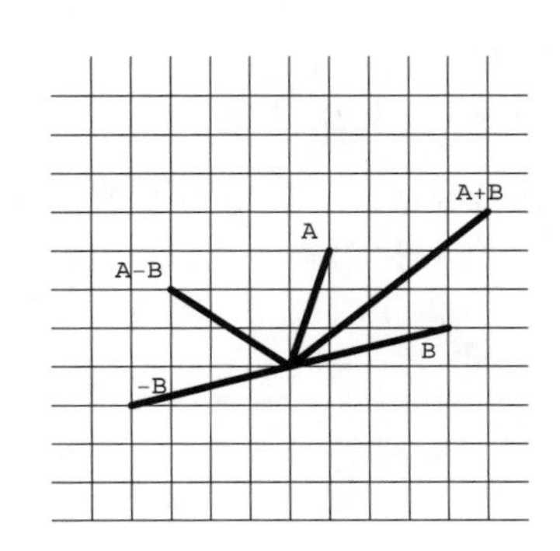

5. 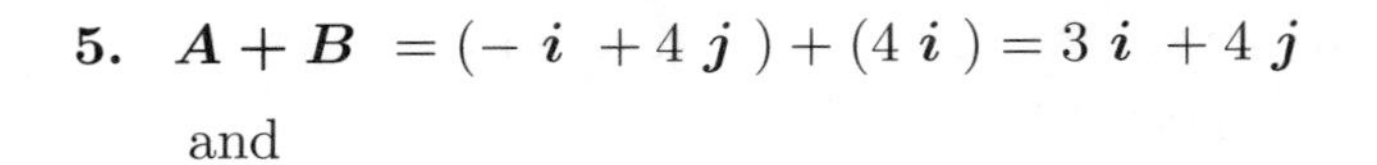
$\boldsymbol{A}+\boldsymbol{B} = (-\boldsymbol{i} + 4\boldsymbol{j}) + (4\boldsymbol{i}) = 3\boldsymbol{i} + 4\boldsymbol{j}$

and

$$\boldsymbol{A}-\boldsymbol{B} = (-\boldsymbol{i} + 4\boldsymbol{j}) - (4\boldsymbol{i}) = -5\boldsymbol{i} + 4\boldsymbol{j}$$

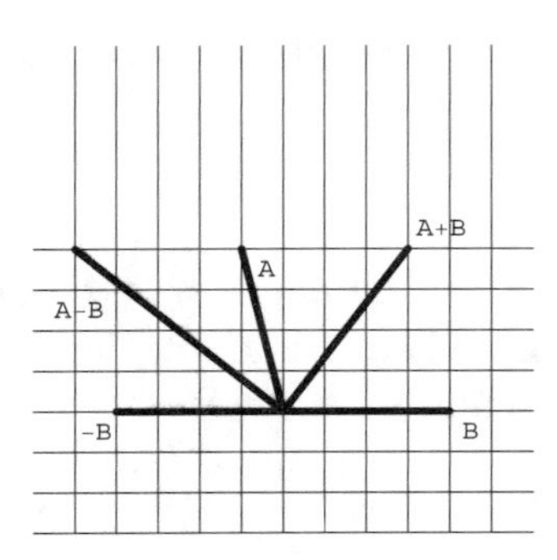

7. D **9.** E **11.** B

13. $|\boldsymbol{v_x}| = |4.5\cos 65.2°| = 1.9$,
$|\boldsymbol{v_y}| = |4.5\sin 65.2°| = 4.1$

15. $|\boldsymbol{v_x}| = |8000\cos 155.1°| \approx 7256.4$,
$|\boldsymbol{v_y}| = |8000\sin 155.1°| \approx 3368.3$

17. $|\boldsymbol{v_x}| = |234\cos 248°| \approx 87.7$,
$|\boldsymbol{v_y}| = |234\sin 248°| \approx 217.0$

19. The magnitude is $\sqrt{\sqrt{3}^2 + 1^2} = 2$.

Since $\tan\alpha = 1/\sqrt{3}$, the direction angle is $\alpha = 30°$.

21. The magnitude is $\sqrt{(-\sqrt{2})^2 + \sqrt{2}^2} = 2$.

Since $\tan\alpha = -\sqrt{2}/\sqrt{2} = -1$, the direction angle is $\alpha = 135°$.

23. The magnitude is $\sqrt{8^2 + (-8\sqrt{3})^2} = 16$.

Since $\tan\alpha = -8\sqrt{3}/8 = -\sqrt{3}$, the direction angle is $\alpha = 300°$.

25. The magnitude is $\sqrt{5^2 + 0^2} = 5$.
Since the terminal point is on the positive x-axis, the direction angle is $0°$.

27. The magnitude is $\sqrt{(-3)^2 + 2^2} = \sqrt{13}$.
Since $\tan^{-1}(-2/3) \approx -33.7°$, the direction angle is $180° - 33.7° = 146.3°$.

29. The magnitude is $\sqrt{3^2 + (-1)^2} = \sqrt{10}$.
Since $\tan^{-1}(-1/3) \approx -18.4°$, the direction angle is $360° - 18.4° = 341.6°$.

31. $\langle 8\cos 45°, 8\sin 45°\rangle = \langle 8(\sqrt{2}/2), 8(\sqrt{2}/2)\rangle$
$= \langle 4\sqrt{2}, 4\sqrt{2}\rangle$

33. $\langle 290\cos 145°, 290\sin 145°\rangle = \langle -237.6, 166.3\rangle$

35. $\langle 18\cos 347°, 18\sin 347°\rangle = \langle 17.5, -4.0\rangle$

37. $\langle 15, -10\rangle$

39. $\langle 6, -4\rangle + \langle 12, -18\rangle = \langle 18, -22\rangle$

41. $\langle -1, 5\rangle + \langle 12, -18\rangle = \langle 11, -13\rangle$

43. $\langle 3, -2\rangle - \langle 3, -1\rangle = \langle 0, -1\rangle$

45. $(3)(-1) + (-2)(5) = -13$

47. If $\boldsymbol{A} = \langle 2, 1\rangle$ and $\boldsymbol{B} = \langle 3, 5\rangle$, then the angle between these vectors is given by

$$\cos^{-1}\left(\frac{\boldsymbol{A}\cdot\boldsymbol{B}}{|\boldsymbol{A}|\cdot|\boldsymbol{B}|}\right) = \cos^{-1}\left(\frac{11}{\sqrt{5}\sqrt{34}}\right) \approx 32.5°$$

49. If $\boldsymbol{A} = \langle -1, 5\rangle$ and $\boldsymbol{B} = \langle 2, 7\rangle$, then the angle between these vectors is given by

$$\cos^{-1}\left(\frac{\boldsymbol{A}\cdot\boldsymbol{B}}{|\boldsymbol{A}|\cdot|\boldsymbol{B}|}\right) = \cos^{-1}\left(\frac{33}{\sqrt{26}\sqrt{53}}\right) \approx 27.3$$

51. Since $\langle -6, 5\rangle \cdot \langle 5, 6\rangle = 0$, the angle between them is $90°$.

53. Perpendicular since their dot product is zero

55. Parallel since $-2\langle 1, 7\rangle = \langle -2, -14\rangle$

57. Neither

59. $\boldsymbol{2i + j}$ **61.** $\boldsymbol{-3i + \sqrt{2}j}$

63. $-9\boldsymbol{j}$ **65.** $-7\boldsymbol{i} - \boldsymbol{j}$

67. The magnitude of $\boldsymbol{A}+\boldsymbol{B} = \langle 1, 4\rangle$ is

$$\sqrt{1^2 + 4^2} = \sqrt{17}$$

and the direction angle is

$$\tan^{-1}(4/1) \approx 76.0°$$

69. The magnitude of $\boldsymbol{-3A} = \langle -9, -3\rangle$ is

$$\sqrt{(-9)^2 + (-3)^2} = \sqrt{90} = 3\sqrt{10}.$$

Since $\tan^{-1}(3/9) \approx 18.4°$, the direction angle is

$$180° + 18.4° = 198.4°.$$

71. The magnitude of $\boldsymbol{B}-\boldsymbol{A}=\langle -5,2\rangle$ is

$$\sqrt{(-5)^2+2^2}=\sqrt{29}.$$

Since $\tan^{-1}(-2/5)\approx -21.8°$, the direction angle is

$$180°-21.8°=158.2°.$$

73. Note $-\boldsymbol{A}+\frac{\mathbf{1}}{\mathbf{2}}\boldsymbol{B}=\langle -3-1,-1+3/2\rangle$ $=\langle -4,1/2\rangle$. The magnitude is

$$\sqrt{(-4)^2+(1/2)^2}=\sqrt{65}/2.$$

Since $\tan^{-1}\left(\frac{1/2}{-4}\right)\approx -7.1°$, the direction angle is $180°-7.1°=172.9°$.

75. The resultant is $\langle 2+6,3+2\rangle=\langle 8,5\rangle$. Then the magnitude is

$$\sqrt{8^2+5^2}=\sqrt{89}$$

and direction angle is

$$\tan^{-1}(5/8)=32.0°.$$

77. The resultant is $\langle -6+4,4+2\rangle=\langle -2,6\rangle$ and its magnitude is

$$\sqrt{(-2)^2+6^2}=2\sqrt{10}.$$

Since $\tan^{-1}(-6/2)\approx -71.6°$, the direction angle is

$$180°-71.6°=108.4°.$$

79. The resultant is $\langle -4+3,4-6\rangle=\langle -1,-2\rangle$ and its magnitude is

$$\sqrt{(-1)^2+(-2)^2}=\sqrt{5}.$$

Since $\tan^{-1}(2/1)\approx 63.4°$, the direction angle is

$$180°+63.4°=243.4°.$$

81. The magnitudes of the horizontal and vertical components are $|520\cos 30°|\approx 450.3$ mph and $|520\sin 30°|=260$ mph, respectively.

For Thought

1. True, since the force required is

$$99\sin 88°\approx 98.9 \text{ kg}.$$

2. True

3. False, the weight of an object is modelled by a vertical vector.

4. True

5. True

6. True

7. False, the bearing of the wind is 45°.

8. False, the airplane's ground speed is slower than 400 mph since the airplane is flying against the wind.

9. False, the bearing of the course is 185°.

10. True, for the bearing of the plane's course is

$$135°+3°=138°.$$

5.5 Exercises

1. Draw two perpendicular vectors whose magnitudes are 3 and 8.

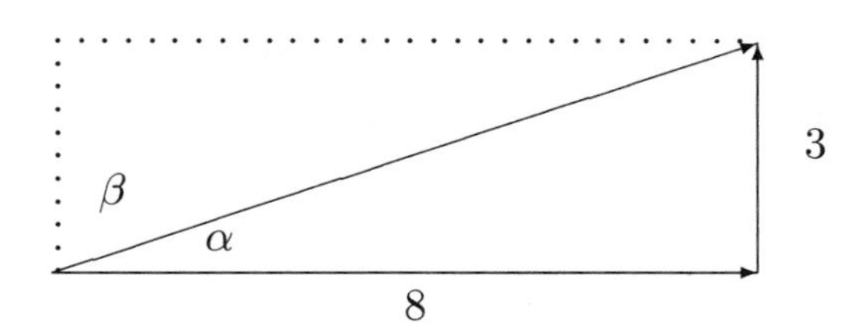

The magnitude of the resultant force is

$$\sqrt{8^2+3^2}=\sqrt{73}\approx 8.5 \text{ lb}$$

by the Pythagorean Theorem.

The angles between the resultant and each force are

$$\tan^{-1}(3/8)\approx 20.6°$$

and

$$\beta=90°-20.6°=69.4°.$$

3. Draw two vectors with magnitudes 10.3 and 4.2 that act at an angle of 130° with each other.

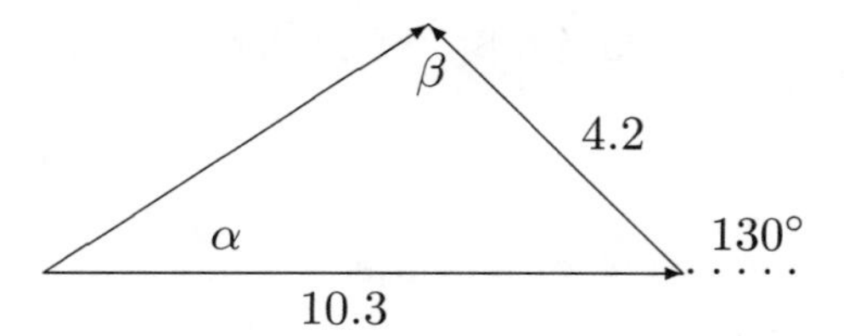

By using the cosine law, the magnitude of the resultant force is

$$\begin{aligned} r &= \sqrt{10.3^2 + 4.2^2 - 2(10.3)(4.2)\cos 50°} \\ &\approx 8.3 \text{ newtons.} \end{aligned}$$

By the sine law, we find

$$\begin{aligned} \frac{4.2}{\sin\alpha} &= \frac{r}{\sin 50°} \\ \sin\alpha &= \frac{4.2\sin 50°}{r} \\ \sin\alpha &\approx 0.3898 \\ \alpha \approx \sin^{-1}(0.3898) &\approx 22.9°. \end{aligned}$$

The angles between the resultant and each force are 22.9° and

$$\beta = 180° - 22.9° - 50° = 107.1°.$$

5. Draw two vectors with magnitudes 10 & 12.3 and whose angle between them is 23.4°.

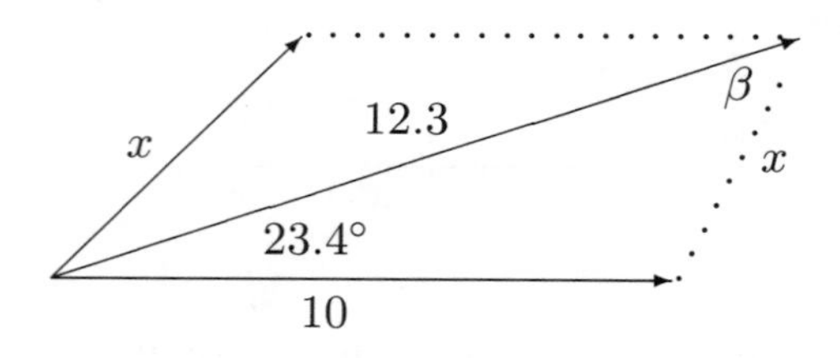

By the cosine law, the magnitude of the other force is

$$\begin{aligned} x &= \sqrt{10^2 + 12.3^2 - 2(10)(12.3)\cos 23.4°} \\ &\approx 5.051 \\ &\approx 5.1 \text{ lb.} \end{aligned}$$

By the sine law, we obtain

$$\begin{aligned} \frac{10}{\sin\beta} &= \frac{5.051}{\sin 23.4°} \\ \sin\beta &= \frac{10\sin 23.4°}{5.051} \\ \sin\beta &\approx 0.7863 \\ \beta \approx \sin^{-1}(0.7863) &\approx 51.8°. \end{aligned}$$

The angle between the two forces is

$$51.8° + 23.4° = 75.2°.$$

7. Since the angles in a parallelogram must add up to 360°, the angle formed by the two forces is

$$\frac{360° - 2(25°)}{2} = 155°.$$

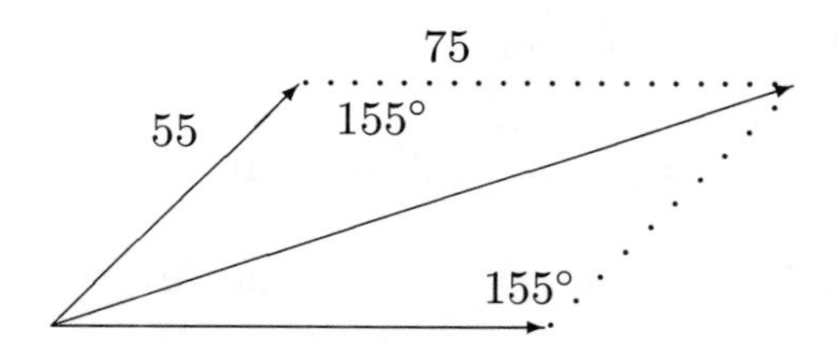

By the cosine law, the magnitude of the resultant force is

$$\sqrt{55^2 + 75^2 - 2(55)(75)\cos 155°} \approx 127.0 \text{ pounds.}$$

Then the donkey must pull a force of 127 pounds in the direction opposite that of the resultant's direction.

9. If x is the amount of force required as shown below, then

$$\begin{aligned} \frac{x}{3000} &= \sin 20^\circ \\ x &= 3000 \sin 20^\circ \\ x &\approx 1026.1 \text{ lb} \end{aligned}$$

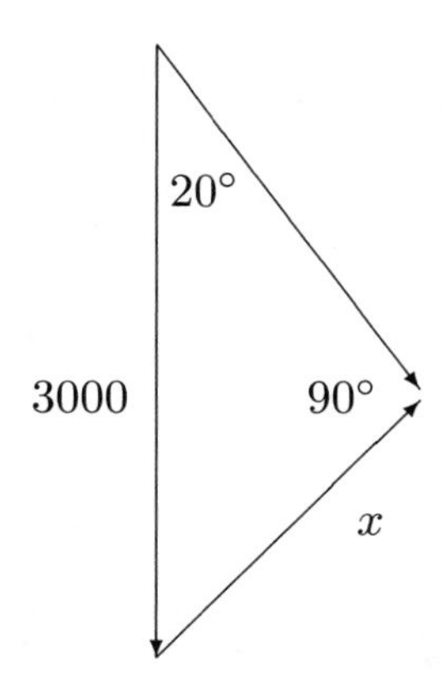

11. If w is the weight of the block of ice as shown below, then

$$\begin{aligned} \sin 25^\circ &= \frac{100}{w} \\ w &= \frac{100}{\sin 25^\circ} \\ w &\approx 236.6 \text{ lb} \end{aligned}$$

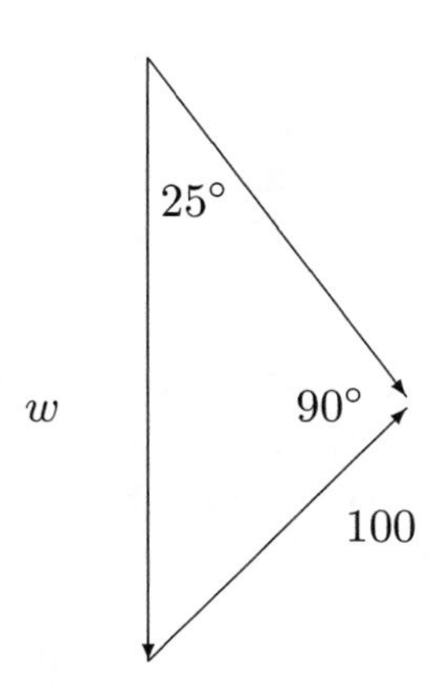

13. If α is the angle of inclination of the hill as shown below, then

$$\begin{aligned} \sin \alpha &= \frac{1000}{5000} \\ \alpha &= \sin^{-1} \frac{1}{5} \\ \alpha &\approx 11.5^\circ \end{aligned}$$

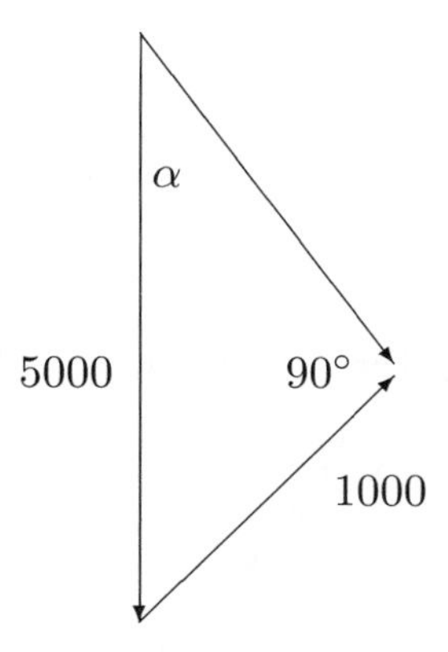

15. Let x be the ground speed and let α be drift angle as shown below.

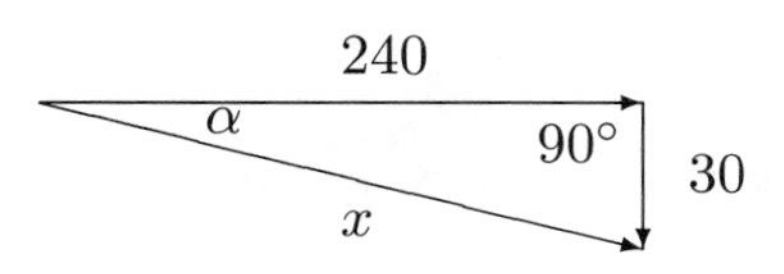

By the Pythagorean Theorem, we obtain

$$x = \sqrt{240^2 + 30^2} \approx 241.9 \text{ mph}$$

Using right triangle trigonometry, we obtain

$$\alpha = \tan^{-1} \frac{30}{240} \approx 7.1^\circ.$$

Thus, the bearing of the course is

$$90^\circ + \alpha \approx 97.1^\circ.$$

17. Let x be the ground speed and let α be drift angle as shown below.

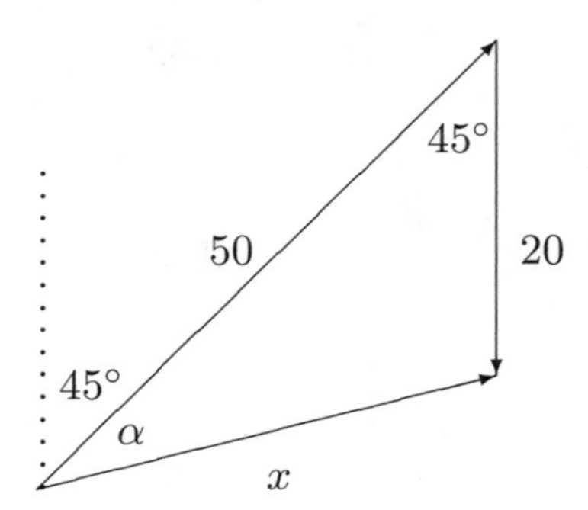

Applying the cosine law, we obtain

$$\begin{aligned} x &= \sqrt{20^2 + 50^2 - 2(20)(50)\cos 45°} \\ &\approx 38.5 \text{ mph.} \end{aligned}$$

Using the sine law, we find

$$\begin{aligned} \frac{\sin\alpha}{20} &= \frac{\sin 45°}{x} \\ \alpha &= \sin^{-1}\left(\frac{20\sin 45°}{x}\right) \\ \alpha &\approx 21.5°. \end{aligned}$$

Thus, the bearing of the course is

$$45° + \alpha \approx 66.5°.$$

19. Let x be the ground speed and let α be drift angle as shown below.

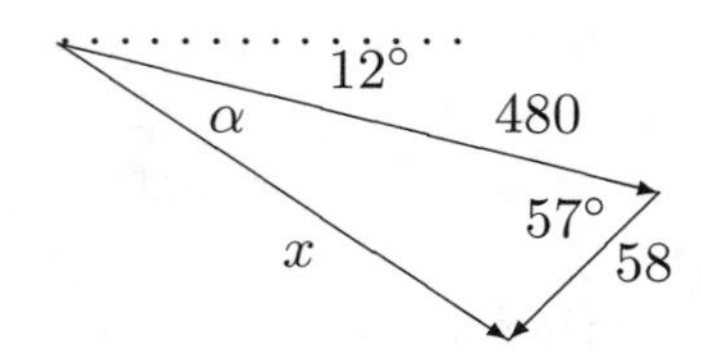

Note, the angle between the vectors representing the airplane and the wind is

$$57° = 12° + 45°.$$

By the cosine law, the ground speed is

$$\begin{aligned} x &= \sqrt{480^2 + 58^2 - 2(480)(58)\cos 57°} \\ &\approx 451.0 \text{ mph.} \end{aligned}$$

By the sine law, we get

$$\begin{aligned} \frac{\sin\alpha}{58} &= \frac{\sin 57°}{x} \\ \sin\alpha &= \frac{58\sin 57°}{x} \\ \alpha &= \sin^{-1}\left(\frac{58\sin 57°}{x}\right) \\ \alpha &\approx 6.2°. \end{aligned}$$

The bearing of the airplane is

$$102° + \alpha \approx 108.2°.$$

21. Draw two vectors representing the canoe and river current; the magnitudes of these vectors are 2 and 6, respectively.

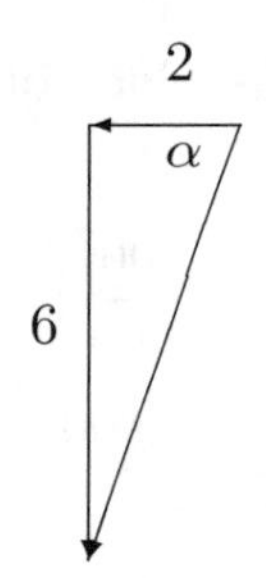

Since $\alpha = \tan^{-1}(6/2) \approx 71.6°$, the direction measured from the north is

$$270° - 71.6° = 198.4°.$$

Also, if d is the distance downstream from a point directly across the river to the point where she will land, then $\tan\alpha = d/2000$. Since $\tan\alpha = 6/2 = 3$, we get

$$d = 2000 \cdot 3 = 6000 \text{ ft.}$$

23. a) Assume we have a coordinate system where the origin is the point where the boat will start.

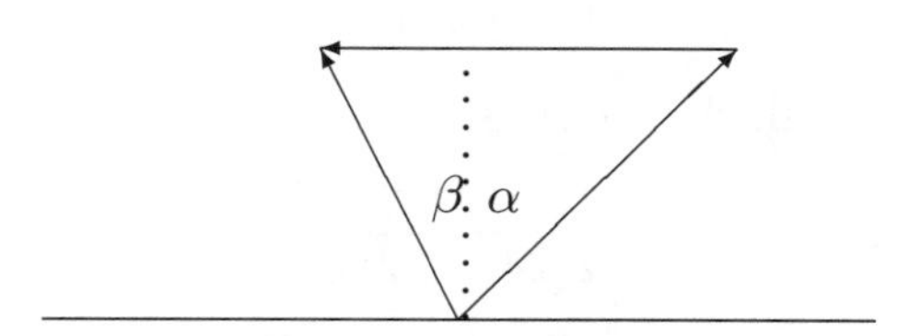

The intended direction and speed of the boat that goes 3 mph in still water is defined by the vector $3\sin\alpha\ \boldsymbol{i} + 3\cos\alpha\ \boldsymbol{j}$ and its actual direction and speed is determined by the vector

$$\boldsymbol{v} = (3\sin\alpha - 1)\ \boldsymbol{i} + 3\cos\alpha\ \boldsymbol{j}\ .$$

The number t of hours it takes the boat to cross the river is given by

$$t = \frac{0.2}{3\cos\alpha},$$

the solution to $3t\cos\alpha = 0.2$.
Suppose $\beta > 0$ if $3\sin\alpha - 1 < 0$ and $\beta < 0$ if $3\sin\alpha - 1 > 0$. Using right triangle trigonometry, we find

$$\tan\beta = \frac{|3\sin\alpha - 1|}{3\cos\alpha}.$$

The distance d the boat travels as a function of β is given by

$$\begin{aligned} d &= |t\ \boldsymbol{v}\ | \\ &= \frac{0.2}{3\cos\alpha}|\ \boldsymbol{v}\ | \\ &= \frac{0.2}{3\cos\alpha}\sqrt{(3\sin\alpha - 1)^2 + (3\cos\alpha)^2} \\ &= 0.2\sqrt{\tan^2\beta + 1} \\ d &= 0.2|\sec\beta|. \end{aligned}$$

b) Since speed is distance divided by time, then by using the answer from part a) the speed r as a function of α and β is

$$\begin{aligned} r &= \frac{d}{t} \\ &= \frac{0.2|\sec\beta|}{0.2/(3\cos\alpha)} \\ r &= 3\cos(\alpha)|\sec\beta|. \end{aligned}$$

25. Let the forces exerted by the papa, mama, and baby elephant be represented by the vectors
$\boldsymbol{v_p} = 800\cos 30°\ \boldsymbol{i} + 800\sin 30°\ \boldsymbol{j}$,
$\boldsymbol{v_m} = 500\ \boldsymbol{i}$, and
$\boldsymbol{v_b} = 200\cos 20°\ \boldsymbol{i} - 200\sin 20°\ \boldsymbol{j}$,
respectively. With a calculator, we find

$$\begin{aligned} \boldsymbol{F} &= \boldsymbol{v_p} + \boldsymbol{v_m} + \boldsymbol{v_b} \\ &\approx 1380.76\ \boldsymbol{i} + 331.60\ \boldsymbol{j}\ . \end{aligned}$$

The magnitude of the resultant of the three forces is

$$|\ \boldsymbol{F}\ | \approx \sqrt{1380.76^2 + 331.60^2} \approx 1420.0 \text{ lb}$$

and the direction is

$$\tan^{-1}\left(\frac{331.60}{1380.76}\right) \approx 13.5°$$

or E13.5°N.

Chapter 5 Review Exercises

1. Draw a triangle with $\gamma = 48°$, $a = 3.4$, $b = 2.6$.

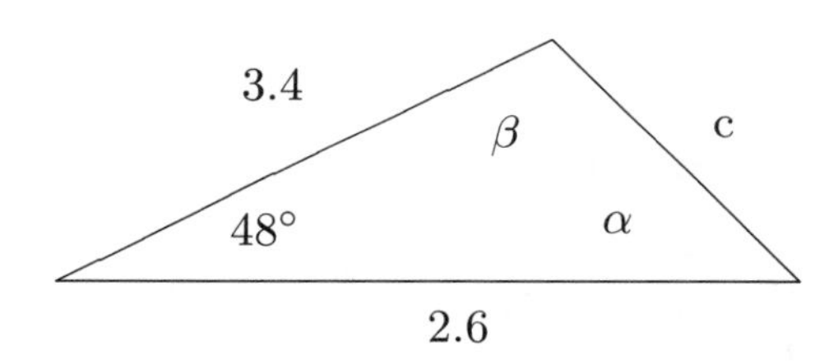

By the cosine law, we obtain
$c = \sqrt{2.6^2 + 3.4^2 - 2(2.6)(3.4)\cos 48°} \approx 2.5475 \approx 2.5$. By the sine law, we find

$$\begin{aligned} \frac{2.5475}{\sin 48°} &= \frac{2.6}{\sin\beta} \\ \sin\beta &= \frac{2.6\sin 48°}{2.5475} \\ \sin\beta &\approx 0.75846 \\ \beta &\approx \sin^{-1}(0.75846) \\ \beta &\approx 49.3°. \end{aligned}$$

Also, $\alpha = 180° - (49.3° + 48°) = 82.7°$.

3. Draw a triangle with $\alpha = 13°$, $\beta = 64°$, $c = 20$.

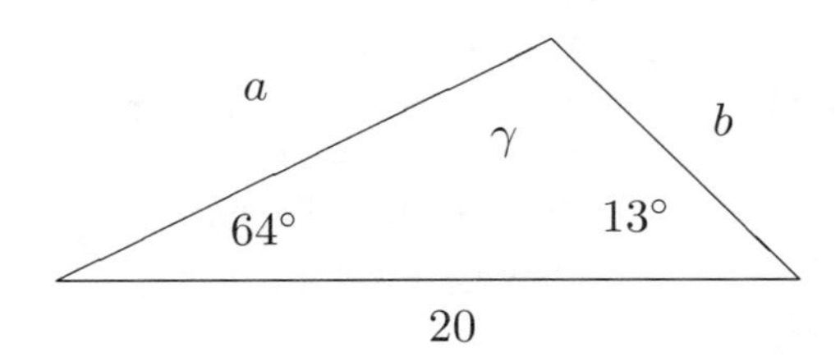

Note $\gamma = 180° - (64° + 13°) = 103°$.

By the sine law, we get $\dfrac{20}{\sin 103°} = \dfrac{a}{\sin 13°}$

and $\dfrac{20}{\sin 103°} = \dfrac{b}{\sin 64°}$.

So $a = \dfrac{20}{\sin 103°} \sin 13° \approx 4.6$

and $b = \dfrac{20}{\sin 103°} \sin 64° \approx 18.4$.

5. Draw a triangle with $a = 3.6$, $b = 10.2$, $c = 5.9$.

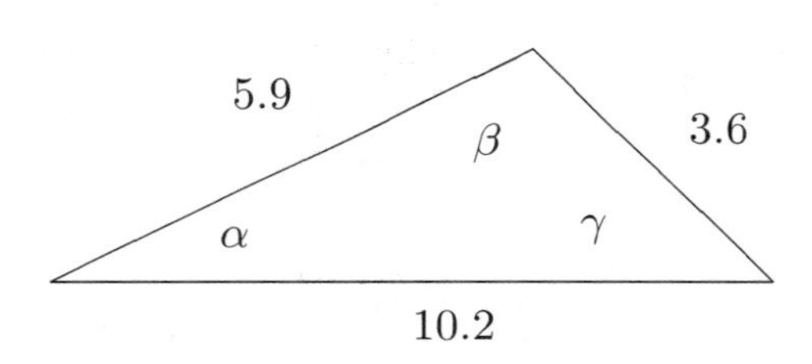

By the cosine law one gets

$$\cos\beta = \frac{5.9^2 + 3.6^2 - 10.2^2}{2(5.9)(3.6)} \approx -1.3.$$

This is a contradiction since the range of cosine is $[-1, 1]$. No triangle exists.

7. Draw a triangle with sides $a = 30.6$, $b = 12.9$, and $c = 24.1$.

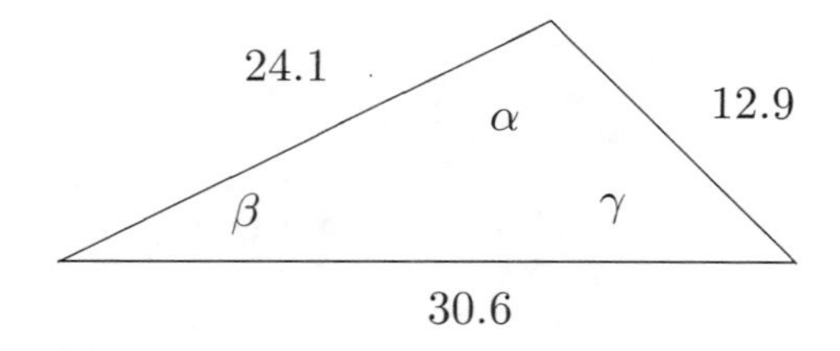

By the cosine law, we get

$\cos\alpha = \dfrac{24.1^2 + 12.9^2 - 30.6^2}{2(24.1)(12.9)} \approx -0.3042.$

So $\alpha = \cos^{-1}(-0.3042) \approx 107.7°$.

Similarly, we find

$\cos\beta = \dfrac{24.1^2 + 30.6^2 - 12.9^2}{2(24.1)(30.6)} \approx 0.9158.$

So $\beta = \cos^{-1}(0.9158) \approx 23.7°$.
Also, $\gamma = 180° - (107.7° + 23.7°) = 48.6°$.

9. Draw angle $\beta = 22°$ and let h be the height.

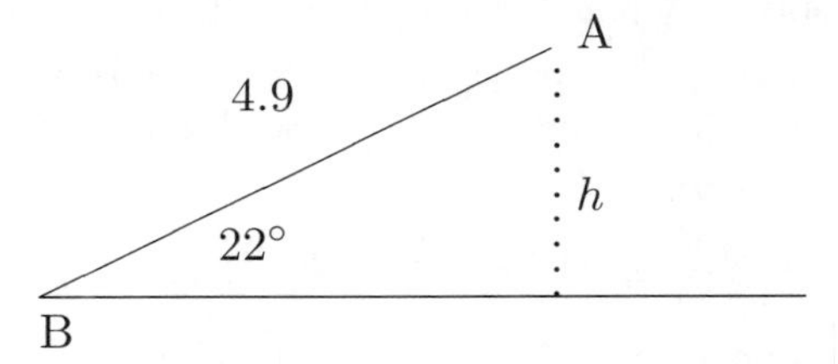

Since $h = 4.9 \sin 22° \approx 1.8$ and $1.8 < b < 4.9$, we have two triangles and they are given by

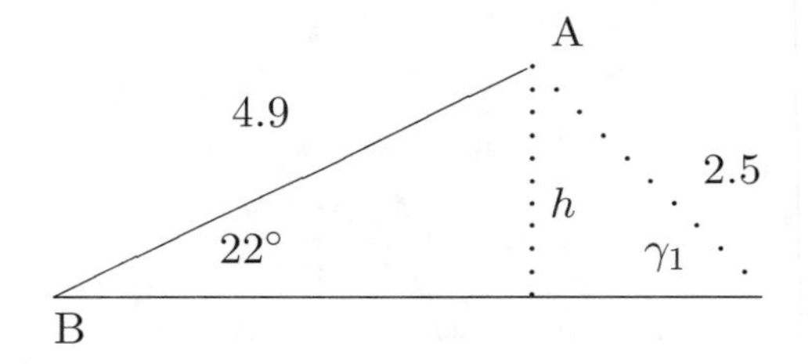

and

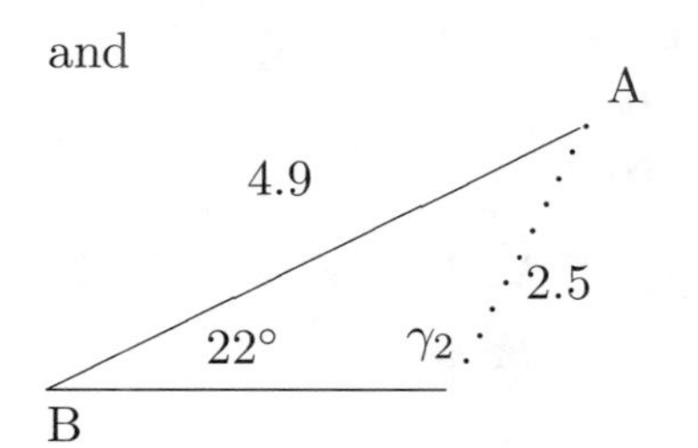

Apply the sine law to case 1.

$$\begin{aligned} \frac{4.9}{\sin\gamma_1} &= \frac{2.5}{\sin 22°} \\ \sin\gamma_1 &= \frac{4.9 \sin 22°}{2.5} \\ \sin\gamma_1 &\approx 0.7342 \\ \gamma_1 = \sin^{-1}(0.7342) &\approx 47.2° \end{aligned}$$

So $\alpha_1 = 180° - (22° + 47.2°) = 110.8°$.

By the sine law, $a_1 = \dfrac{2.5}{\sin 22°} \sin 110.8° \approx 6.2$.

In case 2, $\gamma_2 = 180° - \gamma_1 = 132.8°$
and $\alpha_2 = 180° - (22° + 132.8°) = 25.2°$.

By the sine law, $a_2 = \dfrac{2.5}{\sin 22°} \sin 25.2° \approx 2.8$.

11. Area is $A = \dfrac{1}{2}(12.2)(24.6)\sin 38° \approx 92.4 \text{ ft}^2$.

13. Since $S = \dfrac{5.4 + 12.3 + 9.2}{2} = 13.45$, the area is

$\sqrt{13.45(13.45 - 5.4)(13.45 - 12.3)(13.45 - 9.2)}$

$\approx 23.0 \text{ km}^2$.

15. $|\,\boldsymbol{v_x}\,| = |6\cos 23.3°| \approx 5.5$,
$|\,\boldsymbol{v_y}\,| = |6\sin 23.3°| \approx 2.4$

17. $|\,\boldsymbol{v_x}\,| = |3.2\cos 231.4°| \approx 2.0$,
$|\,\boldsymbol{v_y}\,| = |3.2\sin 231.4°| \approx 2.5$

19. magnitude $\sqrt{2^2 + 3^2} = \sqrt{13}$, direction angle $\tan^{-1}(3/2) \approx 56.3°$

21. The magnitude is $\sqrt{(-3.2)^2 + (-5.1)^2} \approx 6.0$. Since $\tan^{-1}(5.1/3.2) \approx 57.9°$, the direction angle is $180° + 57.9° = 237.9°$.

23. $\langle \sqrt{2}\cos 45°, \sqrt{2}\sin 45° \rangle = \langle 1, 1 \rangle$

25. $\langle 9.1\cos 109.3°, 9.1\sin 109.3° \rangle \approx \langle -3.0, 8.6 \rangle$

27. $\langle -6, 8 \rangle$

29. $\langle 2 - 2, -5 - 12 \rangle = \langle 0, -17 \rangle$

31. $\langle -1, 5 \rangle \cdot \langle 4, 2 \rangle = -4 + 10 = 6$

33. $-4\,\boldsymbol{i} + 8\,\boldsymbol{j}$

35. $(7.2\cos 30°)\,\boldsymbol{i} + (7.2\sin 30°)\,\boldsymbol{j} \approx 3.6\sqrt{3}\,\boldsymbol{i} + 3.6\,\boldsymbol{j}$

37. Parallel since $2\langle 2, 6 \rangle = \langle 4, 12 \rangle$

39. Perpendicular since their dot product is zero

41. Parallel since $-3\langle -3, 8 \rangle = \langle 9, -24 \rangle$

43. Draw two vectors with magnitudes 7 and 12 that act at an angle of 30° with each other.

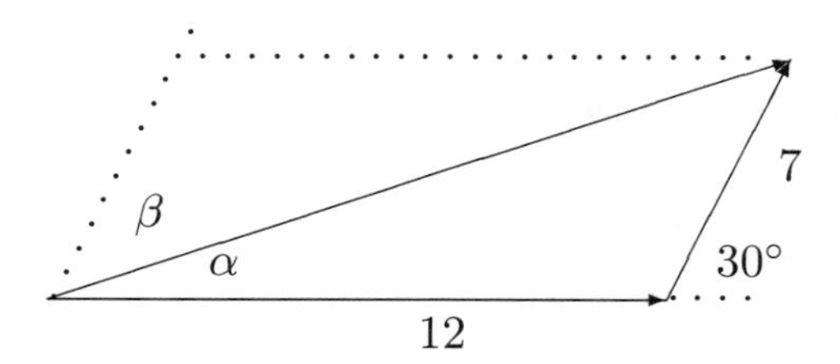

By the cosine law, the magnitude of the resultant force is

$$\sqrt{12^2 + 7^2 - 2(12)(7)\cos 150°} \approx 18.4 \text{ lb}.$$

By the sine law, we find

$$\begin{aligned} \frac{7}{\sin\alpha} &= \frac{18.4}{\sin 150°} \\ \sin\alpha &= \frac{7\sin 150°}{18.4} \\ \sin\alpha &\approx 0.19 \\ \alpha \approx \sin^{-1}(0.19) &\approx 11.0°. \end{aligned}$$

The angles between the resultant and the two forces are 11.0° and $\beta = 180° - 150° - 11° = 19.0°$.

45. Let x be the ground speed and let α be the drift angle, as shown below.

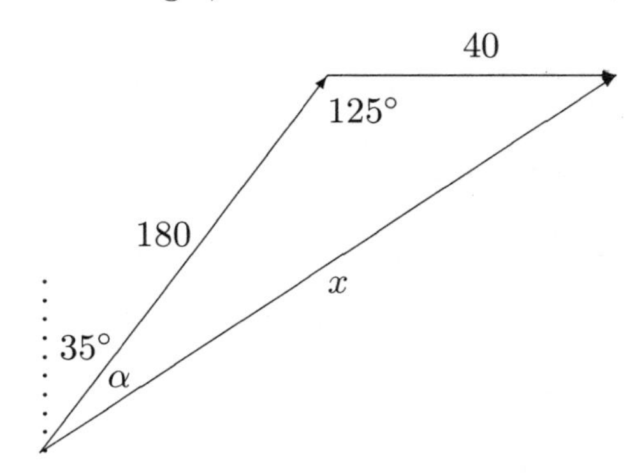

Note, the angle between the vectors of the plane and the wind is

$$125° = 90° + 35°.$$

Applying the cosine law, we obtain

$$\begin{aligned} x &= \sqrt{180^2 + 40^2 - 2(180)(40)\cos 135^\circ} \\ x &\approx 205.6 \text{ mph.} \end{aligned}$$

Applying the sine law, we find

$$\begin{aligned} \frac{\sin\alpha}{40} &= \frac{\sin 125^\circ}{x} \\ \alpha &= \sin^{-1}\left(\frac{40\sin 125^\circ}{x}\right) \\ \alpha &\approx 9.2^\circ. \end{aligned}$$

The bearing of the plane's course is

$$35^\circ + \alpha \approx 44.2^\circ.$$

47. Using Heron's formula and since

$$\frac{482 + 364 + 241}{2} = 543.5,$$

the area of Susan's lot is

$$\sqrt{543.5(543.5 - 482)(543.5 - 364)(543.5 - 241)}$$

which is approximately

$$42{,}602 \text{ ft}^2.$$

Similarly, since

$$\frac{482 + 369 + 238}{2} = 544.5,$$

the area of Seth's lot is

$$\sqrt{544.5(544.5 - 482)(544.5 - 369)(544.5 - 238)}$$

or approximately

$$42{,}785 \text{ ft}^2.$$

Then Seth got the larger piece.

49. Consider triangle below.

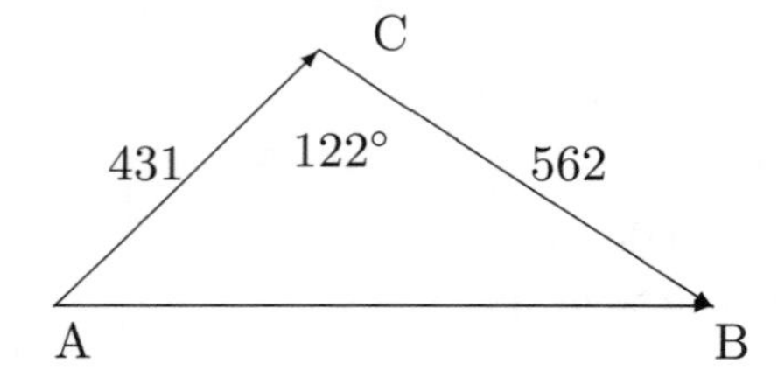

The distance between A and B is

$$\sqrt{431^2 + 562^2 - 2(431)(562)\cos 122^\circ} \approx 870.82 \text{ ft.}$$

The extra amount spent is

$$(431 + 562 - 870.82)(\$21.60) \approx \$2639.$$

51.

a) By the cosine law and by the method of completing the square, one derives

$$\begin{aligned} c^2 &= a^2 + r^2 - 2ar\cos\theta \\ c^2 - r^2 &= a^2 - 2ar\cos\theta \\ c^2 - r^2 + r^2\cos^2\theta &= (a - r\cos\theta)^2 \\ c^2 - r^2(1 - \cos^2\theta) &= (a - r\cos\theta)^2 \\ c^2 - r^2\sin^2\theta &= (a - r\cos\theta)^2 \\ \sqrt{c^2 - r^2\sin^2\theta} &= a - r\cos\theta. \end{aligned}$$

Then $a = \sqrt{c^2 - r^2\sin^2\theta} + r\cos\theta$.

b) When $t = 0.1$ minute, the number of revolutions is 42.6. Then

$$\theta = (0.6)(360^\circ) = 216^\circ$$

and we obtain

$$\begin{aligned} a &= \sqrt{12^2 - 2^2\sin^2 216^\circ} + 2\cos 216^\circ \\ &\approx 10.3 \text{ in.} \end{aligned}$$

Chapter 5 Test

1. Draw a triangle with $\alpha = 30^\circ$, $b = 4$, $a = 2$.

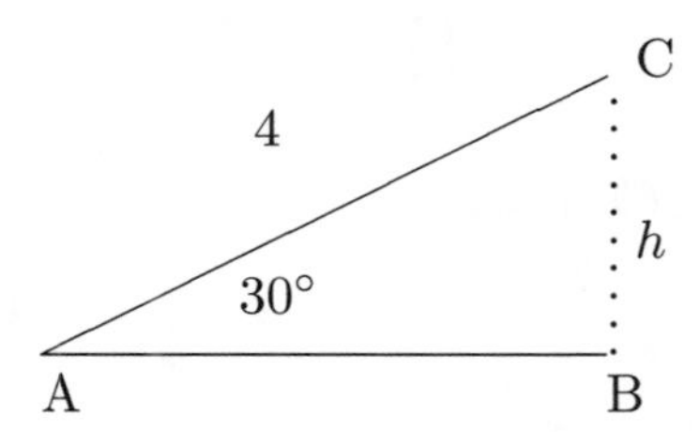

Since $h = 4\sin 30^\circ = 2$ and $a = 2$, there is only one triangle and $\beta = 90^\circ$. Then $\gamma = 90^\circ - 30^\circ = 60^\circ$. Since $c^2 + 2^2 = 4^2$, we get $c = \sqrt{12} = 2\sqrt{3}$.

2. Draw angle $\alpha = 60°$ and let h be the height.

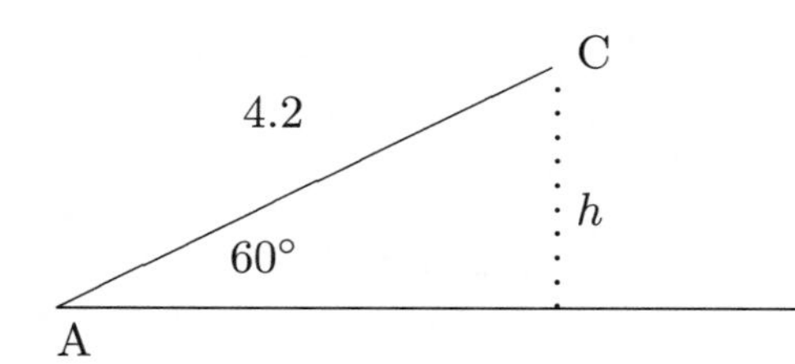

Since $h = 4.2 \sin 60° \approx 3.6$ and $3.6 < a < 4.2$, there are two triangles and they are given by

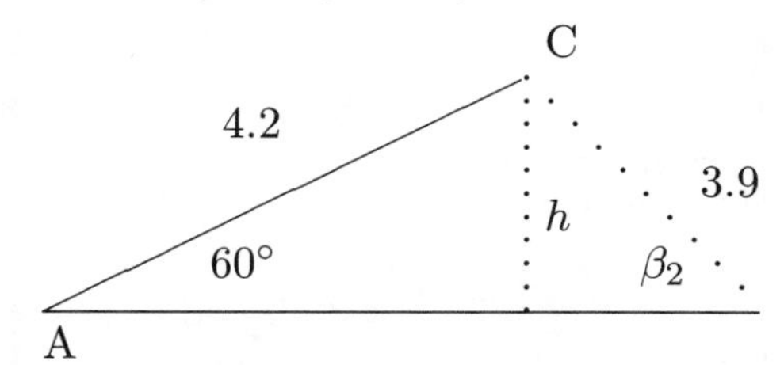

and

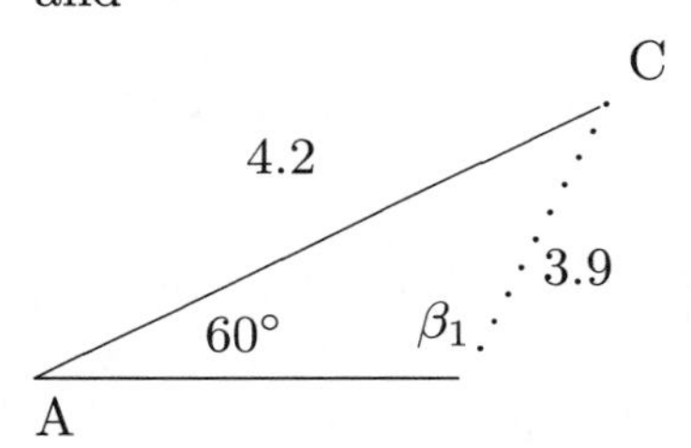

Apply the sine law to the acute triangle.

$$\frac{3.9}{\sin 60°} = \frac{4.2}{\sin \beta_2}$$

$$\sin \beta_2 = \frac{4.2 \sin 60°}{3.9}$$

$$\sin \beta_2 \approx 0.93264$$

$$\beta_2 = \sin^{-1}(0.93264) \approx 68.9°$$

So $\gamma_2 = 180° - (\beta_2 + 60°) = 51.1°$.

By the sine law, $c_2 = \dfrac{3.9}{\sin 60°} \sin 51.1° \approx 3.5$.

In the obtuse triangle, $\beta_1 = 180° - \beta_2 = 111.1°$ and $\gamma_1 = 180° - (\beta_1 + 60°) = 8.9°$.

By the sine law, $c_1 = \dfrac{3.9}{\sin 60°} \sin 8.9° \approx 0.7$.

3. Draw the only triangle with $a = 3.6$, $\alpha = 20.3°$, and $\beta = 14.1°$.

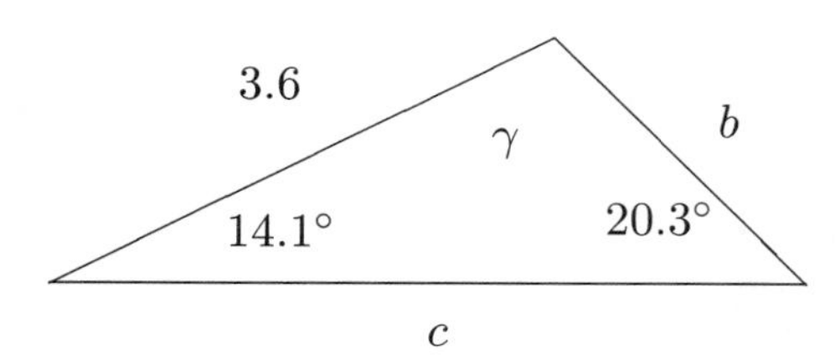

Note, $\gamma = 180° - 14.1° - 20.3° = 145.6°$. Using the sine law, we find

$$\frac{b}{\sin 14.1°} = \frac{3.6}{\sin 20.3°} \text{ and } \frac{c}{\sin 145.6°} = \frac{3.6}{\sin 20.3°}.$$

Then

$$b = \frac{3.6}{\sin 20.3°} \sin 14.1° \approx 2.5$$

and

$$c = \frac{3.6}{\sin 20.3°} \sin 145.6° \approx 5.9.$$

4. Draw the only triangle with $a = 2.8$, $b = 3.9$, and $\gamma = 17°$.

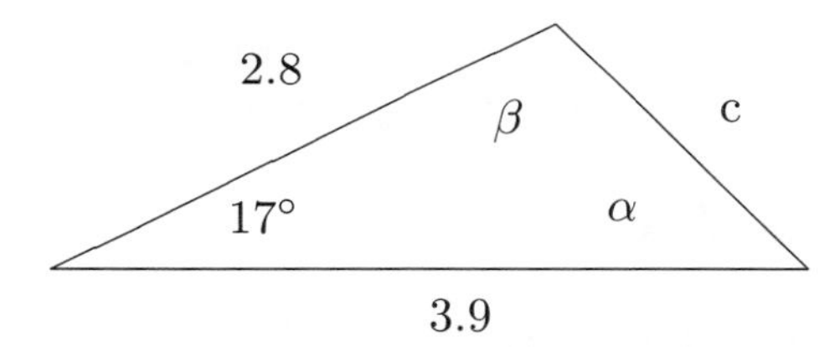

By the cosine law, we get $c = \sqrt{3.9^2 + 2.8^2 - 2(3.9)(2.8)\cos 17°} \approx 1.47 \approx 1.5$. By the sine law,

$$\frac{1.47}{\sin 17°} = \frac{2.8}{\sin \alpha}$$

$$\sin \alpha = \frac{2.8 \sin 17°}{1.47}$$

$$\sin \alpha \approx 0.5569$$

$$\alpha \approx \sin^{-1}(0.5569) \approx 33.8°.$$

Also, $\beta = 180° - (33.8° + 17°) = 129.2°$.

5. Draw the only triangle with the given sides $a = 4.1$, $b = 8.6$, and $c = 7.3$.

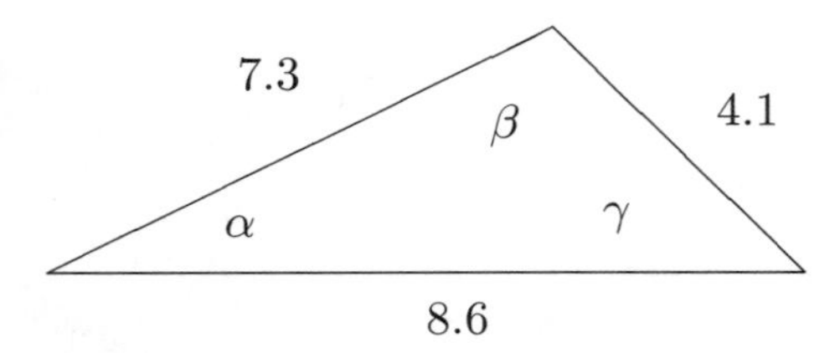

First, find the largest angle β by the cosine law.

$$\begin{aligned}\cos\beta &= \frac{7.3^2 + 4.1^2 - 8.6^2}{2(7.3)(4.1)} \\ \cos\beta &\approx -0.06448 \\ \beta &\approx \cos^{-1}(-0.06448) \\ \beta &\approx 93.7^\circ.\end{aligned}$$

By the sine law,

$$\begin{aligned}\frac{8.6}{\sin 93.7^\circ} &= \frac{7.3}{\sin\gamma} \\ \sin\gamma &= \frac{7.3\sin 93.7^\circ}{8.6} \\ \sin\gamma &\approx 0.8471 \\ \gamma \approx \sin^{-1}(0.8471) &\approx 57.9^\circ.\end{aligned}$$

Also, we obtain

$$\alpha = 180^\circ - (57.9^\circ + 93.7^\circ) = 28.4^\circ.$$

6. The magnitude of $\boldsymbol{A} + \boldsymbol{B} = \langle -2, 6\rangle$ is

$$\sqrt{(-2)^2 + 6^2} = \sqrt{40} = 2\sqrt{10}.$$

The direction angle is

$$\cos^{-1}(-2/\sqrt{40}) \approx 108.4^\circ.$$

7. The magnitude of $\boldsymbol{A} - \boldsymbol{B} = \langle -4, -2\rangle$ is

$$\sqrt{(-4)^2 + (-2)^2} = \sqrt{20} = 2\sqrt{5}.$$

Since $\tan^{-1}(2/4) \approx 26.6^\circ$, the direction angle is

$$180^\circ + 26.6^\circ = 206.6^\circ.$$

8. The magnitude of $\boldsymbol{3B} = \langle 3, 12\rangle$ is

$$\sqrt{3^2 + 12^2} = \sqrt{153} = 3\sqrt{17}.$$

The direction angle is

$$\tan^{-1}(12/3) \approx 76.0^\circ.$$

9. The area is $\frac{1}{2}(12)(10)\sin(22^\circ) \approx 22.5$ ft^2.

10. Using Heron's formula and if

$$s = \frac{4.1 + 6.8 + 9.5}{2} = 10.2$$

then the area is

$$\sqrt{s(s-4.1)(s-6.8)(s-9.5)} \approx 12.2 \text{ m}^2.$$

11. Since $a_1 = 4.6\cos 37.2^\circ \approx 3.66$ and $a_2 = 4.6\sin 37.2^\circ \approx 2.78$, we have

$$\boldsymbol{v} \approx 3.66\,\boldsymbol{i} + 2.78\,\boldsymbol{j}\ .$$

12. Perpendicular since their dot product is zero. That is,

$$\langle -3, 5\rangle \cdot \langle 5, 3\rangle = (-3)(5) + (5)(3) = 0.$$

13. If x is the force required to push the riding lawnmower as shown below, then

$$\begin{aligned}\frac{x}{1000} &= \sin 40^\circ \\ x &= 1000\sin 40^\circ \\ x &\approx 642.8 \text{ lb}\end{aligned}$$

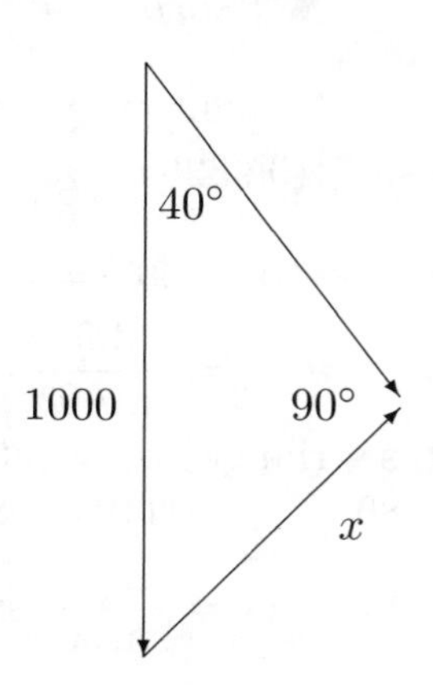

14. Let x be the ground speed and let α be drift angle as shown below.

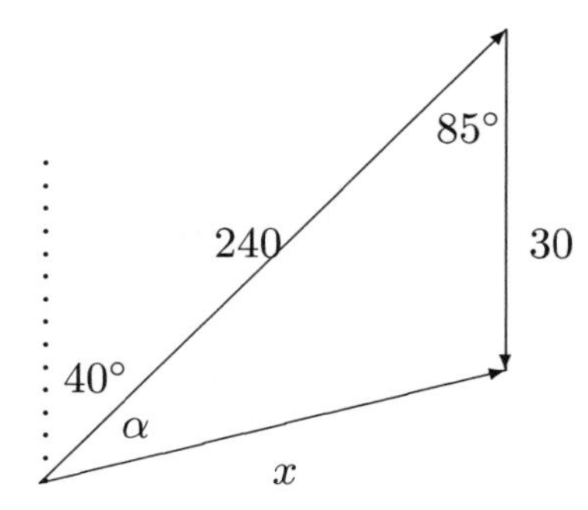

Note, the angle between the vector of the airplane and the vector of the wind is

$$85° = 40° + 45°.$$

Applying the cosine law, we obtain

$$\begin{aligned} x &= \sqrt{240^2 + 30^2 - 2(240)(30)\cos 85°} \\ &\approx 239.3 \text{ mph.} \end{aligned}$$

Using the sine law, we find

$$\begin{aligned} \frac{\sin\alpha}{30} &= \frac{\sin 85°}{x} \\ \alpha &= \sin^{-1}\left(\frac{30\sin 85°}{x}\right) \\ \alpha &\approx 7.2°. \end{aligned}$$

Thus, the bearing of the course is

$$40° + \alpha \approx 47.2°.$$

Tying It All Together

1. $\sin(\pi/6) = 1/2$, $\cos(\pi/6) = \sqrt{3}/2$, $\tan(\pi/6) = \sqrt{3}/3$, $\csc(\pi/6) = 2$, $\sec(\pi/6) = 2\sqrt{3}/3$, and $\cot(\pi/6) = \sqrt{3}$

2. $\sin(\pi/4) = \sqrt{2}/2$, $\cos(\pi/4) = \sqrt{2}/2$, $\tan(\pi/4) = 1$, $\csc(\pi/4) = \sqrt{2}$, $\sec(\pi/4) = \sqrt{2}$, and $\cot(\pi/4) = 1$

3. $\sin(\pi/3) = \sqrt{3}/2$, $\cos(\pi/3) = 1/2$, $\tan(\pi/3) = \sqrt{3}$, $\csc(\pi/3) = 2\sqrt{3}/3$, $\sec(\pi/3) = 2$, and $\cot(\pi/3) = \sqrt{3}/3$

4. $\sin(\pi/2) = 1$, $\cos(\pi/2) = 0$, $\tan(\pi/2)$ is undefined, $\csc(\pi/2) = 1$, $\sec(\pi/2)$ is undefined, and $\cot(\pi/2) = 0$

5. $\pi/2$ **6.** $-\pi/2$ **7.** $-\pi/6$ **8.** $\pi/6$

9. π **10.** 0 **11.** $5\pi/6$ **12.** $\pi/6$

13. 0 **14.** $\pi/4$ **15.** $-\pi/4$ **16.** $\pi/6$

17. $\{x \mid x = k\pi \text{ where } k \text{ is an integer}\}$

18. Factoring, we get

$$\sin(x)(\sin(x) - 1) = 0.$$

Then

$$\sin(x) = 0 \text{ or } \sin(x) = 1.$$

Thus, the solution set is

$$\{x \mid x = k\pi \text{ or } x = \frac{\pi}{2} + 2k\pi\}.$$

19. Factoring, we obtain

$$\begin{aligned} \sin^2 x - \sin x - 2 &= 0 \\ (\sin x + 1)(\sin x - 2) &= 0. \end{aligned}$$

Then

$$\sin x = -1 \text{ or } \sin x = 2.$$

Since $\sin x = 2$ is impossible, we have

$$\sin x = -1.$$

The solution set is

$$\left\{x \mid x = \frac{3\pi}{2} + 2k\pi\right\}.$$

20. Factoring, we find

$$\begin{aligned} 4\sin x\cos x - 2\cos x + 2\sin x - 1 &= 0 \\ 2\cos x(2\sin x - 1) + (2\sin x - 1) &= 0 \\ (2\cos x + 1)(2\sin x - 1) &= 0. \end{aligned}$$

Then

$$\cos x = -\frac{1}{2} \text{ or } \sin x = \frac{1}{2}.$$

The solution set is $\left\{x \mid x = \frac{\pi}{6} + 2k\pi,\right.$

$$\left. x = \frac{5\pi}{6} + 2k\pi, x = \frac{2\pi}{3} + 2k\pi, x = \frac{4\pi}{3} + 2k\pi\right\}.$$

21. Factoring, we find

$$\begin{aligned} 4x \sin x + 2\sin x - 2x - 1 &= 0 \\ 2\sin x(2x+1) - (2x+1) &= 0 \\ (2x+1)(2\sin x - 1) &= 0. \end{aligned}$$

Then

$$x = -\frac{1}{2} \text{ or } \sin x = \frac{1}{2}.$$

The solution set is

$$\left\{x \mid x = -\frac{1}{2}, x = \frac{\pi}{6} + 2k\pi, x = \frac{5\pi}{6} + 2k\pi\right\}.$$

22. Since $\sin 2x = 1/2$, we obtain

$$2x = \frac{\pi}{6} + 2k\pi \text{ or } 2x = \frac{5\pi}{6} + 2k\pi$$

where k is an integer. Then the solution set is

$$\left\{x \mid x = \frac{\pi}{12} + k\pi \text{ or } x = \frac{5\pi}{12} + k\pi\right\}.$$

23. Since $\tan 4x = 1/\sqrt{3}$, we obtain

$$4x = \frac{\pi}{6} + k\pi$$

where k is an integer. Then the solution set is

$$\left\{x \mid x = \frac{\pi}{24} + \frac{k\pi}{4}\right\}.$$

24. Since

$$\sin^2 x + \cos^2 x = 1$$

is an identity, the solution set is the set of all real numbers.

25. Amplitude 1, period $2\pi/3$, phase shift 0, domain $(-\infty, \infty)$, and range $[-1, 1]$

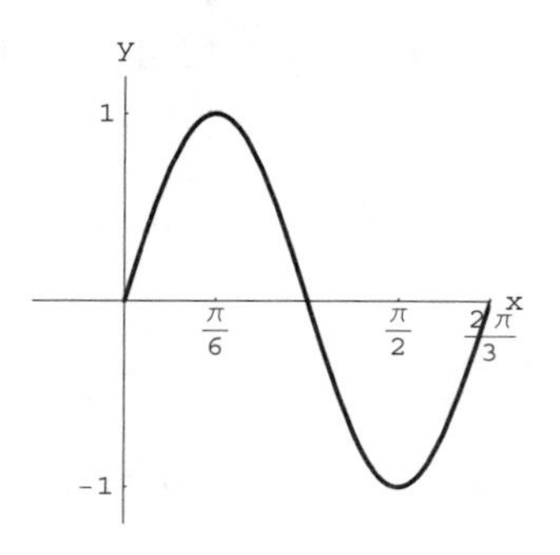

26. Amplitude 3, period π, phase shift 0, domain $(-\infty, \infty)$, and range $[-3, 3]$

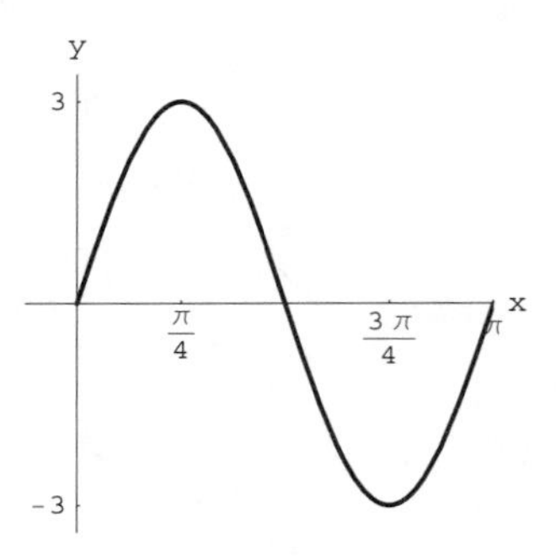

27. Rewriting, we find

$$y = 2\cos(\pi(x-1)).$$

Thus, we have the following: amplitude 2, period $2\pi/\pi$ or 2, phase shift 1, domain $(-\infty, \infty)$, and range $[-2, 2]$.

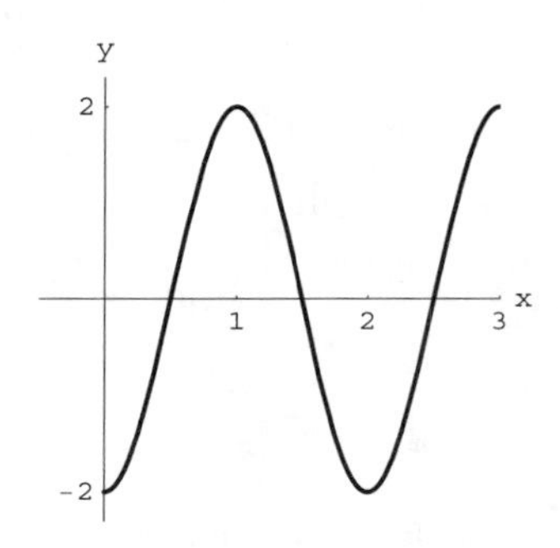

28. Rewriting, we find

$$y = \cos\left(2\left(x - \frac{\pi}{4}\right)\right) + 1.$$

Then we have the following: amplitude 1, period $2\pi/2$ or π, phase shift $\pi/4$, domain $(-\infty, \infty)$, and range is $[0, 2]$.

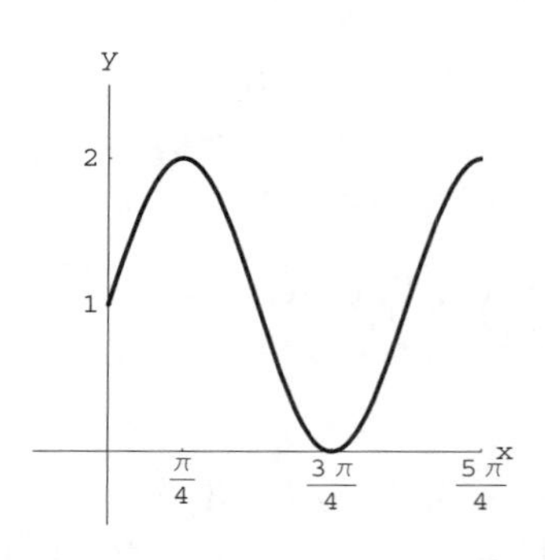

29. The period of

$$y = \tan(x - \pi/2)$$

is π and the phase shift is $\pi/2$. If

$$\cos(x - \pi/2) = 0$$

then

$$x - \frac{\pi}{2} = \frac{\pi}{2} + k\pi$$

or equivalently,

$$x = k\pi$$

where k is an integer. Thus, the domain is

$$\{x : x \neq k\pi\}$$

and the range is $(-\infty, \infty)$.

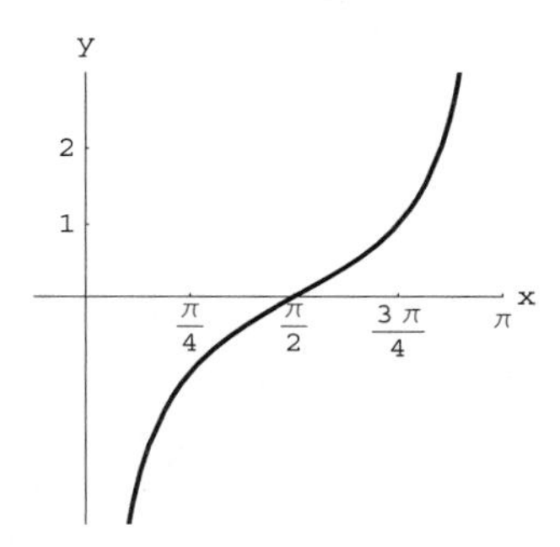

30. The period of

$$y = \tan(\pi x) - 3$$

is π/π or 1, and the phase shift is 0. If

$$\cos(\pi x) = 0$$

then

$$\pi x = \frac{\pi}{2} + k\pi \quad \text{or } x = \frac{1}{2} + k$$

where k is an integer. Thus, the domain is

$$\{x : x \neq \frac{1}{2} + k\}$$

and the range is $(-\infty, \infty)$.

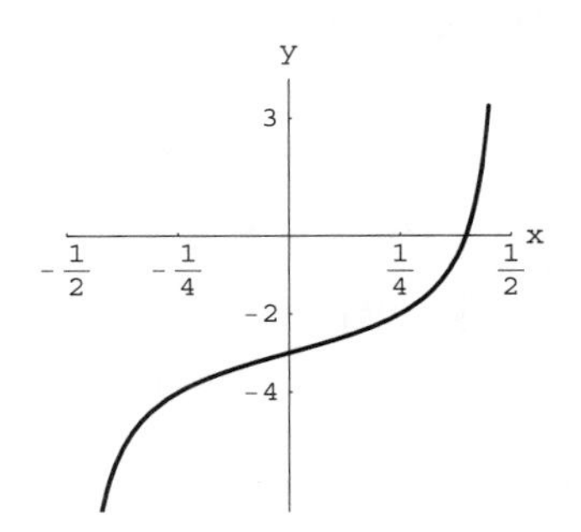

For Thought

1. True, since $i \cdot (-i) = 1$.

2. True, since $\overline{0+i} = 0 - i = -i$.

3. False, the set of real numbers is a subset of the complex numbers.

4. True, $(\sqrt{3} - i\sqrt{2})(\sqrt{3} + i\sqrt{2}) = 3 + 2 = 5$.

5. True, since $(2+5i)(2-5i) = 2^2 - (5i)^2 = 4 - (-25) = 4 + 25$.

6. False, $5 - \sqrt{-9} = 5 - 3i$.

7. True, since $P(3i) = (3i)^2 + 9 = -9 + 9 = 0$.

8. True, since $(-3i)^2 + 9 = -9 + 9 = 0$.

9. True, since $i^4 = i^2 \cdot i^2 = (-1)(-1) = 1$.

10. False, $i^{18} = (i^4)^4 i^2 = (1)^4(-1) = -1$.

6.1 Exercises

1. $0 + 6i$, imaginary

3. $\frac{1}{3} + \frac{1}{3}i$, imaginary

5. $\sqrt{7} + 0i$, real

7. $\frac{\pi}{2} + 0i$, real

9. $7 + 2i$

11. $1 - i - 3 - 2i = -2 - 3i$

13. $-18i + 12i^2 = -12 - 18i$

15. $8 + 12i - 12i - 18i^2 = 26$

17. $(5-2i)(5+2i) = 25 - 4i^2 = 25 - 4(-1) = 29$

19. $(\sqrt{3} - i)(\sqrt{3} + i) = 3 - i^2 = 3 - (-1) = 4$

21. $9 + 24i + 16i^2 = -7 + 24i$

23. $5 - 4i\sqrt{5} + 4i^2 = 1 - 4\sqrt{5}i$

25. $(i^4)^4 \cdot i = (1)^4 \cdot i = i$

27. $(i^4)^{24} i^2 = 1^{24}(-1) = -1$

29. $(i^4)^{-1} = 1^{-1} = 1$

31. Since $i^4 = 1$, we get $i^{-1} = i^{-1}i^4 = i^3 = -i$.

33. $(3-9i)(3+9i) = 9 - 81i^2 = 90$

35. $\left(\frac{1}{2} + 2i\right)\left(\frac{1}{2} - 2i\right) = \frac{1}{4} - 4i^2 = \frac{1}{4} + 4 = \frac{17}{4}$

37. $i(-i) = -i^2 = 1$

39. $(3 - i\sqrt{3})(3 + i\sqrt{3}) = 9 - 3i^2 = 9 - 3(-1) = 12$

41. $\frac{1}{2-i} \cdot \frac{2+i}{2+i} = \frac{2+i}{5} = \frac{2}{5} + \frac{1}{5}i$

43. $\frac{-3i}{1-i} \cdot \frac{1+i}{1+i} = \frac{-3i+3}{2} = \frac{3}{2} - \frac{3}{2}i$

45. $\frac{-2+6i}{2} = \frac{-2}{2} + \frac{6i}{2} = -1 + 3i$

47. $$\frac{-3+3i}{i} \cdot \frac{-i}{-i} = \frac{3i - 3i^2}{1} = 3i - 3(-1) = 3 + 3i$$

49. $$\frac{1-i}{3+2i} \cdot \frac{3-2i}{3-2i} = \frac{3 - 5i - 2}{13} = \frac{1}{13} - \frac{5}{13}i$$

51. $$\frac{\sqrt{2} - i\sqrt{3}}{\sqrt{3} + i\sqrt{2}} \cdot \frac{\sqrt{3} - i\sqrt{2}}{\sqrt{3} - i\sqrt{2}} = \frac{-5i}{5} = -i$$

53. $2i - 3i = -i$

55. $-4 + 2i$

57. $\left(i\sqrt{6}\right)^2 = -6$

59. $(i\sqrt{2})(i\sqrt{50}) = i^2\sqrt{2}{\cdot}5\sqrt{2} = (-1)(2)(5) = -10$

61. $$\frac{-2}{2} + \frac{i\sqrt{20}}{2} = -1 + i\frac{2\sqrt{5}}{2} = -1 + \sqrt{5}i$$

63. $-3 + \sqrt{9 - 20} = -3 + i\sqrt{11}$

65. $2i\sqrt{2}\left(i\sqrt{2} + 2\sqrt{2}\right) = 4i^2 + 8i = -4 + 8i$

67. $(2i + 2\sqrt{2})(i - \sqrt{2}) = 2(i + \sqrt{2})(i - \sqrt{2}) = 2(i^2 - \sqrt{2}^2) = 2(-1-2) = -6$

69. $(-2+i)^2 + 4(-2+i) + 5 = (3 - 4i) + (-8 + 4i) + 5 = 0$

71. $(-2-i)^2 + 4(-2-i) + 5 = (3 + 4i) + (-8 - 4i) + 5 = 0$

73. $(1+i)^2+4(1+i)+5=$
$(2i)+(4+4i)+5=9+6i$

75. $\left(3+i\sqrt{5}\right)^2-6(3+i\sqrt{5})+14=$
$\left(4+6i\sqrt{5}\right)-4-6i\sqrt{5}=0$

77. $(-1)^2-6(-1)+14=1+6+14=21$

79. Yes, since $(2i)^2+4=-4+4=0$

81. No, since $(1-i)^2+2(1-i)-2=-2i-2i=-4i$

83. Yes, since $(3-2i)^2-6(3-2i)+13=$
$5-12i-5+12i=0$

85. Yes, since $3\left(\dfrac{i\sqrt{3}}{3}\right)^2+1=3\left(\dfrac{-3}{9}\right)+1=0$

87. Yes, $\left(2+i\sqrt{3}\right)^2-4(2+i\sqrt{3})+7=$
$\left(1+4i\sqrt{3}\right)-1-4i\sqrt{3}=0$

89. If r is the remainder when n is divided by 4, then $i^n=i^r$. The possible values of r are $0,1,2,3$ and for i^r they are $1,i,-1,-i$, respectively.

91. Yes, since the product of $a+bi$ with its conjugate is $(a+bi)(a-bi)=a^2+b^2$, which is a real number.

For Thought

1. True, since $(3,-4)$ lies in quadrant 4.

2. False, the absolute value is

$$\sqrt{(-2)^2+(-5)^2}=\sqrt{29}.$$

3. True, since $r=\sqrt{1^1+(-3)^2}=\sqrt{10}$ and $\cos\theta=x/r=1/\sqrt{10}$.

4. False, $\tan\theta=3/2$.

5. False, $i=1(\cos 90°+i\sin 90°)$.

6. True, since $\cos 30°=\sqrt{3}/2$ and $30°$ lies in quadrant 1.

7. True, since $|2-5i|=\sqrt{2^2+(-5)^2}=\sqrt{29}$.

8. True, since $2-4i$ lies in quadrant 4 and

$$\cos\theta=x/r=2/\sqrt{20}=1/\sqrt{5}.$$

9. True, since $\dfrac{\pi}{4}+\dfrac{\pi}{2}=\dfrac{3\pi}{4}$.

10. False, since
$\dfrac{3(\cos\pi/4+i\sin\pi/4)}{3(\cos\pi/2+i\sin\pi/2)}=$
$1.5(\cos(-\pi/4)+i\sin(-\pi/4))=$
$1.5(\cos\pi/4-i\sin\pi/4)=$

6.2 Exercises

1. $\sqrt{2^2+(-6)^2}=\sqrt{40}=2\sqrt{10}$

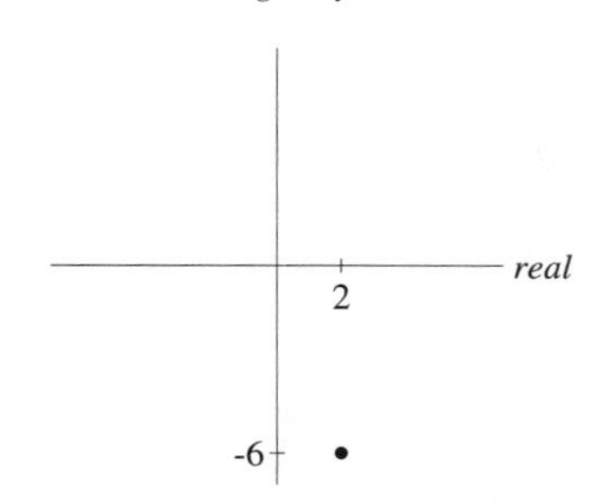

3. $\sqrt{(-2)^2+(2\sqrt{3})^2}=\sqrt{4+12}=4$

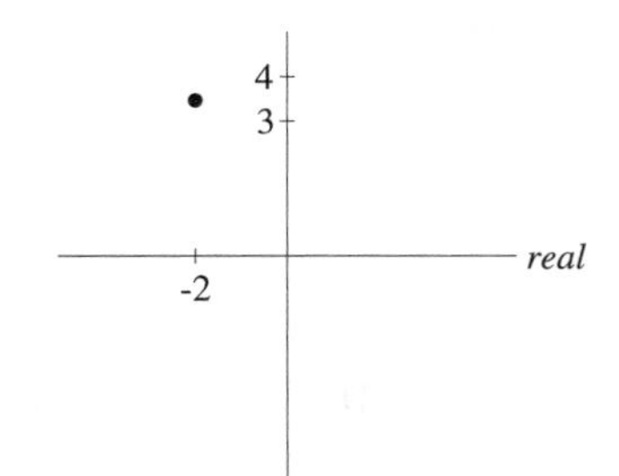

5. $\sqrt{0^2+8^2}=8$

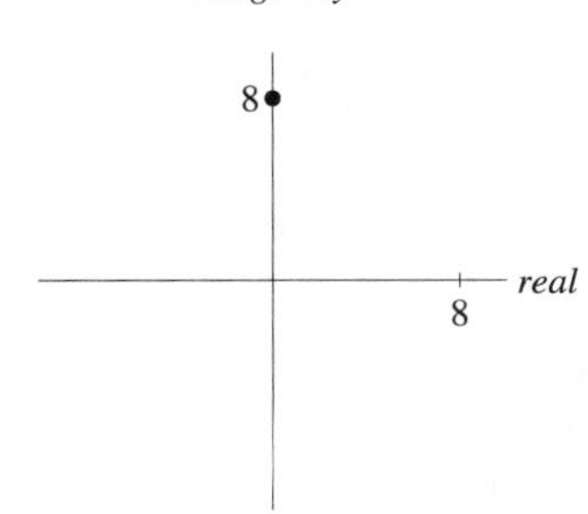

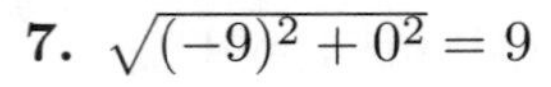

7. $\sqrt{(-9)^2 + 0^2} = 9$

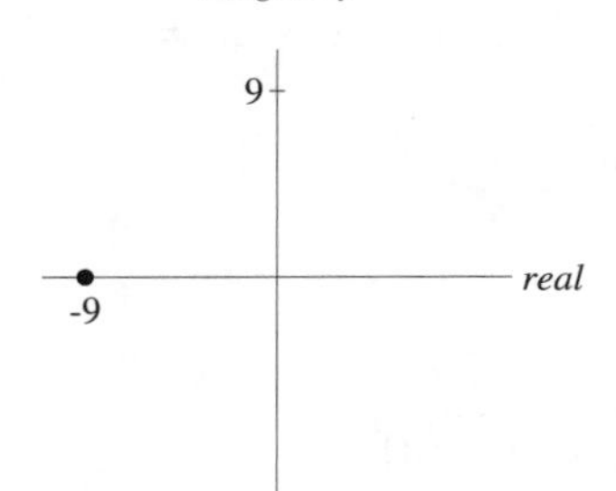

9. $\sqrt{(1/\sqrt{2})^2 + (-1/\sqrt{2})^2} = \sqrt{1/2 + 1/2} = 1$

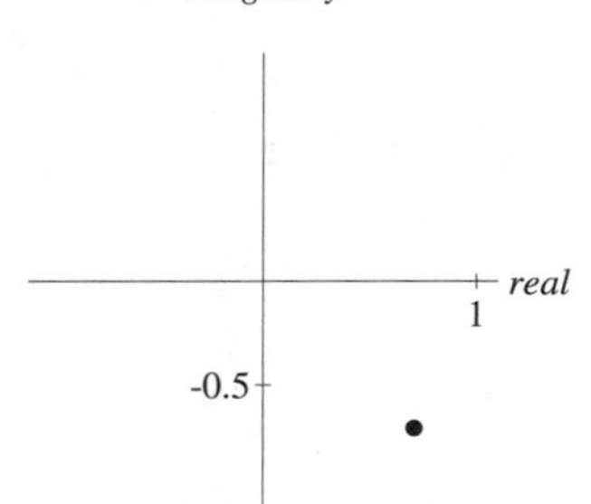

11. $\sqrt{3^2 + 3^2} = \sqrt{18} = 3\sqrt{2}$

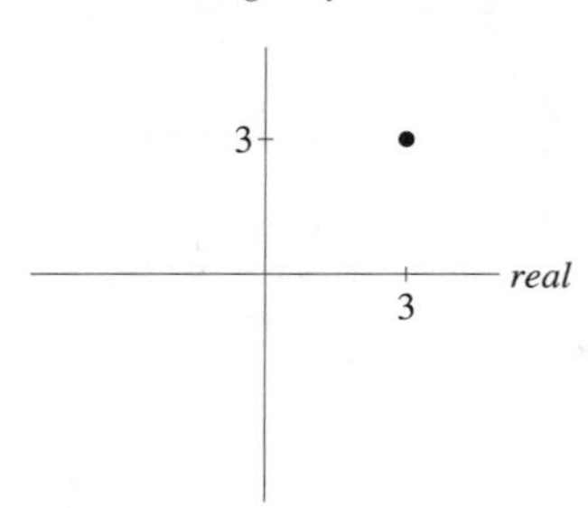

13. Since $|-3 + 3i| = \sqrt{(-3)^2 + 3^2} = 3\sqrt{2}$ and if the terminal side of θ goes through $(-3, 3)$ then $\cos\theta = -3/(3\sqrt{2}) = -1/\sqrt{2}$. One can choose $\theta = 135°$ and the trigonometric form is

$$3\sqrt{2}\left(\cos 135° + i \sin 135°\right).$$

15. Since $\sqrt{(-3/\sqrt{2})^2 + (3/\sqrt{2})^2} = 3$ and if the terminal side of θ goes through $(-3/\sqrt{2}, 3/\sqrt{2})$ then $\cos\theta = (-3/\sqrt{2})/3 = -1/\sqrt{2}$. One can choose $\theta = 135°$. The trigonometric form is

$$3\left(\cos 135° + i \sin 135°\right).$$

17. Since terminal side of $0°$ goes through $(8, 0)$, the trigonomeric form is

$$8\left(\cos 0° + i \sin 0°\right).$$

19. Since terminal side of $90°$ goes through $(0, \sqrt{3})$, the trigonomeric form is

$$\sqrt{3}\left(\cos 90° + i \sin 90°\right).$$

21. Since $|-\sqrt{3} + i| = \sqrt{(-\sqrt{3})^2 + 1^2} = 2$ and if the terminal side of θ goes through $(-\sqrt{3}, 1)$ then $\cos\theta = -\sqrt{3}/2$. One can choose $\theta = 150°$. The trigonometric form is

$$2\left(\cos 150° + i \sin 150°\right).$$

23. Since $|3 + 4i| = \sqrt{3^2 + 4^2} = 5$ and if the terminal side of θ goes through $(3, 4)$ then $\cos\theta = 3/5$. One can choose $\theta = \cos^{-1}(3/5) \approx 53.1°$. The trigonometric form is

$$5\left(\cos 53.1° + i \sin 53.1°\right).$$

25. Since $|-3 + 5i| = \sqrt{(-3)^2 + 5^2} = \sqrt{34}$ and if the terminal side of θ goes through $(-3, 5)$ then $\cos\theta = -3/\sqrt{34}$. One can choose $\theta = \cos^{-1}(-3/\sqrt{34}) \approx 121.0°$. The trigonometric form is

$$\sqrt{34}\left(\cos 121.0° + i \sin 121.0°\right).$$

27. Note $|3 - 6i| = \sqrt{3^2 + (-6)^2} = \sqrt{45} = 3\sqrt{5}$. If the terminal side of θ goes through $(3, -6)$ then $\tan\theta = -6/3 = -2$. Since $\tan^{-1}(-2) \approx -63.4°$, one can choose $\theta = 360° - 63.4° = 296.6°$. The trigonometric form is

$$3\sqrt{5}\left(\cos 296.6° + i \sin 296.6°\right).$$

29. $\sqrt{2}\left(\dfrac{1}{\sqrt{2}} + i\dfrac{1}{\sqrt{2}}\right) = 1 + i$

31. $\dfrac{\sqrt{3}}{2}\left(-\dfrac{\sqrt{3}}{2} + i\dfrac{1}{2}\right) = -\dfrac{3}{4} + i\dfrac{\sqrt{3}}{4}$

33. $\dfrac{1}{2}(-0.848 - 0.53i) \approx -0.42 - 0.26i$

35. $3(0+i) = 3i$

37. $\sqrt{3}(0-i) = -i\sqrt{3}$

39. $\sqrt{6}\left(\frac{1}{2} + i\frac{\sqrt{3}}{2}\right) =$

$\frac{\sqrt{6}}{2} + i\frac{\sqrt{18}}{2} = \frac{\sqrt{6}}{2} + \frac{3\sqrt{2}}{2}i$

41. $6(\cos 450^\circ + i\sin 450^\circ) =$
$6(\cos 90^\circ + i\sin 90^\circ) = 6(0+i) = 6i$

43. $\sqrt{6}(\cos 30^\circ + i\sin 30^\circ) =$

$\sqrt{6}\left(\frac{\sqrt{3}}{2} + i\frac{1}{2}\right) =$

$\frac{\sqrt{18}}{2} + i\frac{\sqrt{6}}{2} = \frac{3\sqrt{2}}{2} + \frac{\sqrt{6}}{2}i$

45. $9(\cos 90^\circ + i\sin 90^\circ) =$
$9(0+i) = 9i$

47. $2(\cos(\pi/6) + i\sin(\pi/6)) =$

$2\left(\frac{\sqrt{3}}{2} + i\frac{1}{2}\right) = \sqrt{3} + i$

49. $0.5(\cos(-47.5^\circ) + i\sin(-47.5^\circ)) \approx$
$0.5(0.6756 - i \cdot 0.7373) \approx 0.34 - 0.37i$

51. Since $z_1 = 4\sqrt{2}(\cos 45^\circ + i\sin 45^\circ)$ and $z_2 = 5\sqrt{2}(\cos 225^\circ + i\sin 225^\circ)$, we have $z_1z_2 = 40(\cos 270^\circ + i\sin 270^\circ) =$ $40(0-i) = -40i$ and
$\frac{z_1}{z_2} = 0.8(\cos(-180^\circ) + i\sin(-180^\circ)) =$
$0.8(-1 + i \cdot 0) = -0.8$

53. Since $z_1 = 2(\cos 30^\circ + i\sin 30^\circ)$ and $z_2 = 4(\cos 60^\circ + i\sin 60^\circ)$, we have $z_1z_2 = 8(\cos 90^\circ + i\sin 90^\circ) =$ $8(0+i) = 8i$ and
$\frac{z_1}{z_2} = \frac{1}{2}(\cos(-30^\circ) + i\sin(-30^\circ)) =$
$\frac{1}{2}\left(\frac{\sqrt{3}}{2} - i\frac{1}{2}\right) = \frac{\sqrt{3}}{4} - \frac{1}{4}i.$

55. Since $z_1 = 2\sqrt{2}(\cos 45^\circ + i\sin 45^\circ)$ and $z_2 = 2(\cos 315^\circ + i\sin 315^\circ)$, we have $z_1z_2 = 4\sqrt{2}(\cos 360^\circ + i\sin 360^\circ) =$ $4\sqrt{2}(1 + i\cdot 0) = 4\sqrt{2}$ and
$\frac{z_1}{z_2} = \sqrt{2}(\cos(-270^\circ) + i\sin(-270^\circ)) =$
$\sqrt{2}(0+i) = \sqrt{2}i.$

57. Let α and β be angles whose terminal side goes through $(3,4)$ and $(-5,-2)$, respectively. Since $|3+4i| = 5$ and $|-5-2i| = \sqrt{29}$, we have $\cos\alpha = 3/5$, $\sin\alpha = 4/5$, $\cos\beta = -5/\sqrt{29}$, and $\sin\beta = -2/\sqrt{29}$. From the sum and difference identities, we find

$$\cos(\alpha+\beta) = -\frac{7}{5\sqrt{29}},$$
$$\sin(\alpha+\beta) = -\frac{26}{5\sqrt{29}},$$
$$\cos(\alpha-\beta) = -\frac{23}{5\sqrt{29}},$$
$$\sin(\alpha-\beta) = -\frac{14}{5\sqrt{29}}.$$

Note $z_1 = 5(\cos\alpha + i\sin\alpha)$ and $z_2 = \sqrt{29}(\cos\beta + i\sin\beta)$. Then

$$\begin{aligned} z_1z_2 &= 5\sqrt{29}(\cos(\alpha+\beta) + i\sin(\alpha+\beta)) \\ &= 5\sqrt{29}\left(-\frac{7}{5\sqrt{29}} - i\frac{26}{5\sqrt{29}}\right) \\ z_1z_2 &= -7 - 26i \end{aligned}$$

and

$$\begin{aligned} \frac{z_1}{z_2} &= \frac{5}{\sqrt{29}}(\cos(\alpha-\beta) + i\sin(\alpha-\beta)) \\ &= \frac{5}{\sqrt{29}}\left(-\frac{23}{5\sqrt{29}} - i\frac{14}{5\sqrt{29}}\right) \\ \frac{z_1}{z_2} &= -\frac{23}{29} - \frac{14}{29}i. \end{aligned}$$

59. Let α and β be angles whose terminal sides go through $(2,-6)$ and $(-3,-2)$, respectively. Since $|2-6i| = 2\sqrt{10}$ and $|-3-2i| = \sqrt{13}$, we have $\cos\alpha = 1/\sqrt{10}$, $\sin\alpha = -3/\sqrt{10}$, $\cos\beta = -3/\sqrt{13}$, and $\sin\beta = -2/\sqrt{13}$. By using the sum and difference identities, we obtain

$$\cos(\alpha+\beta) = -\frac{9}{\sqrt{130}},$$
$$\sin(\alpha+\beta) = \frac{7}{\sqrt{130}},$$
$$\cos(\alpha-\beta) = \frac{3}{\sqrt{130}},$$

$\sin(\alpha-\beta)=\dfrac{11}{\sqrt{130}}$.

Note $z_1=2\sqrt{10}(\cos\alpha+i\sin\alpha)$ and $z_2=\sqrt{13}(\cos\beta+i\sin\beta)$. Thus,

$$\begin{aligned} z_1z_2 &= 2\sqrt{130}\,(\cos(\alpha+\beta)+i\sin(\alpha+\beta)) \\ &= 2\sqrt{130}\left(-\frac{9}{\sqrt{130}}+i\frac{7}{\sqrt{130}}\right) \\ z_1z_2 &= -18+14i \end{aligned}$$

and

$$\begin{aligned} \frac{z_1}{z_2} &= \frac{2\sqrt{10}}{\sqrt{13}}\,(\cos(\alpha-\beta)+i\sin(\alpha-\beta)) \\ &= \frac{2\sqrt{10}}{\sqrt{13}}\left(\frac{3}{\sqrt{130}}+i\frac{11}{\sqrt{130}}\right) \\ \frac{z_1}{z_2} &= \frac{6}{13}+\frac{22}{13}i. \end{aligned}$$

61. Note $3i=3(\cos 90^\circ+i\sin 90^\circ)$ and $1+i=\sqrt{2}(\cos 45^\circ+i\sin 45^\circ)$. Then we get

$$\begin{aligned} (3i)(1+i) &= 3\sqrt{2}\,(\cos 135^\circ+i\sin 135^\circ) \\ &= 3\sqrt{2}\left(-\frac{\sqrt{2}}{2}+i\cdot\frac{\sqrt{2}}{2}\right) \\ (3i)(1+i) &= -3+3i \end{aligned}$$

and

$$\begin{aligned} \frac{3i}{1+i} &= \frac{3}{\sqrt{2}}\,(\cos 45^\circ+i\sin 45^\circ) \\ &= \frac{3}{\sqrt{2}}\left(\frac{\sqrt{2}}{2}+i\cdot\frac{\sqrt{2}}{2}\right) \\ (3i)(1+i) &= 1.5+1.5i. \end{aligned}$$

63. $\left[3\left(\cos\frac{\pi}{6}+i\sin\frac{\pi}{6}\right)\right]\left[3\left(\cos\frac{\pi}{6}-i\sin\frac{\pi}{6}\right)\right]$

$=9\left(\cos^2\frac{\pi}{6}+\sin^2\frac{\pi}{6}\right)=9(1)=9$

65. $[2(\cos 7^\circ+i\sin 7^\circ)]\,[2(\cos 7^\circ-i\sin 7^\circ)]$ $=4(\cos^2 7^\circ+\sin^2 7^\circ)=4(1)=4$

67. Since $3+3i=3\sqrt{2}\,(\cos 45^\circ+i\sin 45^\circ)$, we get

$(3+3i)^3=$

$(3\sqrt{2})^3\,(\cos(3\cdot 45^\circ)+i\sin(3\cdot 45^\circ))=$

$54\sqrt{2}\,(\cos 135^\circ+i\sin 135^\circ)=$

$54\sqrt{2}\left(-\frac{1}{\sqrt{2}}+i\frac{1}{\sqrt{2}}\right)=-54+54i.$

69. The reciprocal of z is $\dfrac{1}{z}=\dfrac{\cos 0+i\sin 0}{r\,[\cos\theta+i\sin\theta]}=$

$r^{-1}\,[\cos(0-\theta)+i\sin(0-\theta)]=$

$r^{-1}\,[\cos\theta-i\sin\theta]$ provided $r\neq 0$.

71. Using polar form, we find $6\,(\cos 9^\circ+i\sin 9^\circ)+3\,(\cos 5^\circ+i\sin 5^\circ)=$ $6\cos 9^\circ+3\cos 5^\circ+i\,(6\sin 9^\circ+3\sin 5^\circ)$.

Since

$$(1+3i)+(5-7i)=6-4i$$

it is easier to add complex numbers in standard form.

For Thought

1. False, $(2+3i)^2=4+12i+9i^2$.

2. False, $z^3=8(\cos 360^\circ+i\sin 360^\circ)=8$.

3. True **4.** False, the argument is 4θ.

5. False, since $\left[\frac{1}{2}+i\cdot\frac{1}{2}\right]^2=\frac{i}{2}$ and $\cos 2\pi/3+i\sin\pi/3=-\frac{1}{2}+i\cdot\frac{\sqrt{3}}{2}$.

6. False, since $\cos 5\pi/6=\cos 7\pi/6$ and $5\pi/6\neq 7\pi/6+2k\pi$ for any integer k. It is possible that $\alpha=2k\pi-\beta$.

7. True, since $360/5=72$. **8.** True, since $|x|=1$.

9. True, $x=i$ is a solution not on $y=\pm x$.

10. False, it has four imaginary solutions.

6.3 Exercises

1. $3^3(\cos 90^\circ+i\sin 90^\circ)=27(0+i)=27i$

3. $(\sqrt{2})^4(\cos 480^\circ+i\sin 480^\circ)=$ $4(\cos 120^\circ+i\sin 120^\circ)=$ $4\left(-\frac{1}{2}+i\cdot\frac{\sqrt{3}}{2}\right)=-2+2i\sqrt{3}$

5. $\cos(8\pi/12)+i\sin(8\pi/12)=$ $\cos(2\pi/3)+i\sin(2\pi/3)=-\frac{1}{2}+i\frac{\sqrt{3}}{2}$

7. $(\sqrt{6})^4\,[\cos(8\pi/3) + i\sin(8\pi/3)] =$
$36\,[\cos(2\pi/3) + i\sin(2\pi/3)] =$
$36\left[-\frac{1}{2} + i\frac{\sqrt{3}}{2}\right] = -18 + 18i\sqrt{3}$

9. $4.3^5\,[\cos 61.5^\circ + i\sin 61.5^\circ] \approx$
$1470.1\,[0.4772 + 0.8788i] \approx 701.5 + 1291.9i$

11. $\left(2\sqrt{2}\,[\cos 45^\circ + i\sin 45^\circ]\right)^3 =$
$16\sqrt{2}\,[\cos 135^\circ + i\sin 135^\circ] =$
$16\sqrt{2}\left[-\frac{1}{\sqrt{2}} + i\frac{1}{\sqrt{2}}\right] = -16 + 16i$

13. $(2\,[\cos(-30^\circ) + i\sin(-30^\circ)])^4 =$
$16\,[\cos(-120^\circ) + i\sin(-120^\circ)] =$
$16\left[-\frac{1}{2} - i\frac{\sqrt{3}}{2}\right] = -8 - 8i\sqrt{3}$

15. $(6\,[\cos 240^\circ + i\sin 240^\circ])^5 =$
$7776\,[\cos 1200^\circ + i\sin 1200^\circ] =$
$7776\,[\cos 120^\circ + i\sin 120^\circ] =$
$7776\left[-\frac{1}{2} + i\frac{\sqrt{3}}{2}\right] = -3888 + 3888i\sqrt{3}$

17. Note $|2+3i| = \sqrt{13}$. If the terminal side of α goes through $(2,3)$ then $\cos\alpha = 2/\sqrt{13}$ and $\sin\alpha = 3/\sqrt{13}$. By using the double-angle identities one can successively obtain
$\cos 2\alpha = -5/13$, $\sin 2\alpha = 12/13$,
$\cos 4\alpha = -119/169$, $\sin 4\alpha = -120/169$.
So $(2+3i)^4 = \left(\sqrt{13}\,[\cos\alpha + i\sin\alpha]\right)^4 =$
$169(\cos 4\alpha + i\sin 4\alpha) =$
$169\,(-119/169 - 120i/169) = -119 - 120i$

19. Note $|2-i| = \sqrt{5}$. If the terminal side of α goes through $(2,-1)$ then $\cos\alpha = 2/\sqrt{5}$ and $\sin\alpha = -1/\sqrt{5}$. By using the double-angle identities one can successively obtain
$\cos 2\alpha = 3/5$, $\sin 2\alpha = -4/5$,
$\cos 4\alpha = -7/25$, $\sin 4\alpha = -24/25$.
So $(2-i)^4 = \left(\sqrt{5}\,[\cos\alpha + i\sin\alpha]\right)^4 =$
$25(\cos 4\alpha + i\sin 4\alpha) =$
$25\,(-7/25 - 24i/25) = -7 - 24i$.

21. Let $\omega = |1.2 + 3.6i|$. If the terminal side of α goes through $(1.2, 3.6)$ then $\cos\alpha = 1.2/\omega$ and $\sin\alpha = 3.6/\omega$. By using the double-angle identities, one obtains
$\cos 2\alpha = -11.52/\omega^2$ and $\sin 2\alpha = 8.64/\omega^2$.
By the sum identities, one gets

$$\cos 3\alpha = \cos(2\alpha + \alpha) = -44.928/\omega^3$$

and

$$\sin 3\alpha = \sin(2\alpha + \alpha) = -31.104/\omega^3.$$

Thus, we obtain
$(1.2 + 3.6i)^3 = (\omega\,[\cos\alpha + i\sin\alpha])^3 =$
$\omega^3(\cos 3\alpha + i\sin 3\alpha) =$
$\omega^3\,(-44.928/\omega^3 - 31.104i/\omega^3) =$
$-44.928 - 31.104i$.

23. The square roots are given by
$2\left[\cos\left(\frac{90^\circ + k360^\circ}{2}\right) + i\sin\left(\frac{90^\circ + k360^\circ}{2}\right)\right] =$
$2\,[\cos(45^\circ + k\cdot 180^\circ) + i\sin(45^\circ + k\cdot 180^\circ)] =$
where k is an integer. If $k = 0, 1$, we get
$2\,(\cos 45^\circ + i\sin 45^\circ)$ and
$2\,(\cos 225^\circ + i\sin 225^\circ)$.

25. The fourth roots are given by
$\cos\left(\frac{120^\circ + k360^\circ}{4}\right) + i\sin\left(\frac{120^\circ + k360^\circ}{4}\right) =$
$\cos(30^\circ + k\cdot 90^\circ) + i\sin(30^\circ + k\cdot 90^\circ)$
where k is an integer. If $k = 0, 1, 2, 3$, we find

$$\cos\alpha + i\sin\alpha$$

where $\alpha = 30^\circ, 120^\circ, 210^\circ, 300^\circ$.

27. The sixth roots are

$$2\left[\cos\left(\frac{\pi + 2k\pi}{6}\right) + i\sin\left(\frac{\pi + 2k\pi}{6}\right)\right]$$

where k is an integer. If $k = 0, 1, 2, 3, 4, 5$, we have

$$2\,(\cos\alpha + i\sin\alpha)$$

where $\alpha = \frac{\pi}{6}, \frac{\pi}{2}, \frac{5\pi}{6}, \frac{7\pi}{6}, \frac{3\pi}{2}, \frac{11\pi}{6}$.

29. The cube roots of 1 are given by

$$\cos\left(\frac{k360^\circ}{3}\right) + i\sin\left(\frac{k360^\circ}{3}\right)$$

where k is an integer. If $k = 0, 1, 2$, we get

$\cos 0^\circ + i\sin 0^\circ = 1$,

$\cos 120^\circ + i\sin 120^\circ = -\frac{1}{2} + i\frac{\sqrt{3}}{2}$, and

$\cos 240^\circ + i\sin 240^\circ = -\frac{1}{2} - i\frac{\sqrt{3}}{2}$.

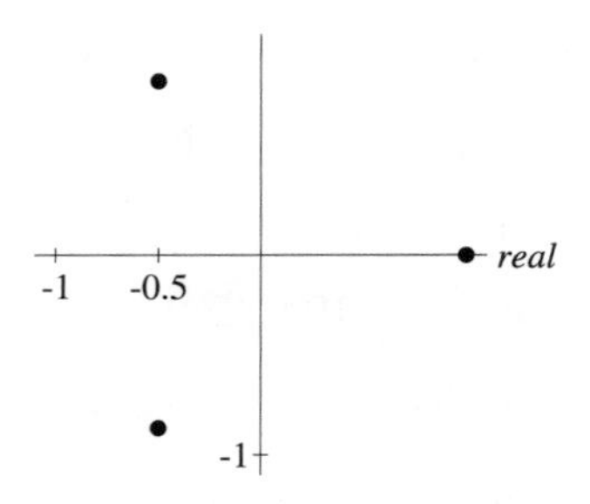

31. The fourth roots of 16 are

$$2\left[\cos\left(\frac{k360^\circ}{4}\right) + i\sin\left(\frac{k360^\circ}{4}\right)\right]$$

where k is an integer. If $k = 0, 1, 2, 3$, we obtain

$2[\cos 0^\circ + i\sin 0^\circ] = 2$,

$2[\cos 90^\circ + i\sin 90^\circ] = 2i$,

$2[\cos 180^\circ + i\sin 180^\circ] = -2$, and

$2[\cos 270^\circ + i\sin 270^\circ] = -2i$.

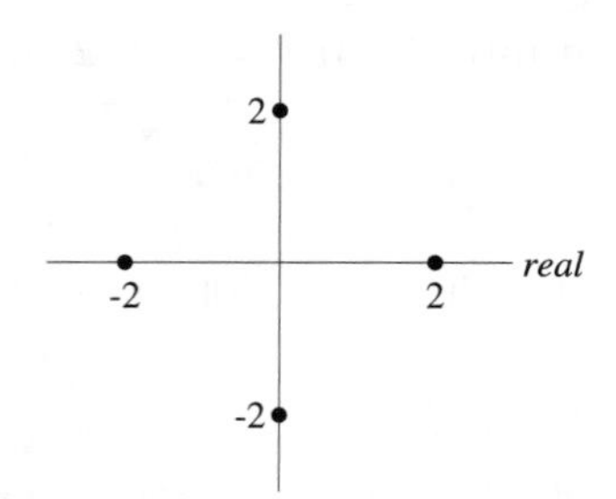

33. The fourth roots of -1 are

$\cos\left(\frac{180^\circ + k360^\circ}{4}\right) + i\sin\left(\frac{180^\circ + k360^\circ}{4}\right) =$

$\cos(45^\circ + k90^\circ) + i\sin(45^\circ + k90^\circ)$

where k is an integer. If $k = 0, 1, 2, 3$, we find

$\cos 45^\circ + i\sin 45^\circ = \frac{\sqrt{2}}{2} + \frac{\sqrt{2}}{2}i$,

$\cos 135^\circ + i\sin 135^\circ = -\frac{\sqrt{2}}{2} + \frac{\sqrt{2}}{2}i$,

$\cos 225^\circ + i\sin 225^\circ = -\frac{\sqrt{2}}{2} - \frac{\sqrt{2}}{2}i$, and

$\cos 315^\circ + i\sin 315^\circ = \frac{\sqrt{2}}{2} - \frac{\sqrt{2}}{2}i$.

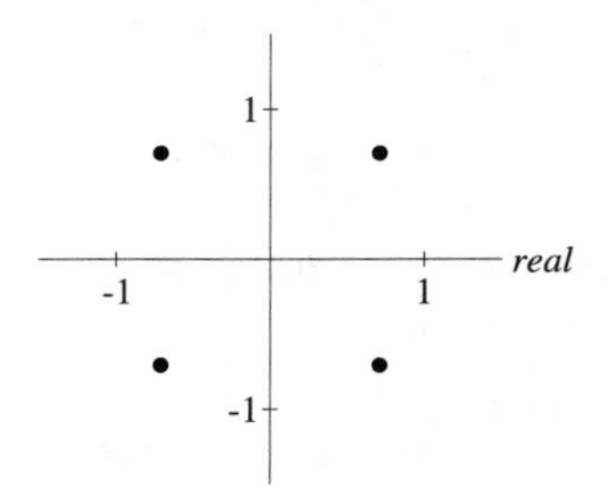

35. The cube roots of i are

$\cos\left(\frac{90^\circ + k360^\circ}{3}\right) + i\sin\left(\frac{90^\circ + k360^\circ}{3}\right) =$

$\cos(30^\circ + k120^\circ) + i\sin(30^\circ + k120^\circ)$

where k is an integer. If $k = 0, 1, 2$, we have

$\cos 30^\circ + i\sin 30^\circ = \frac{\sqrt{3}}{2} + \frac{1}{2}i$,

$\cos 150^\circ + i\sin 150^\circ = -\frac{\sqrt{3}}{2} + \frac{1}{2}i$, and

$\cos 270^\circ + i\sin 270^\circ = -i$.

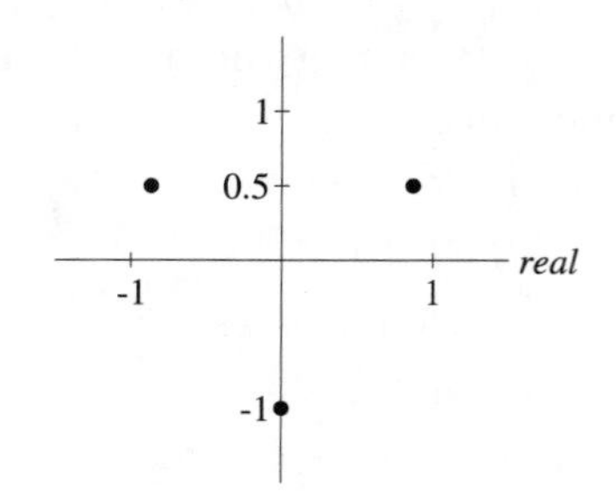

37. Since $|-2+2i\sqrt{3}| = 4$, the square roots are given by
$$2\left[\cos\left(\frac{120^\circ + k360^\circ}{2}\right) + i\sin\left(\frac{120^\circ + k360^\circ}{2}\right)\right]$$
$$= 2\left[\cos\left(60^\circ + k180^\circ\right) + i\sin\left(60^\circ + k180^\circ\right)\right].$$
If $k = 0, 1$ one obtains
$2\left[\cos 60^\circ + i\sin 60^\circ\right] = 1 + i\sqrt{3}$, and
$2\left[\cos 240^\circ + i\sin 240^\circ\right] = -1 - i\sqrt{3}$.

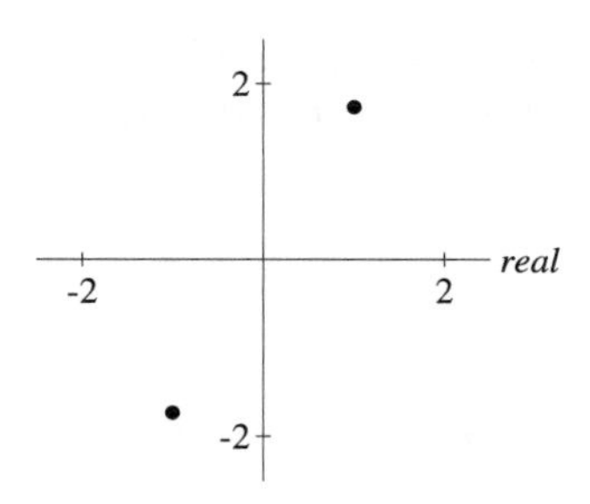

39. Note $|1+2i| = \sqrt{5}$. Since $\tan^{-1} 2 \approx 63.4^\circ$ and
$$\frac{63.4^\circ + k360^\circ}{2} = 31.7^\circ + k180^\circ$$
the square roots are given by
$\sqrt[4]{5}\left[\cos\left(31.7^\circ + k180^\circ\right) + i\sin\left(31.7^\circ + k180^\circ\right)\right]$.
If $k = 0, 1$ one obtains
$\sqrt[4]{5}\left[\cos 31.7^\circ + i\sin 31.7^\circ\right] \approx 1.272 + 0.786i$, and
$\sqrt[4]{5}\left[\cos 211.7^\circ + i\sin 211.7^\circ\right] \approx -1.272 - 0.786i$.

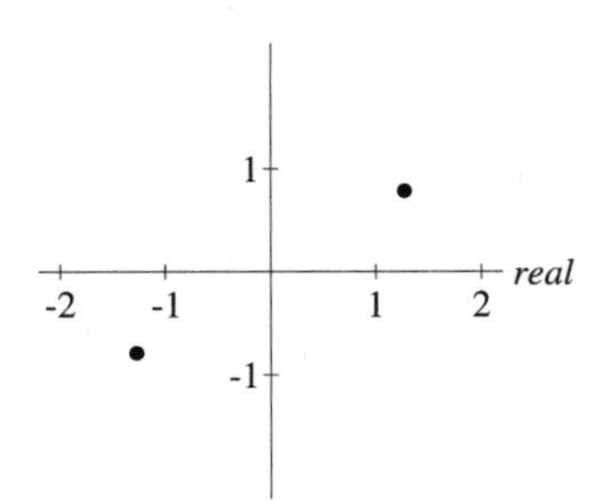

41. Solutions are the cube roots of -1. Namely,
$$\cos\left(\frac{180^\circ + k360^\circ}{3}\right) + i\sin\left(\frac{180^\circ + k360^\circ}{3}\right)$$
$= \cos\left(60^\circ + k120^\circ\right) + i\sin\left(60^\circ + k120^\circ\right)$.
If $k = 0, 1, 2$ one obtains
$\cos 60^\circ + i\sin 60^\circ = \frac{1}{2} + \frac{\sqrt{3}}{2}i$,
$\cos 180^\circ + i\sin 180^\circ = -1$, and
$\cos 300^\circ + i\sin 300^\circ = \frac{1}{2} - \frac{\sqrt{3}}{2}i$.

43. Solutions are the fourth roots of 81. Namely,
$$3\left[\cos\left(\frac{k360^\circ}{4}\right) + i\sin\left(\frac{k360^\circ}{4}\right)\right]$$
$= 3\left[\cos\left(k90^\circ\right) + i\sin\left(k90^\circ\right)\right]$.
If $k = 0, 1, 2, 3$ one obtains
$3\left[\cos 0^\circ + i\sin 0^\circ\right] = 3$,
$3\left[\cos 90^\circ + i\sin 90^\circ\right] = 3i$,
$3\left[\cos 180^\circ + i\sin 180^\circ\right] = -3$ and
$3\left[\cos 270^\circ + i\sin 270^\circ\right] = -3i$.

45. Solutions are the square roots of $-2i$.
If $k = 0, 1$ in $\alpha = \dfrac{-90^\circ + k360^\circ}{2}$ then
$\alpha = -45^\circ, 135^\circ$. These roots are
$\sqrt{2}\left[\cos(-45^\circ) + i\sin(-45^\circ)\right] = 1 - i$ and
$\sqrt{2}\left[\cos 135^\circ + i\sin 135^\circ\right] = -1 + i$.

47. Solutions of $x(x^6 - 64) = 0$ are $x = 0$ and the sixth roots of 64. The sixth roots are given by
$$2\left[\cos\left(\frac{k360^\circ}{6}\right) + i\sin\left(\frac{k360^\circ}{6}\right)\right]$$
$= 2\left[\cos\left(k60^\circ\right) + i\sin\left(k60^\circ\right)\right]$.
If $k = 0, 1, 2, 3, 4, 5$ one obtains
$2\left[\cos 0^\circ + i\sin 0^\circ\right] = 2$,
$2\left[\cos 60^\circ + i\sin 60^\circ\right] = 1 + i\sqrt{3}$,
$2\left[\cos 120^\circ + i\sin 120^\circ\right] = -1 + i\sqrt{3}$,
$2\left[\cos 180^\circ + i\sin 180^\circ\right] = -2$,
$2\left[\cos 240^\circ + i\sin 240^\circ\right] = -1 - i\sqrt{3}$, and
$2\left[\cos 300^\circ + i\sin 300^\circ\right] = 1 - i\sqrt{3}$.

49. Solutions are the fifth roots of 2. Namely,
$$\sqrt[5]{2}\left[\cos\left(\frac{k360^\circ}{5}\right) + i\sin\left(\frac{k360^\circ}{5}\right)\right]$$
$= \sqrt[5]{2}\left[\cos\left(k72^\circ\right) + i\sin\left(k72^\circ\right)\right]$.
Solutions are $x = \sqrt[5]{2}\left[\cos\alpha + i\sin\alpha\right]$ where $\alpha = 0^\circ, 72^\circ, 144^\circ, 216^\circ, 288^\circ$.

51. Solutions are the fourth roots of $-3 + i$. Since $|-3+i| = \sqrt{10}$, an argument of $-3 + i$ is $\cos^{-1}(-3/\sqrt{10}) \approx 161.6^\circ$. Arguments of the fourth roots are given by
$$\frac{161.6^\circ + k360^\circ}{4} = 40.4^\circ + k90^\circ.$$
By choosing $k = 0, 1, 2, 3$, the solutions are $x = \sqrt[8]{10}\left[\cos\alpha + i\sin\alpha\right]$ where $\alpha = 40.4^\circ, 130.4^\circ, 220.4^\circ, 310.4^\circ$.

53. $[\cos \pi/3 + i \sin \pi/6]^3 =$
$[1/2 + i(1/2)]^3 = \left[\frac{1}{2}(1+i)\right]^3 =$
$\frac{1}{8}\left[\sqrt{2}\,(\cos 45^\circ + i \sin 45^\circ)\right]^3 =$
$\frac{1}{8}\left[2\sqrt{2}\,(\cos 135^\circ + i \sin 135^\circ)\right] =$
$\frac{\sqrt{2}}{4}\left(-\frac{\sqrt{2}}{2} + i \cdot \frac{\sqrt{2}}{2}\right) = -\frac{1}{4} + \frac{1}{4}i$

55. By the quadratic formula, we get

$$x = \frac{-(-1+i) \pm \sqrt{(-1+i)^2 - 4(1)(-i)}}{2}$$
$$x = \frac{1 - i \pm \sqrt{(1 - 2i - 1) + 4i}}{2}$$
$$x = \frac{1 - i \pm \sqrt{2i}}{2}.$$

Note the square roots of $2i$ are given by
$\sqrt{2}(\cos 45^\circ + i \sin 45^\circ) = 1 + i$ and
$\sqrt{2}(\cos 225^\circ + i \sin 225^\circ) = -1 - i$.
These two roots differ by a minus sign. So

$$x = \frac{1 - i \pm (1+i)}{2}$$
$$x = \frac{1 - i + 1 + i}{2} \quad \text{or} \quad x = \frac{1 - i - 1 - i}{2}$$
$$x = 1 \quad \text{or} \quad x = -i.$$

The solutions are $x = 1, -i$.

For Thought

1. True, since the distance is $|r|$.

2. False, the distance is $|r|$.

3. False

4. False, $x = r\cos\theta$, $y = r\sin\theta$, and $x^2 + y^2 = r^2$.

5. True, since $x = -4\cos 225^\circ = 2\sqrt{2}$ and
$y = -4\sin 225^\circ = 2\sqrt{2}$.

6. True, $\theta = \pi/4$ is a straight line through the origin which makes an angle of $\pi/4$ with the positive x-axis.

7. True, since each circle is centered at the origin with radius 5.

8. False, since upon substitution one gets
$\cos 2\pi/3 = -1/2$ while $r^2 = 1/2$.

9. False, for $r = 1/\sin\theta$ is undefined when $\theta = k\pi$ for any integer k.

10. False, since $r = \theta$ is a reflection of $r = -\theta$ about the origin.

6.4 Exercises

1. Since $r = \sqrt{3^2 + 3^2} = 3\sqrt{2}$ and
$\theta = \tan^{-1}(3/3) = \pi/4$, $(r, \theta) = (3\sqrt{2}, \pi/4)$.

3. $(3, \pi/2)$

5. $(3, \pi/6)$

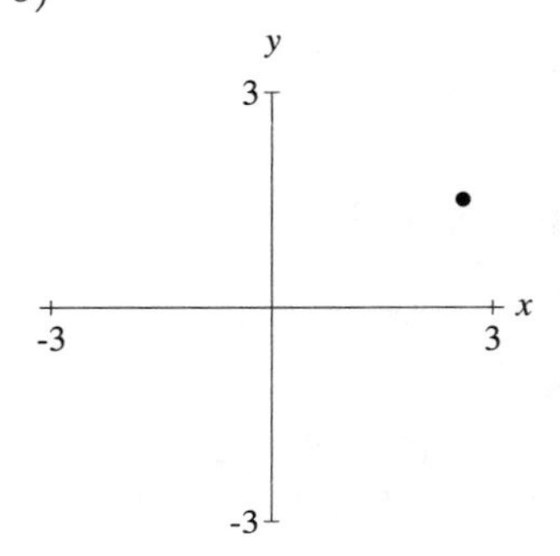

7. $(-2, 2\pi/3)$

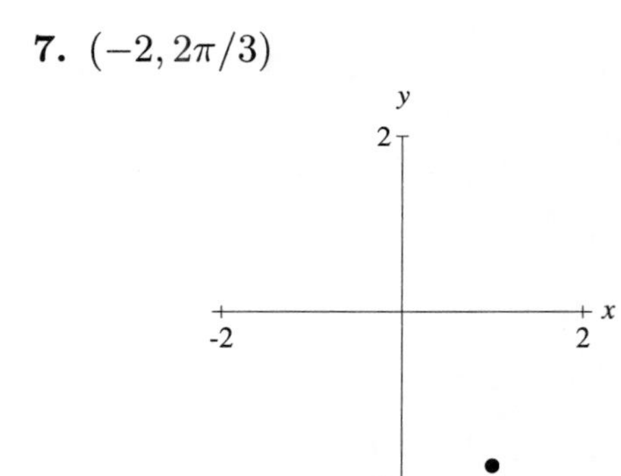

9. $(2, -\pi/4)$

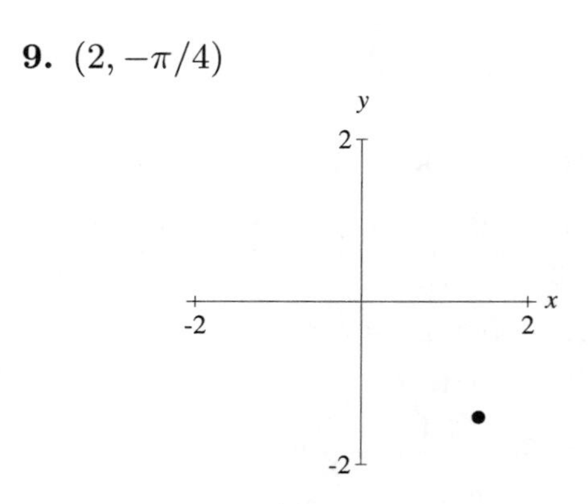

11. $(3, -225°)$

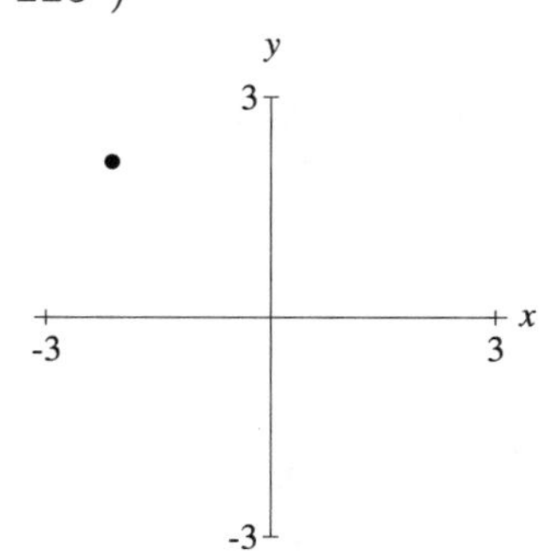

13. $(-2, 45°)$

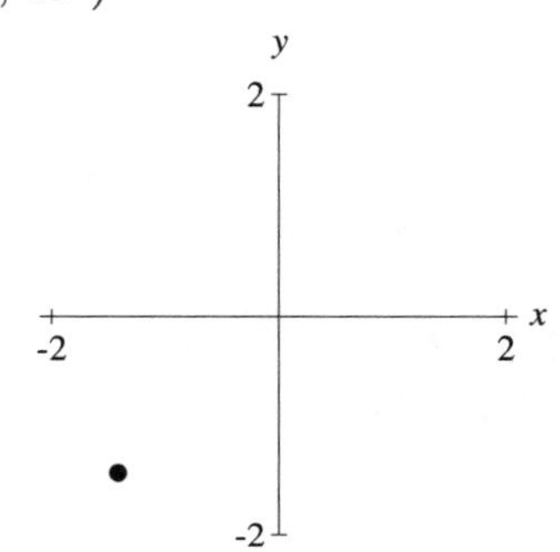

15. $(4, 390°)$

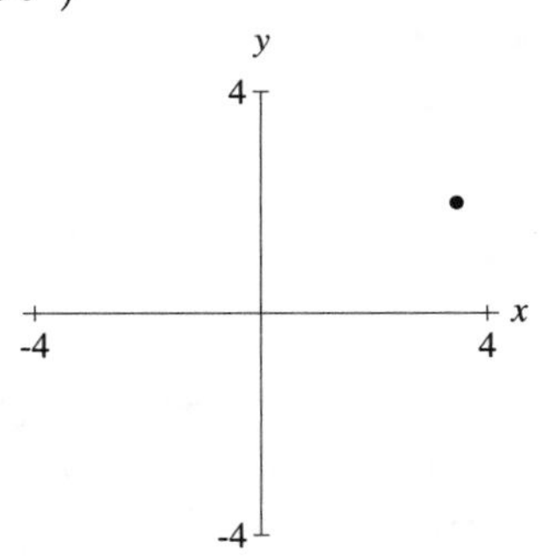

17. $(x, y) = (1 \cdot \cos(\pi/6), 1 \cdot \sin(\pi/6)) = \left(\frac{\sqrt{3}}{2}, \frac{1}{2}\right)$

19. $(x, y) = (-3\cos(3\pi/2), -3\sin(3\pi/2)) = (0, 3)$

21. $(x, y) = \left(\sqrt{2}\cos 135°, \sqrt{2}\sin 135°\right) = (-1, 1)$

23. $(x, y) = \left(-\sqrt{6}\cos(-60°), -\sqrt{6}\sin(-60°)\right) = \left(-\frac{\sqrt{6}}{2}, \frac{3\sqrt{2}}{2}\right)$

25. Since $r = \sqrt{(\sqrt{3})^2 + 3^2} = 2\sqrt{3}$ and $\cos\theta = \frac{x}{r} = \frac{\sqrt{3}}{2\sqrt{3}} = \frac{1}{2}$, one can choose $\theta = 60°$. Then

$$(r, \theta) = (2\sqrt{3}, 60°).$$

27. Since $r = \sqrt{(-2)^2 + 2^2} = 2\sqrt{2}$ and $\cos\theta = \frac{x}{r} = \frac{-2}{2\sqrt{2}} = -\frac{1}{\sqrt{2}}$, one can choose $\theta = 135°$. Then

$$(r, \theta) = (2\sqrt{2}, 135°).$$

29. $(r, \theta) = (2, 90°)$

31. Note $r = \sqrt{(-3)^2 + (-3)^2} = 3\sqrt{2}$.

Since $\tan\theta = \frac{y}{x} = \frac{-3}{-3} = 1$ and $(-3, -3)$ is in quadrant III, one can choose $\theta = 225°$. Then

$$(r, \theta) = (3\sqrt{2}, 225°).$$

33. Since $r = \sqrt{1^2 + 4^2} = \sqrt{17}$ and $\theta = \cos^{-1}\left(\frac{x}{r}\right) = \cos^{-1}\left(\frac{1}{\sqrt{17}}\right) \approx 75.96°$, we find

$$(r, \theta) = (\sqrt{17}, 75.96°).$$

35. Since $r = \sqrt{(\sqrt{2})^2 + (-2)^2} = \sqrt{6}$ and $\theta = \tan^{-1}\left(\frac{y}{x}\right) = \tan^{-1}\left(\frac{-2}{\sqrt{2}}\right) \approx -54.7°$, we find

$$(r, \theta) = (\sqrt{6}, -54.7°).$$

37. $r = 2\sin\theta$ is a circle centered at $(x, y) = (0, 1)$. It goes through the following points in polar coordinates: $(0, 0)$, $(1, \pi/6)$, $(2, \pi/2)$, $(1, 5\pi/6)$, $(0, \pi)$.

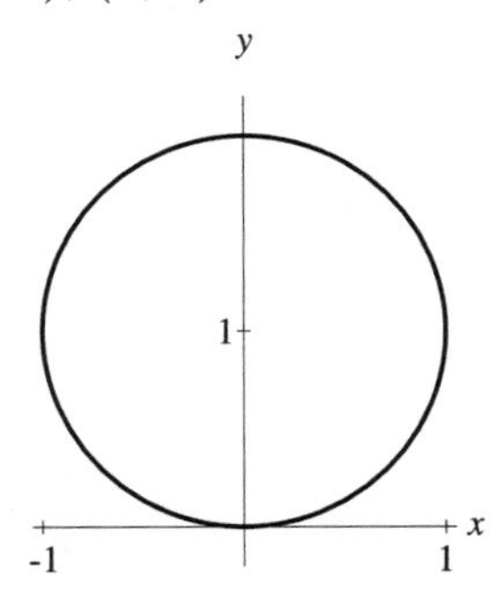

39. $r = 3\cos 2\theta$ is a four-leaf rose that goes through the following points in polar coordinates $(3,0)$, $(3/2,\pi/6)$, $(0,\pi/4)$, $(-3,\pi/2)$, $(0,3\pi/4)$, $(3,\pi)$, $(-3,3\pi/2)$, $(0,7\pi/4)$.

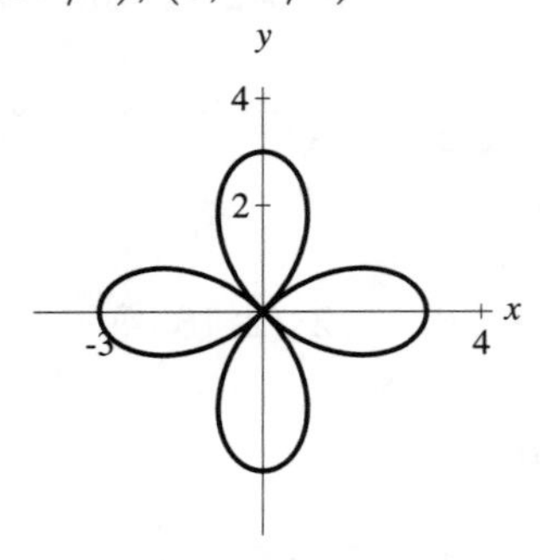

41. $r = 2\theta$ is spiral-shaped and goes through the following points in polar coordinates $(-\pi,-\pi/2)$, $(0,0)$, $(\pi,\pi/2)$, $(2\pi,\pi)$

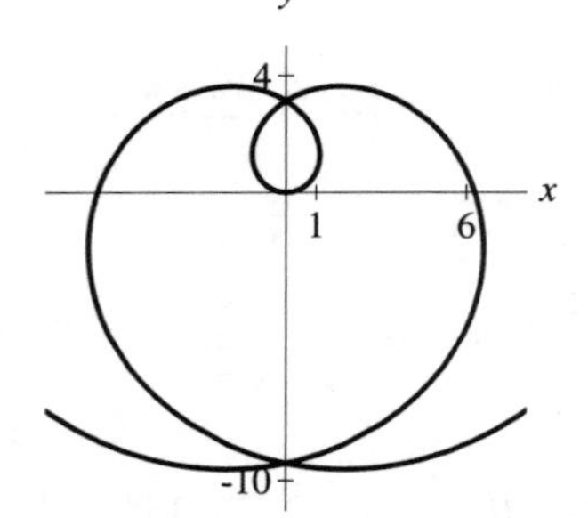

43. $r = 1+\cos\theta$ goes through the following points in polar coordinates $(2,0)$, $(1.5,\pi/3)$, $(1,\pi/2)$, $(0.5,2\pi/3)$, $(0,\pi)$.

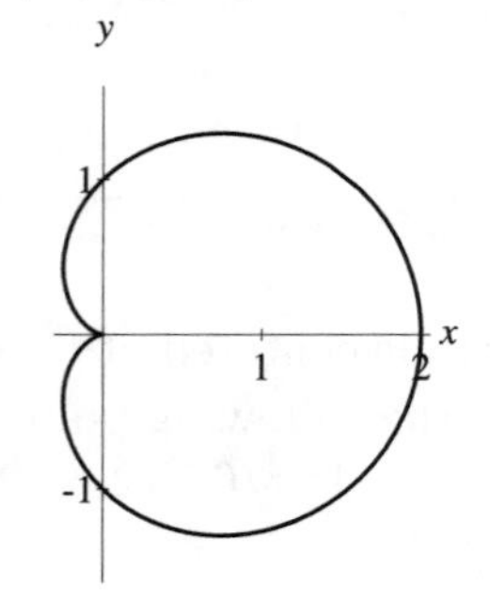

45. $r^2 = 9\cos 2\theta$ goes through the following points in polar coordinates $(0,-\pi/4)$, $(\pm 3\sqrt{2}/2,-\pi/6)$, $(\pm 3,0)$, $(\pm 3\sqrt{2}/2,\pi/6)$, and $(0,\pi/4)$.

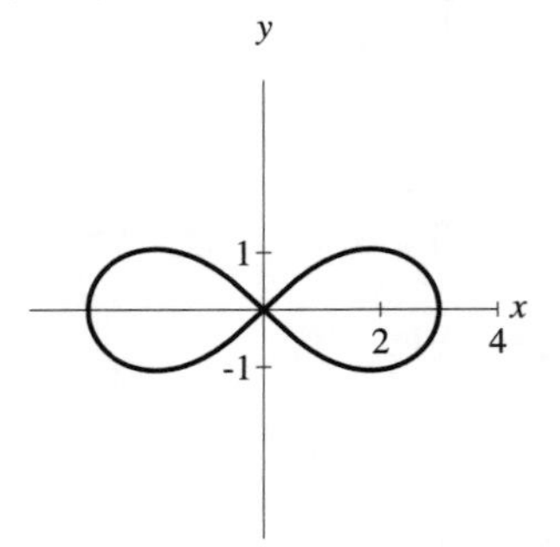

47. $r = 4\cos 2\theta$ is a four-leaf rose that goes through the following points in polar coordinates $(4,0)$, $(2,\pi/6)$, $(0,\pi/4)$, $(-4,\pi/2)$, $(2,5\pi/6)$, $(4,\pi)$, $(0,5\pi/4)$, $(-4,3\pi/2)$, $(0,7\pi/4)$.

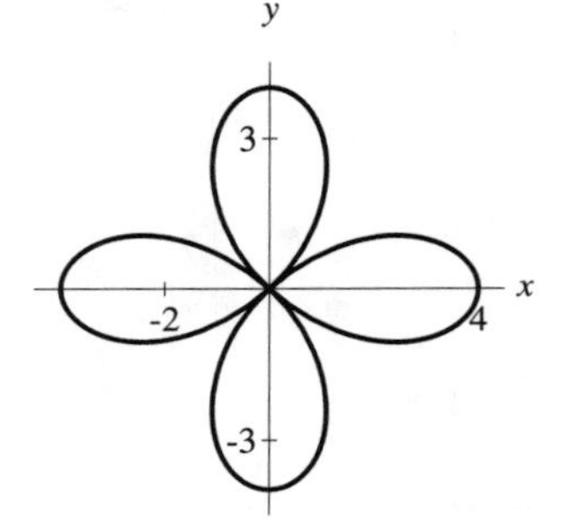

49. $r = 2\sin 3\theta$ is a three-leaf rose that goes through the following points in polar coordinates $(0,0)$, $(2,\pi/6)$, $(-2,\pi/2)$, $(2,5\pi/6)$.

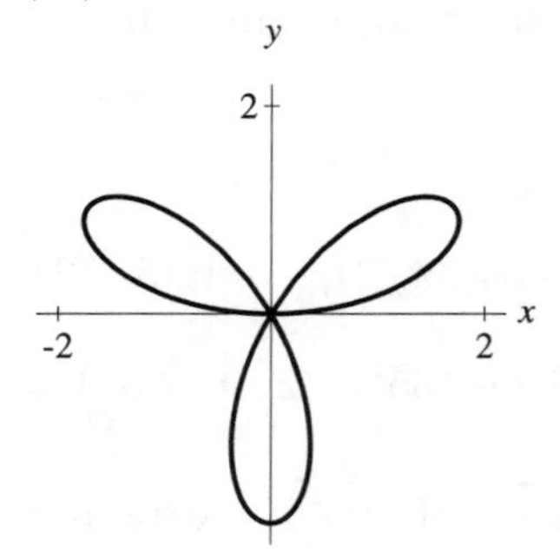

51. $r = 1 + 2\cos\theta$ goes through the following points in polar coordinates $(3, 0)$, $(2, \pi/3)$, $(1, \pi/2)$, $(0, 2\pi/3)$, $(1 - \sqrt{3}, 5\pi/6)$, $(-1, \pi)$.

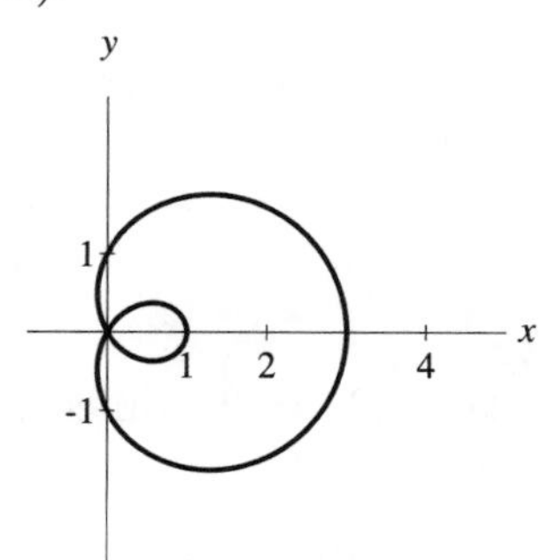

53. $r = 3.5$ is a circle centered at the origin with radius 3.5.

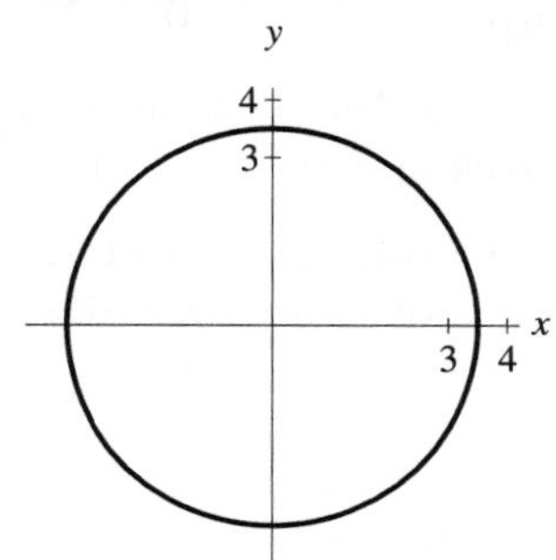

55. $\theta = 30°$ is a line through the origin that makes a 30° angle with the positive x-axis.

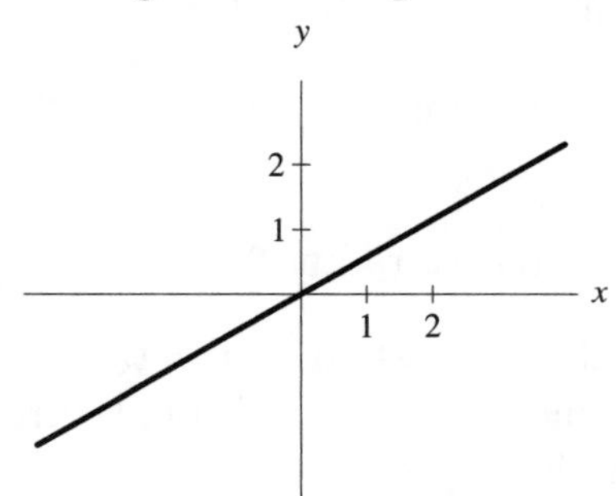

57. Multiply equation by r.

$$\begin{aligned} r^2 &= 4r\cos\theta \\ x^2 + y^2 &= 4x \\ x^2 - 4x + y^2 &= 0 \end{aligned}$$

59. Multiply equation by $\sin\theta$.

$$\begin{aligned} r\sin\theta &= 3 \\ y &= 3 \end{aligned}$$

61. Multiply equation by $\cos\theta$.

$$\begin{aligned} r\cos\theta &= 3 \\ x &= 3 \end{aligned}$$

63. Since $r = 5$, we get $r = \sqrt{x^2 + y^2} = 5$. Squaring, we find $x^2 + y^2 = 25$.

65. Note, $\tan\theta = \dfrac{\sin\theta}{\cos\theta} = \dfrac{y/r}{x/r} = \dfrac{y}{x}$.
Since $\tan\pi/4 = 1$, $\dfrac{y}{x} = 1$ or $y = x$.

67. Multiply equation by $1 - \sin\theta$.

$$\begin{aligned} r(1 - \sin\theta) &= 2 \\ r - r\sin\theta &= 2 \\ r - y &= 2 \\ \pm\sqrt{x^2 + y^2} &= y + 2 \\ x^2 + y^2 &= y^2 + 4y + 4 \\ x^2 - 4y &= 4 \end{aligned}$$

69. $r\cos\theta = 4$

71. Note $\tan\theta = y/x$.

$$\begin{aligned} y &= -x \\ \frac{y}{x} &= -1 \\ \tan\theta &= -1 \\ \theta &= -\pi/4 \end{aligned}$$

73. Note $x = r\cos\theta$ and $y = r\sin\theta$.

$$\begin{aligned} (r\cos\theta)^2 &= 4r\sin\theta \\ r^2\cos^2\theta &= 4r\sin\theta \\ r &= \frac{4\sin\theta}{\cos^2\theta} \\ r &= 4\tan\theta\sec\theta \end{aligned}$$

75. $r = 2$

77. Note $x = r\cos\theta$ and $y = r\sin\theta$.

$$\begin{aligned} y &= 2x - 1 \\ r\sin\theta &= 2r\cos\theta - 1 \\ r(\sin\theta - 2\cos\theta) &= -1 \\ r(2\cos\theta - \sin\theta) &= 1 \\ r &= \frac{1}{2\cos\theta - \sin\theta}. \end{aligned}$$

79. Note that $y = r\sin\theta$ and $x^2 + y^2 = r^2$.

$$\begin{aligned} x^2 + (y^2 - 2y + 1) &= 1 \\ x^2 + y^2 - 2y &= 0 \\ r^2 - 2r\sin\theta &= 0 \\ r^2 &= 2r\sin\theta \\ r &= 2\sin\theta \end{aligned}$$

81. There are six points of intersection and in polar coordinates these are approximately $(1, 0.17)$, $(1, 2.27)$, $(1, 4.36)$, $(1, 0.87)$, $(1, 2.97)$, $(1, 5.06)$

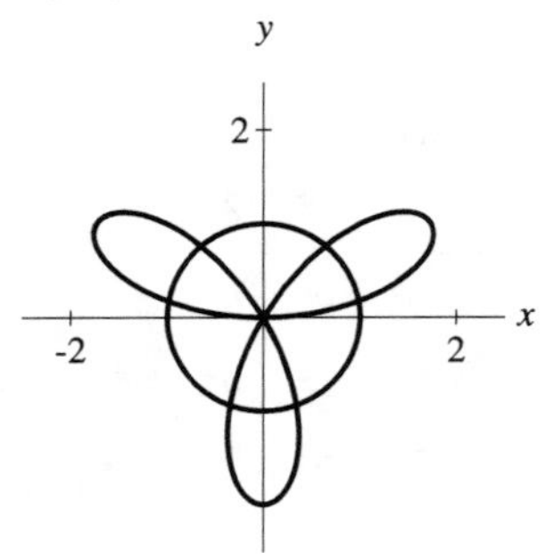

83. There are seven points of intersection and in polar coordinates these are approximately $(0, 0)$, $(0.9, 1.4)$, $(1.2, 1.8)$, $(1.9, 2.8)$, $(1.9, 3.5)$, $(1.2, 4.5)$, $(0.9, 4.9)$

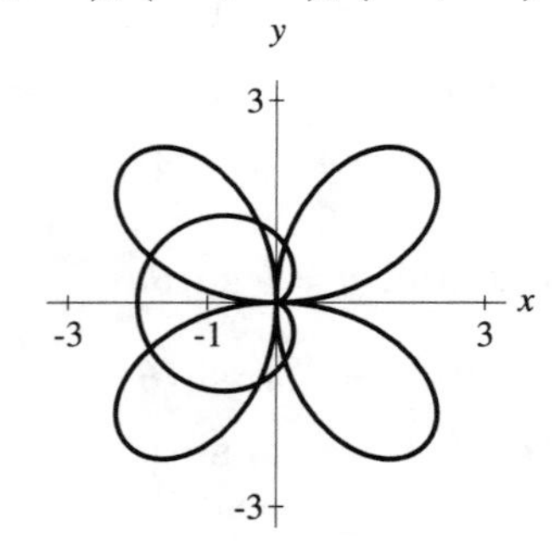

85. If the coordinate system is centered at A, then polar coordinates of B is $(1, 90°)$.

If the coordinate system is centered at B, then polar coordinates of C is $(\sqrt{2}, -45°)$.

If the coordinate system is centered at C, then polar coordinates of A is $(1, 180°)$.

87. Consider the trapezoid below.

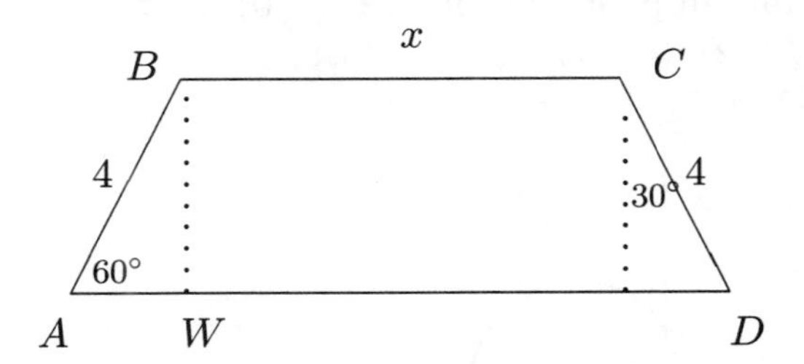

Note, $AW = 4\cos 60° = 2$. Since $AD = 10$, we find $x = 6$.

Thus, if the coordinate system is centered at A, then polar coordinates of B is $(4, 60°)$.

If the coordinate system is centered at B, then polar coordinates of C is $(6, 0°)$.

If the coordinate system is centered at C, then polar coordinates of D is $(4, -60°)$.

If the coordinate system is centered at D, then polar coordinates of A is $(10, 180°)$.

For Thought

1. False, t is the parameter.

2. True, graphs of parametric equations are sketched in a rectangular coordinate system in this book.

3. True, since $2x = t$ and $y = 2t + 1 = 2(2x) + 1 = 4x + 1$.

4. False, it is a circle of radius 1.

5. True, since if one substitutes $x = \tan t$ in $y = \tan^2 t$ one gets $y = x^2$. Note, the range of $y = \tan t$ is (∞, ∞).

6. True, since if $t = \dfrac{1}{3}$ then $x = 3\left(\dfrac{1}{3}\right) + 1 = 2$ and $y = 6\left(\dfrac{1}{3}\right) - 1 = 1$.

7. False, for if $w^2 - 3 = 1$ then $w = \pm 2$ and this does not satisfy $-2 < w < 2$.

8. False, since some of the points lie in the 3rd quadrant.

9. True, since $x = -\sin(t) < 0$ and $y = \cos(t) > 0$ for $0 < t < \pi/2$.

10. True, since $x = r\cos\theta$ and $y = r\sin\theta$ where $r = \cos\theta$ is given.

6.5 Exercises

1. If $t = 0$, then $x = 4(0) + 1 = 1$ and $y = 0 - 2 = -2$. If $t = 1$, then $x = 4(1) + 1 = 5$ and $y = 1 - 2 = -1$.

If $x = 7$, then $7 = 4t + 1$. Solving for t, we get $t = 1.5$. Substitute $t = 1.5$ into $y = t - 2$. Then $y = 1.5 - 2 = -0.5$.

If $y = 1$, then $1 = t - 2$. Solving for t, we get $t = 3$. Consequently $x = 4(3) + 1 = 13$.

We tabulate the results as follows.

t	x	y
0	1	-2
1	5	-1
1.5	7	-0.5
3	13	1

3. If $t = 1$, then $x = 1^2 = 1$ and $y = 3(1) - 1 = 2$. If $t = 2.5$, then $x = (2.5)^2 = 6.25$ and $y = 3(2.5) - 1 = 6.5$.

If $x = 5$, then $5 = t^2$ and $t = \sqrt{5}$. Consequently, $y = 3\sqrt{5} - 1$.

If $y = 11$, then $11 = 3t - 1$. Solving for t, we get $t = 4$. Consequently $x = 4^2 = 16$.

If $x = 25$, then $25 = t^2$ and $t = 5$. Consequently, $y = 3(5) - 1 = 14$.

We tabulate the results as follows.

t	x	y
1	1	2
2.5	6.25	6.5
$\sqrt{5}$	5	$3\sqrt{5} - 1$
4	16	11
5	25	14

5.

Some points are given by

t	x	y
0	-2	3
4	10	7

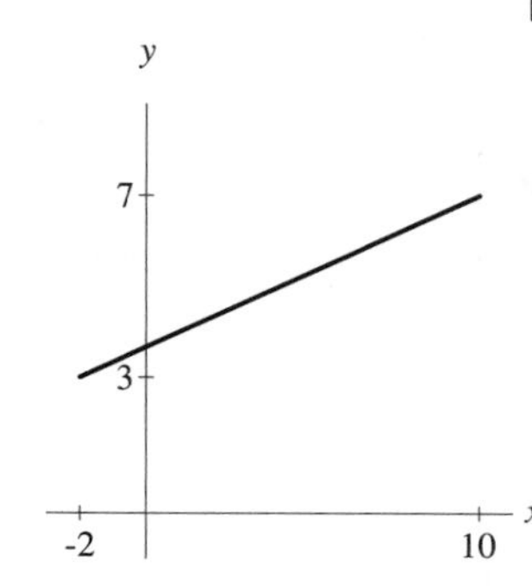

7.

Some points are given by

t	x	y
0	-1	0
2	1	4

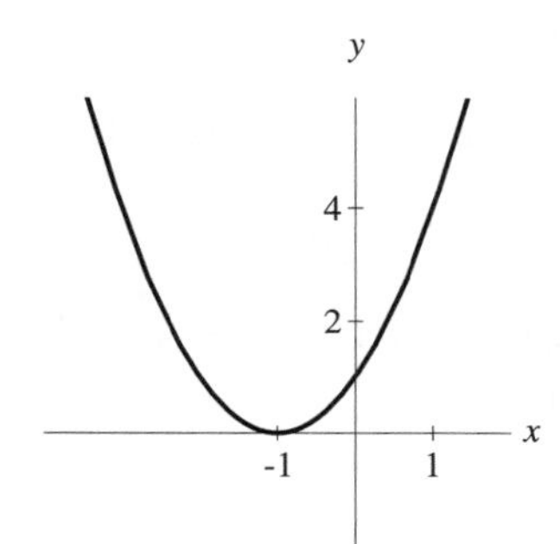

9.

A few points are approximated by

t	x	y
0.2	0.4	0.9
0.8	0.9	0.4

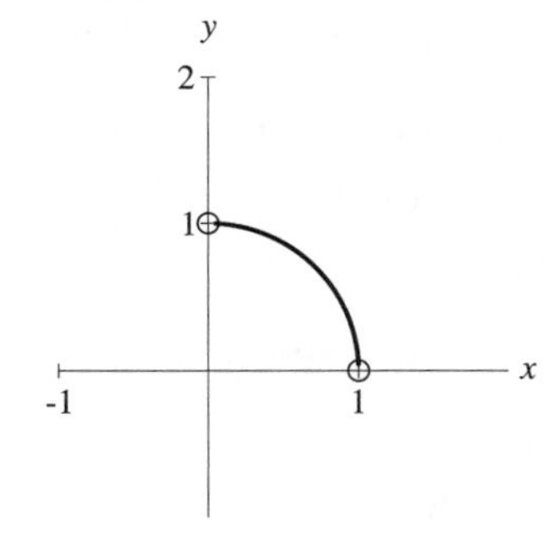

11. A circle of radius 1, centered at the origin

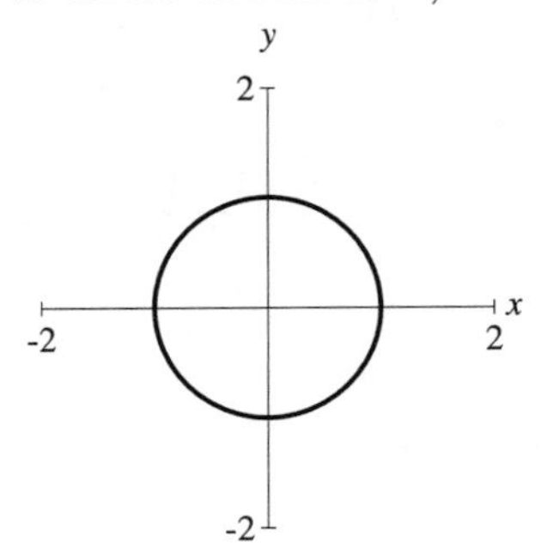

13. Since $t = \dfrac{x+5}{4}$, we obtain

$$y = 3 - 4\left(\frac{x+5}{4}\right) = -x - 2$$

or equivalently

$$x + y = -2.$$

The graph is a straight line.

15. Since $x^2 + y^2 = 16\sin^2(3t) + 16\cos^2(3t) = 16$, we get

$$x^2 + y^2 = 16.$$

The graph is a circle with radius 4 and with center at the origin.

17. Since $t = x - 4$, we find

$$y = \sqrt{(x-4) - 5} = \sqrt{x - 9}.$$

The graph of

$$y = \sqrt{x - 9}$$

is a square root curve.

19. $y = 2x + 3$ is the graph of a straight line

21. An equation (in terms of x and y) of the line through $(2,3)$ and $(5,9)$ is $y = 2x - 1$.

An equation (in terms of t and x) of the line through $(0,2)$ and $(2,5)$ is $x = \dfrac{3}{2}t + 2$.

The parametric equations are

$$x = \frac{3}{2}t + 2$$

and

$$y = 2\left(\frac{3}{2}t + 2\right) - 1 = 3t + 3$$

where $0 \le t \le 2$.

23. $x = 2\cos(t)$, $y = 2\sin(t)$, $\pi < t < \dfrac{3\pi}{2}$

25. $x = 3$, $y = t$, $-\infty < t < \infty$

27. Since $x = r\cos\theta$ and $y = r\sin\theta$, we get $x = 2\sin(t)\cos(t)$ and $y = 2\sin(t)\sin(t)$ where $0 \le t \le 2\pi$.

Equivalently, the parametric equations are

$$x = \sin(2t), \quad y = 2\sin^2(t)$$

where $0 \le t \le 2\pi$.

29. For $-\pi \le t \le \pi$, one obtains the given graph (for a larger range of values for t, more points are filled and the graph would look different)

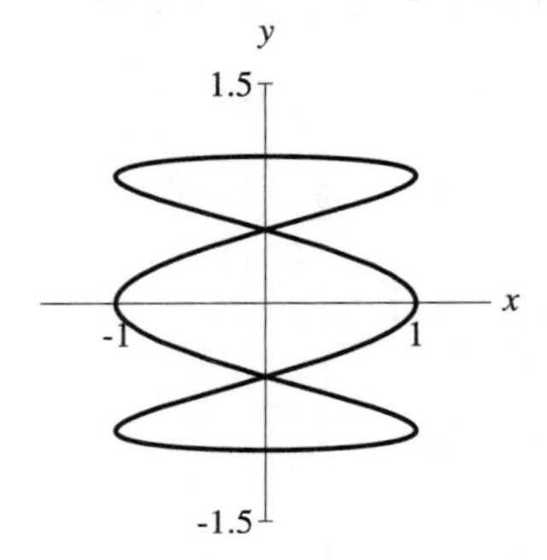

31. For $-15 \le t \le 15$, one finds

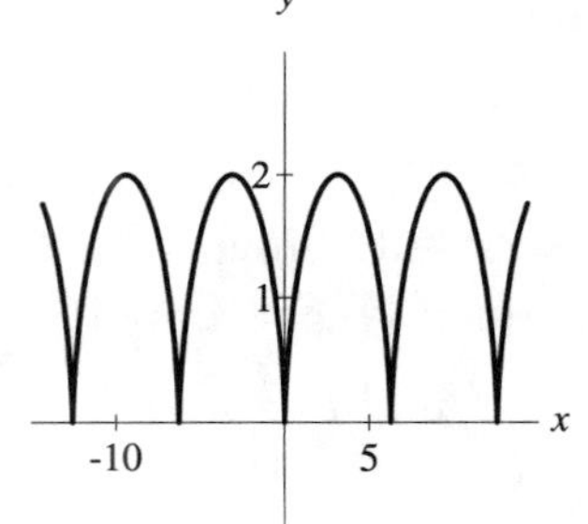

33. For $-10 \le t \le 10$, one obtains

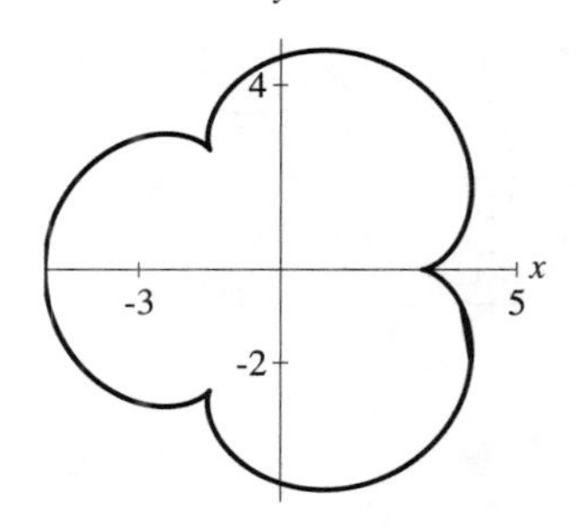

35. The graph of the parametric equations

$$x = 150\sqrt{3}t, \quad y = -16t^2 + 150t + 5$$

for $0 \le t \le 10$ is given below

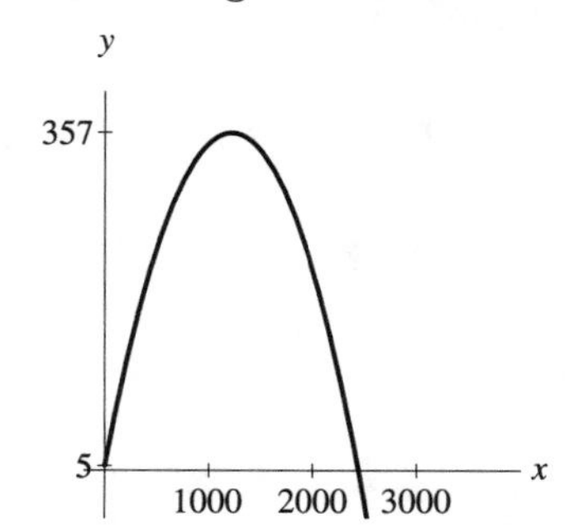

37. Solving $y = -16t^2 + 150t + 5 = 0$, one finds

$$t = \frac{-150 \pm \sqrt{150^2 - 4(-16)(5)}}{-32} \approx 9.41, -0.03$$

The arrow is in the air for 9.4 seconds.

Chapter 6 Review Exercises

1. $-1 - i$

3. $(4-5i)^2 = 16 - 40i + (25i^2) = -9 - 40i$

5. $(1-3i)2-(1-3i)6i = 2-6i-6i-18 = -16-12i$

7. $\frac{2-3i}{i} \cdot \frac{-i}{-i} = \frac{-3-2i}{1} = -3 - 2i$

9. $\frac{1+i}{2-3i} \cdot \frac{2+3i}{2+3i} = \frac{-1+5i}{13} = -\frac{1}{13} + \frac{5}{13}i$

11. $\frac{6+2i\sqrt{2}}{2} = 3 + i\sqrt{2}$

13. $i^{32}i^2 + i^8 i = i^2 + i = -1 + i$

15. $|3-5i| = \sqrt{3^2 + (-5)^2} = \sqrt{34}$

17. $|\sqrt{5} + i\sqrt{3}| = \sqrt{(\sqrt{5})^2 + (\sqrt{3})^2} = \sqrt{8} = 2\sqrt{2}$

19. Note $\sqrt{(-4.2)^2 + (4.2)^2} \approx 5.94$. If the terminal side of α goes through $(-4.2, 4.2)$, then $\tan\alpha = -1$. If $\alpha = 135°$, then

$$-4.2 + 4.2i \approx 5.94\left[\cos 135° + i \sin 135°\right].$$

21. Note $\sqrt{(-2.3)^2 + (-7.2)^2} \approx 7.6$. If the terminal side of α goes through $(-2.3, -7.2)$ and since $\tan^{-1}(7.2/2.3) \approx 72.3°$, then one can choose $\alpha = 180° + 72.3° = 252.3°$. Thus,

$$-2.3 - 7.2i \approx 7.6\left[\cos 252.3° + i \sin 252.3°\right].$$

23. $\sqrt{3}\left(-\frac{\sqrt{3}}{2} + i\frac{1}{2}\right) = -\frac{3}{2} + \frac{\sqrt{3}}{2}i$

25. $6.5[0.8377 + (0.5461)i] \approx 5.4 + 3.5i$

27. Since $z_1 = 2.5\sqrt{2}\left[\cos 45° + i \sin 45°\right]$ and $z_2 = 3\sqrt{2}\left[\cos 225° + i \sin 225°\right]$, we have

$$z_1 z_2 = 15\left[\cos 270° + i \sin 270°\right] = -15i$$

and

$$\frac{z_1}{z_2} = \frac{(5/2)\sqrt{2}}{3\sqrt{2}}\left[\cos(-180°) + i \sin(-180°)\right] = -\frac{5}{6}.$$

29. Let α and β be angles whose terminal sides go through $(2, 1)$ and $(3, -2)$, respectively. Since $|2+i| = \sqrt{5}$ and $|3-2i| = \sqrt{13}$, we get $\cos\alpha = 2/\sqrt{5}$, $\sin\alpha = 1/\sqrt{5}$, $\cos\beta = 3/\sqrt{13}$, and $\sin\beta = -2/\sqrt{13}$. From the sum and difference identities, we obtain

$$\cos(\alpha+\beta) = \frac{8}{\sqrt{65}}, \; \sin(\alpha+\beta) = -\frac{1}{\sqrt{65}},$$

$$\cos(\alpha-\beta) = \frac{4}{\sqrt{65}}, \; \sin(\alpha-\beta) = \frac{7}{\sqrt{65}}.$$

Note $z_1 = \sqrt{5}(\cos\alpha + i\sin\alpha)$ and $z_2 = \sqrt{13}(\cos\beta + i\sin\beta)$. Then

$$\begin{aligned} z_1 z_2 &= \sqrt{65}\left(\cos(\alpha+\beta) + i\sin(\alpha+\beta)\right) \\ &= \sqrt{65}\left(\frac{8}{\sqrt{65}} - i \cdot \frac{1}{\sqrt{65}}\right) \\ z_1 z_2 &= 8 - i \end{aligned}$$

and

$$\begin{aligned} \frac{z_1}{z_2} &= \frac{\sqrt{5}}{\sqrt{13}}\left(\cos(\alpha-\beta) + i\sin(\alpha-\beta)\right) \\ &= \frac{\sqrt{65}}{13}\left(\frac{4}{\sqrt{65}} + i \cdot \frac{7}{\sqrt{65}}\right) \\ \frac{z_1}{z_2} &= \frac{4}{13} + \frac{7}{13}i \approx 0.31 + 0.54i. \end{aligned}$$

31. $2^3[\cos 135° + i \sin 135°] =$
$8\left[-\frac{\sqrt{2}}{2} + i\frac{\sqrt{2}}{2}\right] = -4\sqrt{2} + 4\sqrt{2}i$

33. $(4+4i)^3 = (4\sqrt{2})^3[\cos 45° + i \sin 45°]^3 =$
$128\sqrt{2}[\cos 135° + i \sin 135°] =$
$128\sqrt{2}\left[-\frac{1}{\sqrt{2}} + i\frac{1}{\sqrt{2}}\right] = -128 + 128i$

35. Square roots of i are given by
$\cos\left(\frac{90° + k360°}{2}\right) + i \sin\left(\frac{90° + k360°}{2}\right) =$
$\cos(45° + k \cdot 180°) + i \sin(45° + k \cdot 180°)$
where k is an integer.
If $k = 0, 1$ one gets
$\cos 45° + i \sin 45° = \frac{\sqrt{2}}{2} + i\frac{\sqrt{2}}{2}$ and
$\cos 225° + i \sin 225° = -\frac{\sqrt{2}}{2} - i\frac{\sqrt{2}}{2}$.

37. Since $|\sqrt{3} + i| = 2$, the cube roots are
$\sqrt[3]{2}\left[\cos\left(\frac{30° + k360°}{3}\right) + i \sin\left(\frac{30° + k360°}{3}\right)\right]$
where k is an integer. If $k = 0, 1, 2$, we find

$$\sqrt[3]{2}[\cos\alpha + i \sin\alpha]$$

where $\alpha = 10°, 130°, 250°$.

39. Since $|2 + i| = \sqrt{5}$ and $\tan^{-1}(1/2) \approx 26.6°$, the arguments of the cube roots are

$$\frac{26.6° + k360°}{3} \approx 8.9° + k120°$$

where k is an integer.
If $k = 0, 1, 2$, we obtain

$$\sqrt[6]{5}[\cos\alpha + i \sin\alpha]$$

where $\alpha = 8.9°, 128.9°, 248.9°$.

41. Since $\sqrt[4]{625} = 5$, the fourth roots of $625i$ are
$5\left[\cos\left(\frac{90° + k360°}{4}\right) + i \sin\left(\frac{90° + k360°}{4}\right)\right]$
where k is an integer.
When $k = 0, 1, 2, 3$ the fourth roots are

$$5[\cos\alpha + i \sin\alpha]$$

where $\alpha = 22.5°, 112.5°, 202.5°, 292.5°$.

43. $(5\cos 60°, 5 \sin 60°) = \left(\frac{5}{2}, \frac{5\sqrt{3}}{2}\right)$

45. $(\sqrt{3}\cos 100°, \sqrt{3}\sin 100°) \approx (-0.3, 1.7)$

47. Note, we have

$$r = \sqrt{(-2)^2 + (-2\sqrt{3})^2} = \sqrt{16} = 4.$$

Since $\tan\theta = \sqrt{3}$ and the terminal side of θ goes through $(-2, -2\sqrt{3})$, we may let $\theta = 4\pi/3$. Then

$$(r, \theta) = \left(4, \frac{4\pi}{3}\right).$$

49. Note, the maginitude is

$$r = \sqrt{2^2 + (-3)^2} = \sqrt{13}.$$

Since $\theta = \tan^{-1}(-3/2) \approx -0.98$, we have

$$(r, \theta) \approx \left(\sqrt{13}, -0.98\right).$$

51. Circle centered at $(r, \theta) = (1, -\pi/2)$

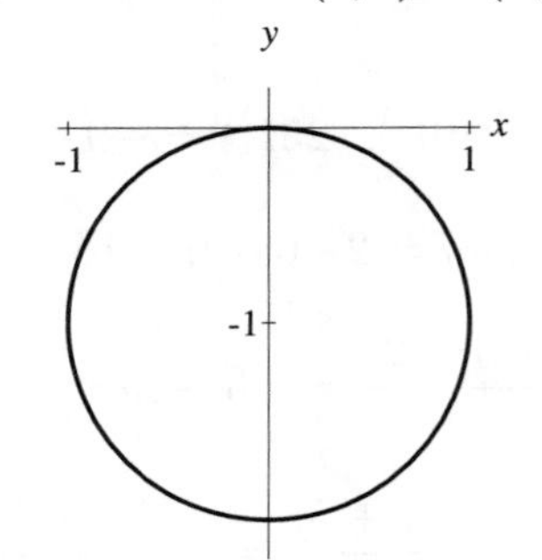

53. four-leaf rose

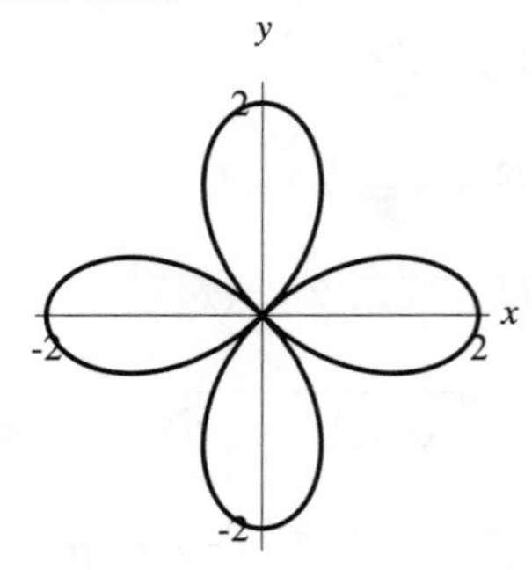

55. Limacon $r = 500 + \cos\theta$

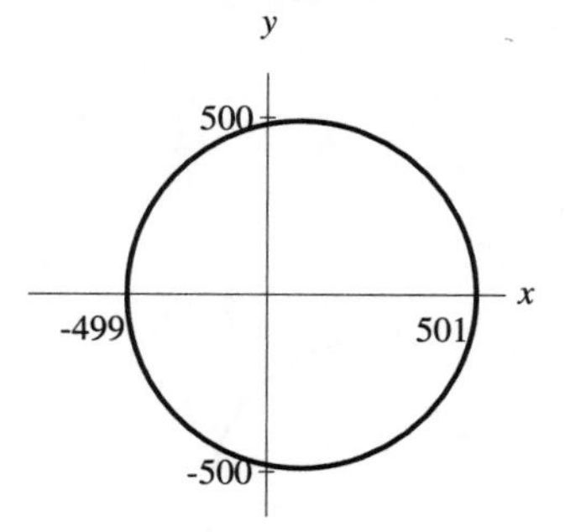

57. Horizontal line $y = 1$

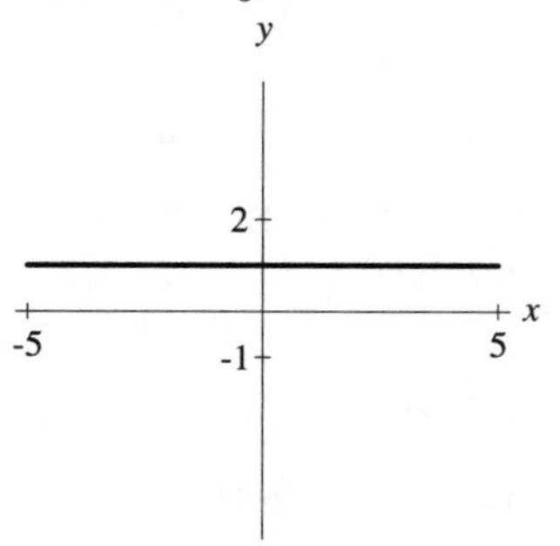

59. Since we have

$$r = \frac{1}{\sin\theta + \cos\theta},$$

we obtain

$$\begin{aligned} r\sin\theta + r\cos\theta &= 1 \\ y + x &= 1. \end{aligned}$$

61. $x^2 + y^2 = 25$

63. Since $y = 3$, we find $r\sin\theta = 3$ and $r = \dfrac{3}{\sin\theta}$.

65. $r = 7$

67. The boundary points are given by

t	x	y
0	0	3
1	3	2

Note, the boundary points do not lie on the graph.

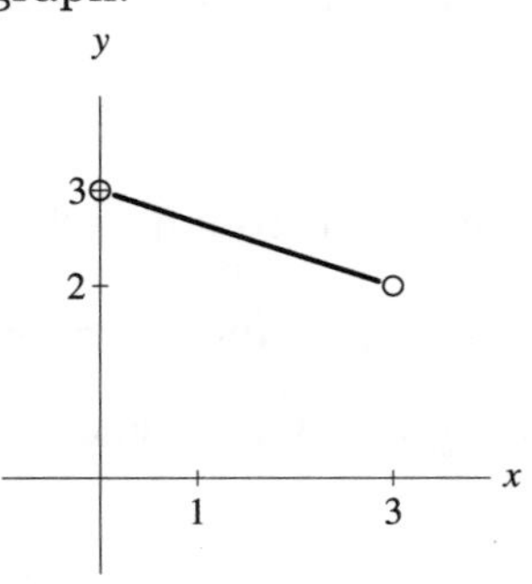

69. The graph is a quarter of a circle.

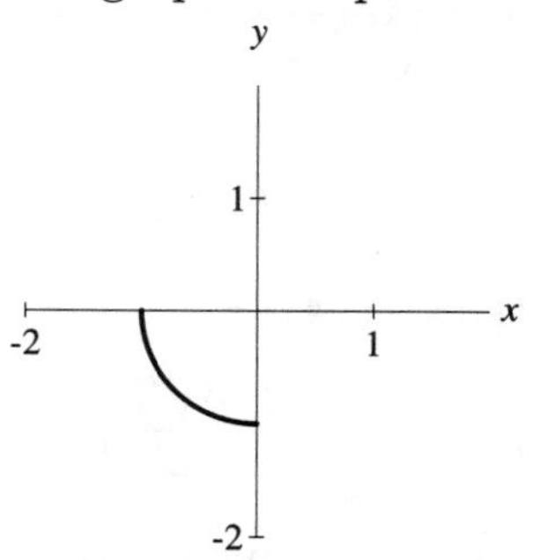

Chapter 6 Test

1. $(4 - 3i)^2 = 16 - 24i + (3i)^2 = 7 - 24i$

2. $\dfrac{2-i}{3+i} \cdot \dfrac{3-i}{3-i} = \dfrac{5-5i}{10} = \dfrac{1}{2} - \dfrac{1}{2}i$

3. $i^4 i^2 - i^{32} i^3 = i^2 - i^3 =$

$-1 - (-i) = -1 + i$

4. $2i\sqrt{2}(i\sqrt{2} + \sqrt{6}) = -4 + 2i\sqrt{12} = -4 + 4i\sqrt{3}$

5. Since $|3 + 3i| = 3\sqrt{2}$ and $\tan^{-1}(3/3) = 45°$, we have

$$3 + 3i = 3\sqrt{2}\left[\cos 45° + i\sin 45°\right].$$

6. Since $|-1 + i\sqrt{3}| = 2$ and $\cos^{-1}(-1/2) = 120°$, we have

$$|-1 + i\sqrt{3}| = 2\left[\cos 120° + i\sin 120°\right].$$

7. Note $|-4 - 2i| = \sqrt{20} = 2\sqrt{5}$.
Since $\tan^{-1}(2/4) = 26.6°$, the direction angle of $-4 - 2i$ is

$$180° + 26.6° = 206.6°.$$

Then we obtain

$$-4 - 2i = 2\sqrt{5}\left[\cos 206.6° + i\sin 206.6°\right].$$

8. $6\left[\cos 45° + i\sin 45°\right] =$

$$6\left[\frac{\sqrt{2}}{2} + i\frac{\sqrt{2}}{2}\right] = 3\sqrt{2} + 3i\sqrt{2}$$

9. $2^9\left[\cos 90° + i\sin 90°\right] = 512\left[0 + i\right] = 512i$

10. $\dfrac{3}{2}\left[\cos 45° + i\sin 45°\right] =$

$$\frac{3}{2}\left[\frac{\sqrt{2}}{2} + i\frac{\sqrt{2}}{2}\right] = \frac{3\sqrt{2}}{4} + i\frac{3\sqrt{2}}{4}$$

11. $(5\cos 30°, 5\sin 30°) = \left(5 \cdot \frac{\sqrt{3}}{2}, 5 \cdot \frac{1}{2}\right) =$
$\left(\frac{5\sqrt{3}}{2}, \frac{5}{2}\right)$

12. $(-3\cos(-\pi/4), -3\sin(-\pi/4)) =$
$\left(-3 \cdot \frac{\sqrt{2}}{2}, 3 \cdot \frac{\sqrt{2}}{2}\right) = \left(-\frac{3\sqrt{2}}{2}, \frac{3\sqrt{2}}{2}\right)$

13. $(33\cos 217°, 33\sin 217°) \approx (-26.4, -19.9)$

14. Circle of radius $5/2$ with center at $(r, \theta) = (5/2, 0)$

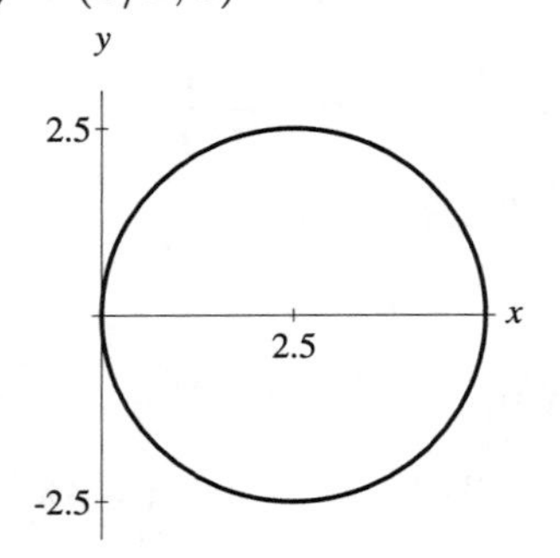

15. Four-leaf rose.

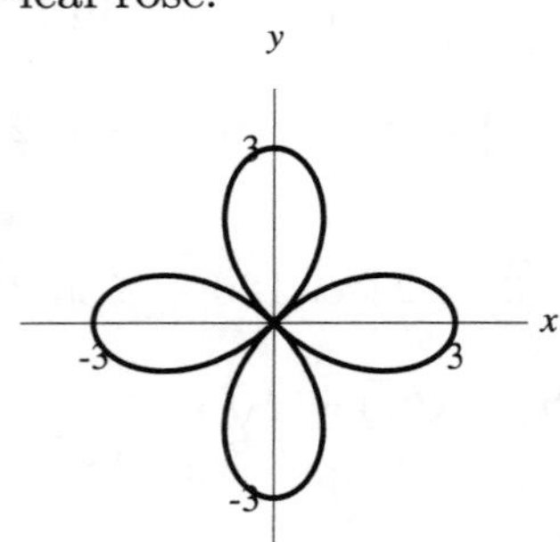

16. Some points on the graph are given by

t	x	y
-1	-2	-10
3	6	6

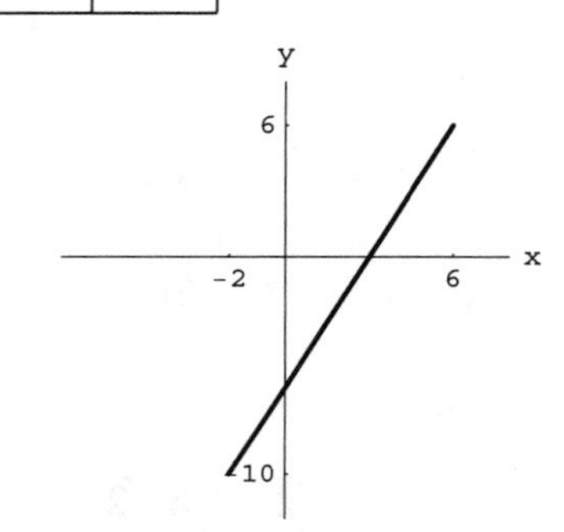

17. Some points on the graph are given by

t	x	y
0	0	3
π	0	-3

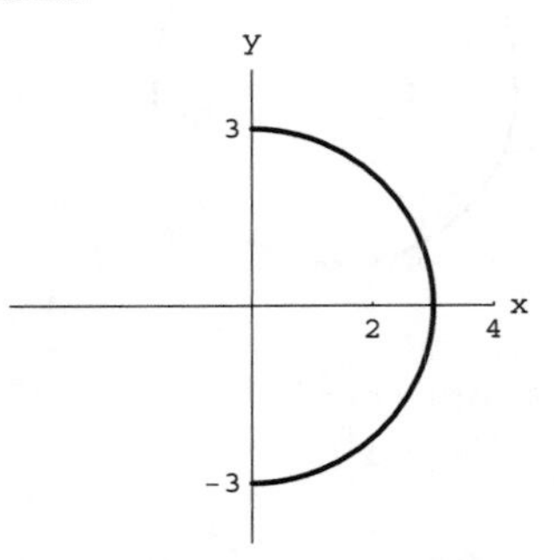

18. Fourth roots of -81 are given by
$3\left[\cos\left(\frac{180° + k360°}{4}\right) + i\sin\left(\frac{180° + k360°}{4}\right)\right]$
$= 3\left[\cos(45° + k90°) + i\sin(45° + k90°)\right].$
When $k = 0, 1, 2, 3$ one gets

$$3[\cos 45° + i\sin 45°] = \frac{3\sqrt{2}}{2} + i\frac{3\sqrt{2}}{2},$$

$$3[\cos 135° + i\sin 135°] = -\frac{3\sqrt{2}}{2} + i\frac{3\sqrt{2}}{2},$$

$$3[\cos 225° + i\sin 225°] = -\frac{3\sqrt{2}}{2} - i\frac{3\sqrt{2}}{2},$$

$$3[\cos 315° + i\sin 315°] = \frac{3\sqrt{2}}{2} - i\frac{3\sqrt{2}}{2}.$$

19. Since $x^2 + y^2 + 5y = 0$, we obtain

$$\begin{aligned} r^2 + 5r\sin\theta &= 0 \\ r + 5\sin\theta &= 0 \\ r &= -5\sin\theta \end{aligned}$$

20. Since $r = 5(2\sin\theta\cos\theta)$, we find

$$\begin{aligned} r^3 &= 10(r\sin\theta)(r\cos\theta) \\ r^3 &= 10yx \\ (x^2 + y^2)^{3/2} &= 10yx. \end{aligned}$$

21. Since the slope of the line through $(-2, -3)$ and $(4, 5)$ is $\frac{5+3}{4+2}$ or $\frac{8}{6}$, parametric equations are $x = -2 + 6t$ and $y = -3 + 8t$ where $0 \le t \le 1$.

Tying It All Together

1. $\frac{\sqrt{2}}{2}$ **2.** $\frac{1}{2}$ **3.** $\frac{\pi}{4}$ **4.** $-\frac{\pi}{4}$

5. $\frac{\pi}{3}$ **6.** $\frac{2\pi}{3}$

7. $\cos\left(\frac{\pi}{4}\right) = \frac{\sqrt{2}}{2}$ **8.** $\sin\left(\frac{-\pi}{4}\right) = -\frac{\sqrt{2}}{2}$

9. Let $\cos\theta = \frac{3}{5} = \frac{x}{r}$. Since $\cos^{-1}(3/5)$ is an angle in the 1st quadrant, we get $x = 3$, $r = 5$ and $y > 0$. Since $5^2 = 3^2 + y^2$, we find $y = 4$. Then

$$\sin\theta = \frac{y}{r} = \frac{4}{5}.$$

10. Let $\sin\theta = -\frac{3}{5} = \frac{y}{r}$. Since $\sin^{-1}\left(-\frac{3}{5}\right)$ is an angle in the 4th quadrant, we get $y = -3$, $r = 5$ and $x > 0$. Since $5^2 = x^2 + (-3)^2$, we get $x = 4$. Then $\tan\theta = \frac{y}{x} = -\frac{3}{4}$.

11. $\left\{x \mid x = \frac{\pi}{2} + 2k\pi \text{ where } k \text{ is an integer}\right\}$

12. $\{x \mid x = 2k\pi \text{ where } k \text{ is an integer}\}$

13. Since $\sin x = \pm 1$, the solution set is

$$\left\{x \mid x = \frac{\pi}{2} + 2k\pi \text{ or } x = \frac{3\pi}{2} + 2k\pi\right\}.$$

14. Since $\cos x = \pm 1$, the solution set is

$$\{x \mid x = k\pi \text{ where } k \text{ is an integer}\}.$$

15. Since $\sin x = \frac{1}{2}$, the solution set is

$$\left\{x \mid x = \frac{\pi}{6} + 2k\pi \text{ or } x = \frac{5\pi}{6} + 2k\pi\right\}.$$

16. Since $\cos x = \frac{1}{2}$, the solution set is

$$\left\{x \mid x = \frac{\pi}{3} + 2k\pi \text{ or } x = \frac{5\pi}{3} + 2k\pi\right\}.$$

17. Since $\sin(2x) = 1$, we obtain $2x = \frac{\pi}{2} + 2k\pi$. The solution set is

$$\left\{x \mid x = \frac{\pi}{4} + k\pi\right\}.$$

18. Since $\sin^2 x + \cos^2 x = 1$ is an identity, the solution set is the set of all real numbers.

19. Note, $x \cdot \frac{\sqrt{3}}{2} = 1$. Solving for x, we get $x = \frac{2}{\sqrt{3}}$. The solution set is $\left\{\frac{2\sqrt{3}}{3}\right\}$.

20. Note, $x \cdot 1 + x \cdot \frac{\sqrt{3}}{2} = 1$. Factoring, we get $x\left(1 + \frac{\sqrt{3}}{2}\right) = 1$ or $x\left(2 + \sqrt{3}\right) = 2$. Solving for x, we find $x = \frac{2}{2 + \sqrt{3}}$. By rationalizing the denominator, we find that the solution set is $\left\{4 - 2\sqrt{3}\right\}$.

21. Note $\sin 2x = 2\sin x\cos x$.

$$\begin{aligned} 4\sin x\cos x - 2\cos x + 2\sin x - 1 &= 0 \\ 2\cos x(2\sin x - 1) + (2\sin x - 1) &= 0 \\ (2\cos x + 1)(2\sin x - 1) &= 0 \end{aligned}$$

Then $\cos x = -\frac{1}{2}$ or $\sin x = \frac{1}{2}$.

The solution set is $\left\{x | x = \frac{\pi}{6} + 2k\pi,\right.$

$$\left. x = \frac{5\pi}{6} + 2k\pi, x = \frac{2\pi}{3} + 2k\pi, x = \frac{4\pi}{3} + 2k\pi\right\}.$$

22. First, we factor by grouping.

$$\begin{aligned} 4x\sin x + 2\sin x - 2x - 1 &= 0 \\ 2\sin x(2x + 1) - (2x + 1) &= 0 \\ (2x + 1)(2\sin x - 1) &= 0 \end{aligned}$$

Then $x = -\frac{1}{2}$ or $\sin x = \frac{1}{2}$.

The solution set is

$$\left\{x | x = -\frac{1}{2}, x = \frac{\pi}{6} + 2k\pi, x = \frac{5\pi}{6} + 2k\pi\right\}.$$

23. $y = \sin x$

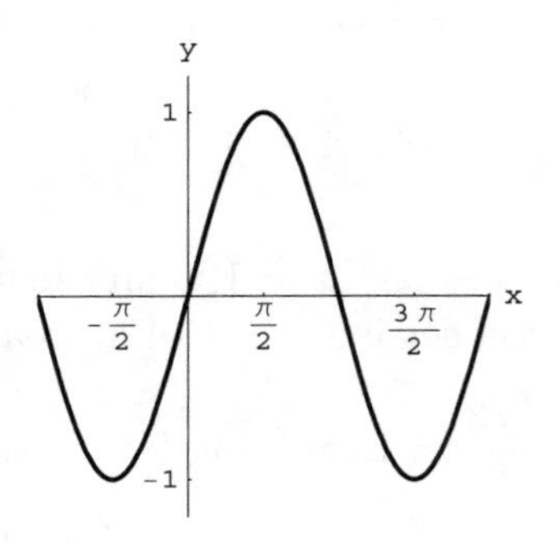

24. $y = \cos x$

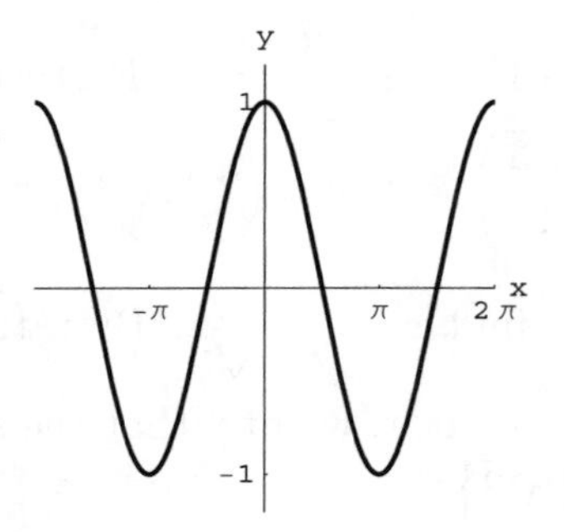

25. $r = \sin\theta$

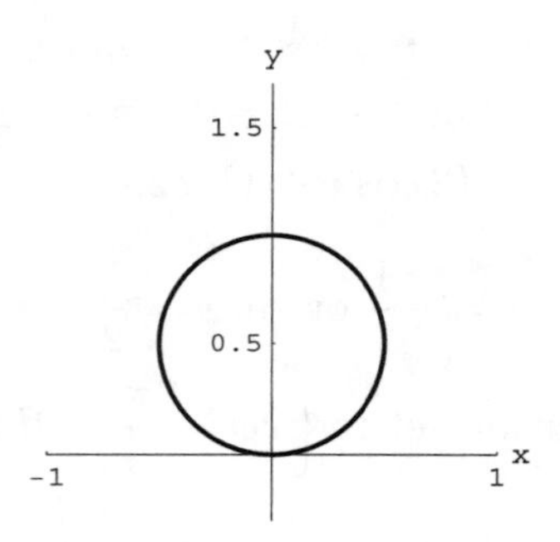

26. $r = \cos\theta$

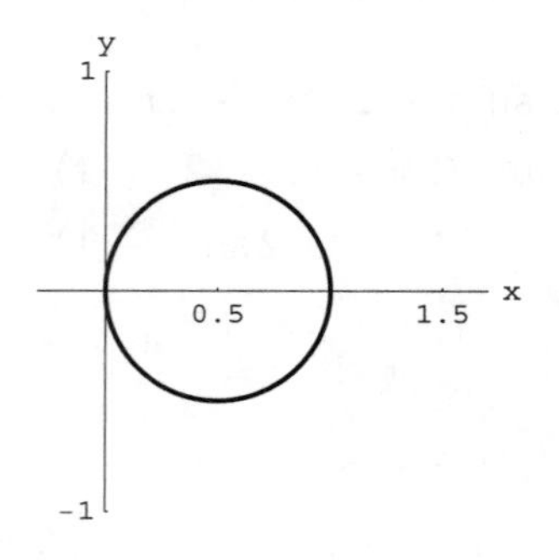

27. $r = \theta$

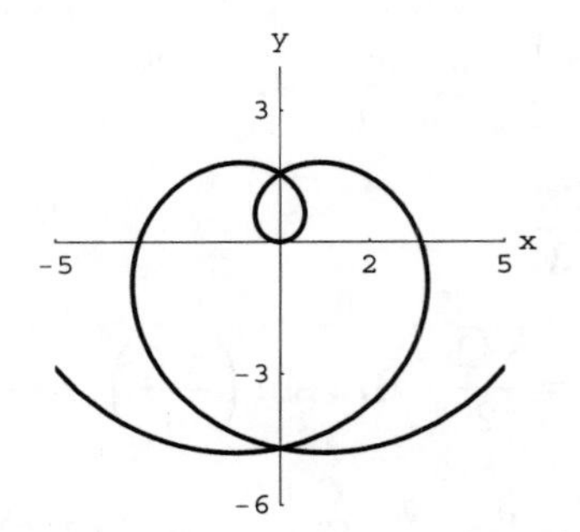

28. $y = x$

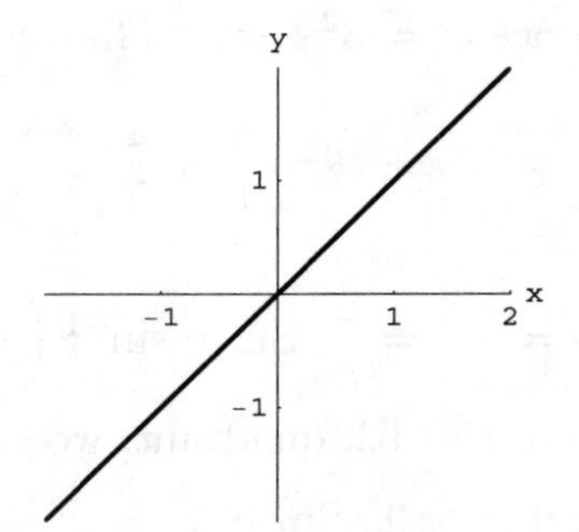

29. $y = \frac{\pi}{2}$

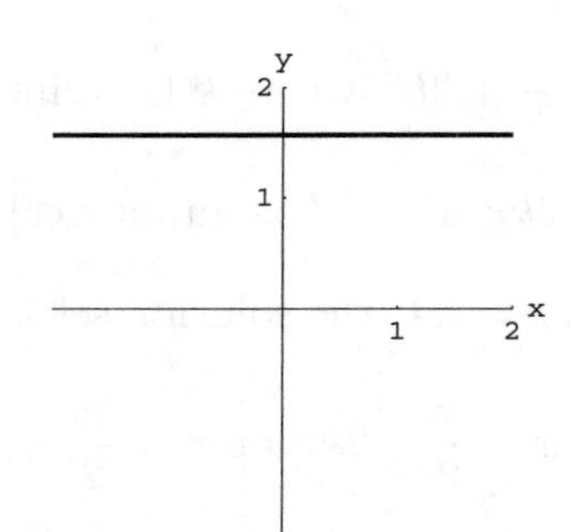

30. $r = \sin\left(\frac{\pi}{3}\right)$

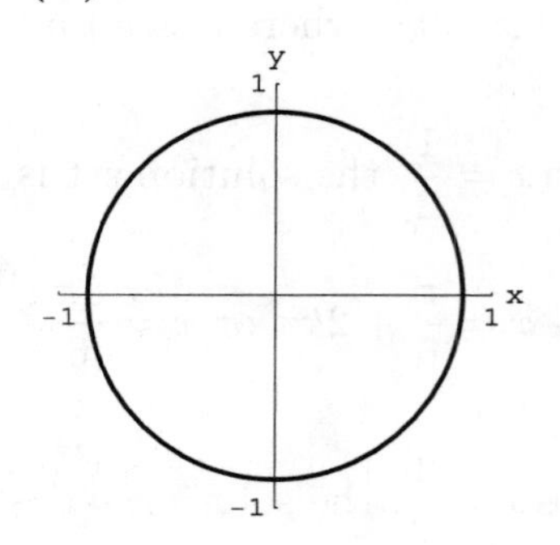

31. $y = \sqrt{\sin x}$

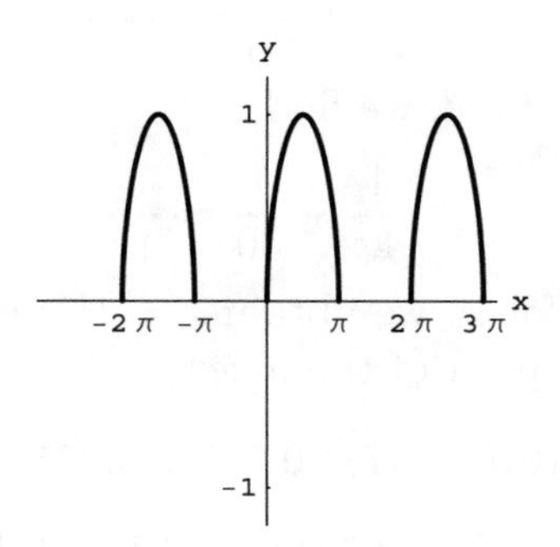

32. $y = \cos^2 x$

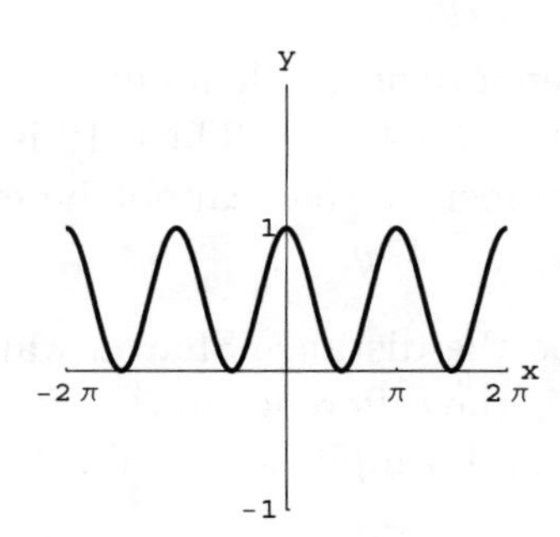

Thinking Outside the Box

I. Let x be the rate of the current in the river, and let $2x$ be the rate that Milo and Bernard can paddle. Since the rate going upstream is x and it takes 21 hours going upstream, the distance going upstream is $21x$.

Note, the rate going downstream is $3x$. Then the time going downstream is

$$\text{time} = \frac{\text{distance}}{\text{rate}} = \frac{21x}{3x} = 7 \text{ hours.}$$

Thus, in order to meet Vince at 5pm, Milo and Bernard must start their return trip at 10 AM.

II. Let x be the uniform width of the swath. When Eugene is half done, we find

$$(300-2x)(400-2x) = \frac{300(400)}{2}.$$

We could rewrite the equation as

$$x^2 - 350x + 15,000 = (x-50)(x-300).$$

Then $x = 50$ or $x = 300$. Since $x = 300$ is not possible, the width of the swath is $x = 50$ feet.

III. 117, since $117 = 13 \times 9$

Note, $x = 7$, $y = 1$, and $z = 1$ satisfies $x + 10y + 100z = 13(x+y+z)$.

IV. Let x be the top speed of the hiker, and let d be the length of the tunnel. The time it takes the hiker to cover one-fourth of the tunnel is

$$\frac{d/4}{x}.$$

Let y be the number of miles that the tunnel is ahead of the train when the hiker spots the train. Since the hiker can return to the entrance of the tunnel just before the train enters the tunnel, we obtain.

$$\frac{y}{30} = \frac{d/4}{x}.$$

Similarly, if the hiker runs towards the other end of the tunnel then

$$\frac{y+d}{30} = \frac{3d/4}{x}.$$

Thus,

$$\frac{d/4}{x} + \frac{d}{30} = \frac{3d/4}{x}.$$

Dividing by d, we find

$$\frac{1}{4x} + \frac{1}{30} = \frac{3}{4x}.$$

Solving for x, we obtain $x = 15$ mph which is the top speed of the hiker.

V. $1 - (0.7)(0.7)^2(0.7)^3(0.7)^4 \approx 0.97175$ or 97.2%

VI. Since $20 = 3(3)+11(1)$ and $1 = 3(4)+11(-1)$, all integers at least 20 can be expressed in the form $3x + 11y$.

Note, there are no whole numbers x and y that satisfy $19 = 3x + 11y$. Thus, 19 is the largest whole number N that cannot be expressed in the form $3x + 11y$.

VII. Let d be the distance Sharon walks. Since 2 minutes is the difference in the times of arrival at 4mph and 5 mph, we obtain

$$\frac{d}{4} - \frac{d}{5} = \frac{2}{60}.$$

The solution of the above equation is $d = 2/3$ mile. Since $d/4 = (2/3)/4 = 10/60$, Sharon must walk to school in 9 minutes to arrive on time. Thus, Sharon's speed in order to arrive on time is

$$r = \frac{d}{t} = \frac{2/3}{9/60} = \frac{40}{9} \text{ mph.}$$

VIII. The vertices of the quadrilateral where the triangle and square overlap are

$A(81/30, 81/40)$, $B(81/30, 19/10)$,

$C(143/40, 19/10)$, and $D(96/30, 24/10)$.

Subdividing the quadrilateral into two triangles and a rectangle, we find that the area of the quadrilateral is $1/4$.

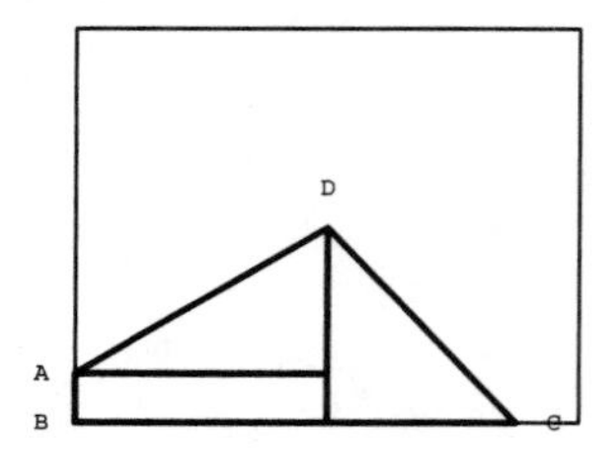

IX. 2178

X. $\frac{1}{2}+\frac{1}{3}+\frac{1}{10}+\frac{1}{15}$, $\frac{1}{2}+\frac{1}{3}+\frac{1}{9}+\frac{1}{18}$
$\frac{1}{2}+\frac{1}{3}+\frac{1}{8}+\frac{1}{24}$, $\frac{1}{2}+\frac{1}{3}+\frac{1}{7}+\frac{1}{42}$
$\frac{1}{2}+\frac{1}{4}+\frac{1}{6}+\frac{1}{12}$, $\frac{1}{2}+\frac{1}{4}+\frac{1}{5}+\frac{1}{20}$

XI. If $x=\sqrt{3}+\sqrt{5}$, then

$$\begin{aligned} x^2 &= 8+2\sqrt{15} \\ x^2-8 &= 2\sqrt{15} \\ (x^2-8)^2 &= 4(15) \\ x^4-16x^2+64 &= 60 \\ x^4-16x^2+4 &= 0. \end{aligned}$$

XII.

$$\begin{array}{r} 85254 \\ +50671 \\ \hline 135925 \end{array} \qquad \begin{array}{r} 63732 \\ +39841 \\ \hline 103573 \end{array}$$

XIII. Let x be the rate of the swimmer who after 40 feet passes the other swimmer. Let y be the rate of the other swimmer. If x is the length of the pool, then

$$\frac{40}{x}=\frac{d-40}{y}$$

and

$$\frac{d+45}{x}=\frac{2d-45}{y}.$$

Since we may solve for the ratio y/x from both equations, we find

$$\frac{y}{x}=\frac{d-40}{40}=\frac{2d-45}{d+45}.$$

Solving for d, we obtain $d=75$ or $d=0$. Thus, the length of the pool is 75 feet.

XIV. Move the second "I" in "VII", and form a square root symbol to obtain the fraction

$$\frac{1}{\sqrt{\text{I}}}$$

which reads as "one divided by the square root of one".

XV. First, \$59 is not a possible total, i.e., for all whole numbers x and y we have

$$7x+11y\neq 59.$$

But for each $60\leq n\leq 69$, there are whole numbers x and y satisfying

$$n=7x+11y.$$

Suppose

$$m=7a+11b\geq 69$$

and a,b are whole numbers. Since

$$m+1=7(a-3)+11(b+2)$$

we see that $m+1$ is a possible total if $a-3\geq 0$. However, if $a-3<0$ then a is either 0, 1, or 2 for $a\geq 0$. Since $m-7a=11b$ and $m\geq 69$, we find

$$\begin{aligned} b-5 &= \frac{m-7a-55}{11} \\ &\geq \frac{69-7a-55}{11} \\ &= \frac{14-7a}{11} \\ &\geq 0 \end{aligned}$$

for $a=0,1,2$. In any case, $m+1$ is a possible total since

$$m+1=7(a+8)+11(b-5).$$

Since $m\geq 69$, all whole numbers greater than 59 is a possible total. Hence, 59 is the largest integer that is not a possible total.

XVI.

a) $\frac{6}{23}=\frac{1}{4}+\frac{1}{92}$
b) $\frac{14}{15}=\frac{1}{2}+\frac{1}{3}+\frac{1}{10}$
c) $\frac{7}{11}=\frac{1}{2}+\frac{1}{11}+\frac{1}{22}$

XVII. Consider the point of intersection between the two left-most circles on the bottom row. The (vertical) distance between this point of intersection and the center of the left-most circle in the middle row is $\sqrt{5}$.

Let x be the radius of the circle at the top row. We draw another right triangle but this time its hypotenuse is the line segment that joins the center of the left-most circle in the middle row and the center of the circle in the top row. The length of the hypotenuse of this triangle is $x + 1$.

From the center of the left-most circle in the middle row, draw a horizontal side that is 2-units long in such a way that the endpoint of this side should be directly below the center of the circle on the top row. Then a right triangle is formed when this endpoint is joined to the center of the top-most circle by a line segment.

Applying the Pythagorean Theorem, we find

$$2^2 + (x + 2 - \sqrt{5})^2 = (x + 1)^2.$$

Solving for x, we find

$$x = \sqrt{5} - 1.$$

XVIII. We begin by noting that

$$\begin{pmatrix} \text{area of} \\ \text{two crescents} \end{pmatrix} + \begin{pmatrix} \text{area of} \\ \text{largest semicircle} \end{pmatrix} =$$

$$\begin{pmatrix} \text{area of two} \\ \text{smaller circles} \end{pmatrix} + \begin{pmatrix} \text{area of the} \\ \text{right triangle} \end{pmatrix} =$$

Let x and y be the sides of the right triangle where crescents A and B intersect the triangle, respectively. Then the hypotenuse of the right triangle is $\sqrt{x^2 + y^2}$.

Recall, the area of a semicircle with diameter d is $\pi d^2/8$. Thus, the sum of the areas of the two smaller semicircles with diameters x and y is equal to the area of the largest semicircle with diameter $\sqrt{x^2 + y^2}$.

If we subtract the areas of the circles from the equation above, then we obtain

$$\begin{pmatrix} \text{area of} \\ \text{two crescents} \end{pmatrix} = \begin{pmatrix} \text{area of the} \\ \text{right triangle} \end{pmatrix}.$$

Hence, the ratio of the total area of the two crescents to the area of the triangle is 1.

XIX. Let p and f be the number of students who passed and failed, respectively. Since the mean score for the passing students is 65, the mean score for the failing students is 53, and the mean score for all students is 53, we obtain

$$\frac{65p + 35f}{p + f} = 53.$$

Solving for f, we find

$$f = \frac{2p}{3}.$$

Then the percentage of passing students in the class is

$$\frac{p}{p + f} \cdot 100 = \frac{p}{p + 2p/3} \cdot 100 = \frac{3}{5} \cdot 100 = 60\%.$$

XX. Factoring the exponent, we find

$$x^3 - 9x^2 + 20x = x(x^2 - 9x + 20) = x(x - 4)(x - 5).$$

The zeros of the exponent are $x = 0, 4, 5$.

Factoring the base, we obtain

$$x^2 + 2x - 24 = (x + 6)(x - 4).$$

Since 0^0 is undefined, we find that $x = 4$ is not a solution of

$$(x^2 + 2x - 24)^{x^3 - 9x^2 + 20x} = 0.$$

If we set the base equal to 1, we find

$$\begin{aligned} x^2 + 2x - 24 &= 1 \\ (x + 1)^2 &= 26 \\ x &= -1 \pm \sqrt{26}. \end{aligned}$$

Thus, the solution set is

$$\{0, 5, -1 \pm \sqrt{26}\}.$$

XXI. Let h, f, and c be the costs of a hamburger, an order of french fries, and a coke, respectively.

$$\begin{aligned} 8h + 5f + 2c &= 14.25 \\ 5h + 3f + c &= 8.51 \end{aligned}$$

If you subtract the 2nd equation from the 1st equation, we obtain

$$3h + 2f + c = 5.74.$$

If you multiply the 2nd equation by -2 and add the result to the first equation, we find

$$-2h - f = -2.77.$$

When we add the last two equations, we obtain

$$h + f + c = \$2.97$$

which is the cost of 1 hamburger, 1 order of french fries, and a coke.

XXII. Since $x + \sqrt{y} = 32$ and $y + \sqrt{x} = 54$, we obtain $\sqrt{y} = 32 - x$ and $x^2 = (54 - y)^2$. Then

$$y = (32 - (54 - y))^2.$$

Using a graphing calculator, we find $y = 49$, $y \approx 47.8$, $y \approx 58.9$, or $y \approx 60.3$

Note, $y \approx 58.9$ and $y \approx 60.3$ cannot satisfy $y + \sqrt{x} = 54$ for any real number x.

If $y \approx 47.8$, then $\sqrt{x} \approx 54 - 47.8 = 6.2$ or $x \approx 38.44$. Note, $x \approx 38.44$ does not satisfy $x + \sqrt{y} = 32$ for any real number y.

However, if $y = 49$ then $\sqrt{x} = 54 - 49 = 5$ or $x = 25$. Thus, the solution is $(x, y) = (25, 49)$.

XXIII. Since $(142,857)(3) = 428,571$ and no number smaller than 142,857 satisfies the condition described in the problem, we find that the smallest such number is 142,857.

XXIV. Open the cylinder to form a 4ft-by-6ft rectangle in such a way that the lizard is on the left side and 1 ft from the base of the rectangle, and the fly is 1 ft from the top of the rectangle and 3 ft from the left side.

If we imagine the rectangle as a piece of paper, then the lizard is on the front page of the paper and the fly in on the back page. Cut the rectangle vertically in the middle into two pieces, then hold the right piece and turn it over. Now, the lizard and fly are on the same side of the pages. Next, move the right piece up by 4 ft.

At this point, we have two 4 ft-by-3 ft rectangles that are joined together at the right top corner of one rectangle (with the lizard) and the left bottom corner of the other rectangle (with the fly). Recall, the lizard and fly are on the same side of the pages.

If the lizard moves 3ft to the right and 4ft up, then the lizard would have reached the fly. However, the shortest path for the lizard is the path along the hypotenuse of a right triangle whose sides are the 3 ft horizontal side and the 4 ft vertical side. The length of the hypotenuse is 5 ft, which is the shortest path from the lizard to the fly.

XXV. A rower can either step over another rower or slide to an adjacent empty seat. The total number of moves is the number of slides plus the number of step overs. Each man in front must step over a woman or be stepped over by a woman. So the minimum number of step overs is $5 \cdot 5$ or 25. Each rower must move a total of 6 spaces. So the total number of spaces moved is $10 \cdot 6$ or 60. Since each step over is two spaces, the number of slides is $60 - 2 \cdot 25$ or 10.

So the minimum number of moves is $10 + 25$ or 35. Represent the five women in back and five men in front as $BBBBB\_\_FFFFF$

B will only move to the right, F will only move to the left, and step overs will only occur between opposite types. The 35 moves can be represented as:

$$B, F, F, B, B, B, F, F, F, F, B, B, B, B, B,$$

$$F, F, F, F, F, B, B, B, B, B, F, F, F, F,$$

$$B, B, B, F, F, B$$

XXVI. Let x be the distance between B and the point P_1 of tangency along side BC of the circle on the left, and let y be the distance between D and P_1. Then the distance between B and D is $\overline{BD} = x + y$.

Let P_2 be the point of tangency along side AB of the circle on the left. Since $\overline{AB} = 7$, we find $\overline{AP_2} = 7 - x$.

Let Q_1 be the point of tangency along side AC of the circle on the right. Since $\overline{BC} = 10$, we find $\overline{CQ_1} = 10 - 2y - x$. Since $\overline{AC} = 12$, we obtain $\overline{AQ_1} = 2 + x + 2y$.

Since $\overline{AP_2} = \overline{AQ_1}$, we find

$$2 + x + 2y = 7 - x.$$

Then $2x + 2y = 5$ and $x + y = 2.5$. Thus, $\overline{BD} = 2.5$.

XXVII. He can get 30,000 miles by rotating the tires as follows:

$$S \to LR \to LF \to RF \to RR \to S$$

where S is the spare tire.

XXVIII. Plant 101 trees on the 1000 foot edge, Then offset the next row so that each tree is the top vertex of an equilateral triangle with side of 10 feet. This gives 174 rows with 101 trees in each row plus 173 rows with 100 trees in each row for a total of 34,874 trees.

This *seems* to be the maximum number of trees that can be planted. I could not find a way of planting more trees.

XXIX. Let x be the amount of water that is already in, y the amount leaking in per hour, and z the amount pumped out per hour by one pump. Then we obtain a system of equations

$$\begin{aligned} 12(3)(z) &= x + 3y \\ 5(10)(z) &= x + 10y \\ n(2)(z) &= x + 2y. \end{aligned}$$

From the first two equations, we find

$$14z = 7y \text{ or } 2z = y.$$

Using the first equation, we get

$$36z = x + 6z \text{ or } 30z = x.$$

Solving for n, we find

$$n = \frac{x + 2y}{2z} = \frac{x + 4z}{2z} = \frac{30z + 4z}{2z} = 17.$$

Thus, 17 pumps are needed.

XXX. Let the ordered triple (x, y, z) represent the amount of antifreeze in the 8-quart radiator, the 5-quart container, and the 3-quart container, respectively.

To leave four quarts of antifreeze in the radiator, perform the sequence of operations defined by the ordered triples:

$$(8,0,0), (5,0,3), (5,3,0), (2,3,3),$$

$$(2,5,1), (7,0,1), (7,1,0), (4,1,3).$$

Afterwards, fill the radiator with water.

XXXI. There were 15 soccer teams in the north section, and 9 teams in the south section. In the south section, undefeated Springville won three games and tied five games for a total of 5.5 points.

XXXII.

a) Substituting $x = 0, 1, 2, ..., 10$ into

$$x(10 - x),$$

we find that the maximum product is 25. This maximum is achieved when $x = 5$. Thus, the two numbers are 5 and 5.

b) First, we enumerate all possible whole numbers satisfying

$$1 \le x \le y \le z \le w \le 10$$

and

$$x + y + z + w = 10.$$

We find that the maximum product for $xyzw$ is 36, and this happens when $x = y = 2$ and $z = w = 3$.

c) The maximum product is 729, and the numbers are six 3's.

XXXIII. Take out one more of type A. Cut them in half one at a time and take half of each pill. Take the other halves the next day.

XXXIV. It makes no sense to say "three times 9 is 27 plus 2 is 29". Rather, three times 9 (total paid by the three boys) is the same as 25 (rent) plus 2 (bellboy).